Simulation of Transport in Nanodevices

Series Editor
Robert Baptist

Simulation of Transport in Nanodevices

Edited by

François Triozon
Philippe Dollfus

iSTE

WILEY

First published 2016 in Great Britain and the United States by ISTE Ltd and John Wiley & Sons, Inc.

ISTE Ltd
27-37 St George's Road
London SW19 4EU
UK

www.iste.co.uk

John Wiley & Sons, Inc.
111 River Street
Hoboken, NJ 07030
USA

www.wiley.com

Library of Congress Control Number: 2016950150

British Library Cataloguing-in-Publication Data
A CIP record for this book is available from the British Library
ISBN 978-1-84821-566-5

Contents

Preface

This book gives an overview of the various methods used to simulate electronic transport in nanoelectronics devices. The miniaturization of transistors to sizes below 20 nm, together with the increasing variety of materials involved in these devices, leads to new challenges for simulation. Quantum mechanical effects are enhanced at such small length scales. Moreover, simulations at the atomic scale are needed to obtain an accurate description of new materials and of their interfaces and defects. Predicting the performances of next generation devices hence requires a multi-scale simulation strategy, from the simulation of materials at the atomic scale to the simulation of electron and heat transport at the device and even circuit scale.

In this book, we focus on the simulation of electronic transport at the device level. In Chapter 1, the field of nanoelectronics is briefly presented and basic notions of solid state physics and of electronic transport are introduced. This chapter serves as an introduction to the other chapters of the book, which are much more specialized and detailed. Before examining the various simulation methods, it is first necessary to introduce some theoretical background concerning solid state physics and the simulation and modelling of materials at the atomic scale. This is the purpose of Chapters 2, 3 and 4. The other chapters then present different methods for the simulation of electronic transport, based on quantum mechanical or semi-classical formalisms.

The book is mainly intended for Master's or PhD students beginning a research project in nanoelectronics.

François TRIOZON
Philippe DOLLFUS
September 2016

List of Symbols

e, q	elementary charge
h	Planck constant
$\hbar$	reduced Planck constant (= $h/2\pi$)
k_B	Boltzmann constant
ψ, ϕ, φ, χ	wavefunction
H	Hamiltonian
V, U	potential energy
m_0	electron mass
m, m^*	effective mass
E, ε	energy
$\mathbf{k}$	electron wavevector
$\mathbf{q}$	phonon wavevector
$\mathbf{p}$	linear momentum
ω	angular frequency
$\mathbf{v}(\mathbf{k})$	group velocity

f(E)	Fermi-Dirac distribution
E, $\mathcal{E}$	electric field
B	magnetic field
T	temperature
n	carrier density
ρ	density matrix
$G, G^r, G^a, G^<, G^>$	Green's functions
j(E)	spectral current density
A(E)	spectral function
Σ	self-energy
$c_k^+, c_k, \psi(r)^+, \psi(r)$	electron creation and annihilation operators
a_q^+, a_q	phonon creation and annihilation operators
f_b	Boltzmann distribution function
f_w	Wigner function
V_w	Wigner potential
D_s	density of states
$s_i(\mathbf{k},\mathbf{k}')$	scattering rate from state **k** to state **k'**
$\Gamma_i(\mathbf{k},\mathbf{k}')$	scattering rate from state **k** to any final state
$\mathcal{R}(\mathbf{k},\mathbf{k}')$	overlap integral
$\mathcal{G}(\mathbf{k},\mathbf{k}')$	overlap factor

List of Abbreviations

CMOS	Complementary Metal-Oxide-Semiconductor
MOSFET	Metal-Oxide-Semiconductor Field Effect Transistor
SPICE	Simulation Program with Integrated Circuit Emphasis
AC	Alternating Current
DC	Direct Current
BTE	Boltzmann Transport Equation
WTE	Wigner Transport Equation
DFT	Density Functional Theory
MC	Monte Carlo
LDA	Local Density Approximation
GGA	Generalized Gradient Approximation
FDSOI	Fully-Depleted Silicon On Insulator
EPM	Empirical Pseudopotential Method
CB	Conduction Band
VB	Valence Band

TB	Tight-Binding
LCAO	Linear Combinations of Atomic Orbitals
SCBA	Self-Consistent Born Approximation
ITRS	International Technology Roadmap for Semiconductors
DM	Density Matrix
GW	Commonly used approximation for many-body effects. G is the Green's functions and W the screened Coulomb potential
MD	Molecular Dynamics
RTA	Relaxation Time Approximation
BZ	Brillouin Zone
PES	Potential Energy Surface
DOS	Density of States

1

Introduction: Nanoelectronics, Quantum Mechanics, and Solid State Physics

1.1. Nanoelectronics

1.1.1. *Evolution of complementary metal–oxide–semiconductor microelectronics toward the nanometer scale*

Current microprocessors are based on the complementary metal–oxide–semi-conductor (CMOS) technology, whose main building blocks are field-effect transistors (MOSFETs). A transistor is made of a semiconducting silicon "channel" connected to "source" and "drain" electrodes. The electrical current through the channel is controlled by a voltage applied to a third electrode, called the "gate" electrode, separated from the channel by a thin insulating layer. Figure 1.1 shows transmission electron microscopy images of MOSFETs. During the past decades, the microelectronics industry has constantly reduced the size of transistors in order to increase the complexity and speed of microprocessors. Current transistors have a channel length L_G of the order of 20 nm, and a channel thickness below 10 nm. Such length scales are close to the typical wavelength of the electrons' wavefunctions propagating through the channel, which enhances quantum effects. The main quantum and atomistic effects occurring in CMOS technology are summarized in Table 1.1.

Chapter written by Philippe DOLLFUS and François TRIOZON.

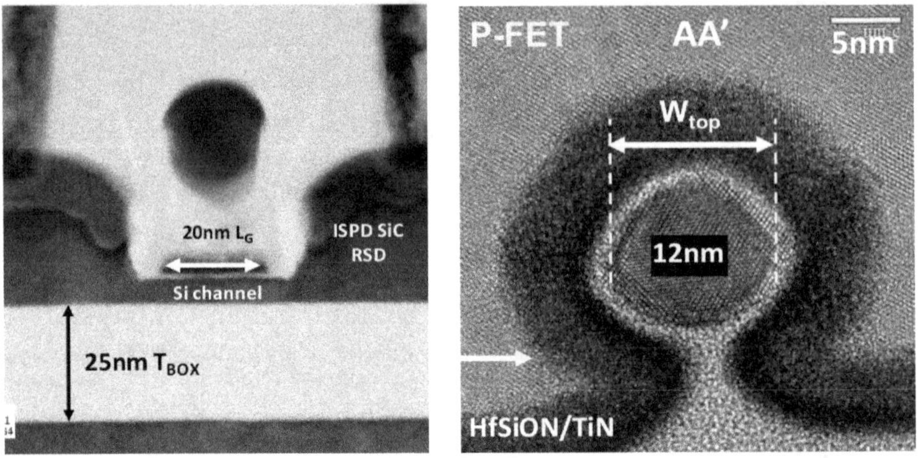

Figure 1.1. *Transmission electron microscopy cross-sections of MOSFETs. Left panel: longitudinal cross-section of a fully-depleted silicon-on-insulator (FDSOI) transistor [LIU 13]. The channel is made of a thin silicon film lying on an oxide layer. Right panel: transverse cross-section of an 'Ω-gate' transistor. The channel is made up of a SiGe nanowire with a diameter of* 12 *nm [NGU 14]*

$L_G > 100$ nm	"Classical" microelectronics
$10\text{ nm} < L_G < 100\text{ nm}$	Quantization of energy levels in the channel Gate leakage by tunneling through the thin oxide Small and uncontrolled number of impurities
$L_G < 10$ nm	Wave phenomena along the transport direction Few electrons in the channel

Table 1.1. *Phenomena occurring at different transistor gate lengths* L_G

When decreasing the gate length, the thickness of the silicon oxide (SiO_2) layer separating the gate electrode from the channel must be reduced accordingly in order to keep a good electrostatic control of conduction inside the channel. However, reducing the SiO_2 thickness below 2 nm leads to detrimental current leakage through the oxide. Hence, materials with higher dielectric constant, such as HfO_2, have been introduced. They allow for a good electrostatic control with larger oxide thickness, hence limiting gate leakage. The transistors shown in Figure 1.1 feature such high-κ gate stacks. This is a first example of a new material introduced in nanoelectronics devices. Other examples are SiGe alloys, used in "p-type" transistors (see the right panel in Figure 1.1) to improve the "hole" mobility, and silicidation of silicon in the source and drain regions to reduce the electrical resistance between the metal contacts and the transistor.

1.1.2. *Post-CMOS nanoelectronics*

While the quantum effects occurring at the nanometer scale tend to limit the performance of CMOS devices, they can be exploited to develop novel types of devices. This is the purpose of post-CMOS nanoelectronics, a research field that has grown considerably during the last two decades. Figure 1.2 shows a tentative classification of these phenomena.

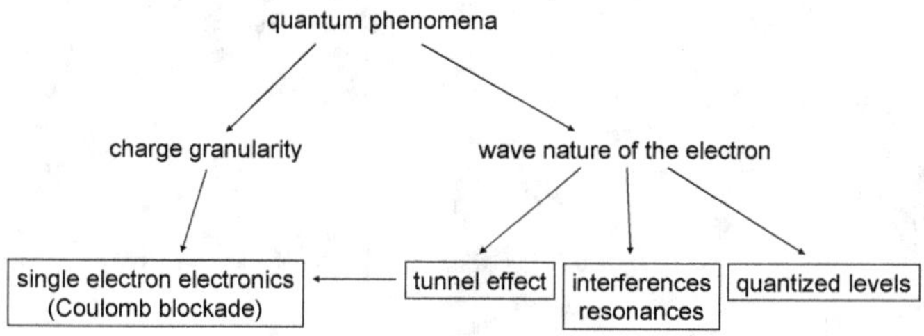

Figure 1.2. *Tentative classification of quantum phenomena that can be exploited in nanoelectronics*

Figures 1.3 shows examples of post-CMOS devices exploiting the wave nature of the electron: a resonant tunneling diode [PAU 00] and a tunnel FET whose channel is made up of a carbon nanotube [APP 04]. Figure 1.4 shows devices exploiting the granularity of the charge: a "flash" memory with silicon nanocrystals [MOL 06] and a single electron transistor [LAV 15]. Such a variety of materials, nanostructures and quantum effects involved in nanoelectronics poses significant challenges to simulation.

1.1.3. *Theory and simulation*

From the very beginning, the development of transistors and microelectronics has been associated with theoretical progress in solid-state physics. Even the most basic properties of solids cannot be explained without quantum mechanics. In particular, their conducting or insulating properties are related to the wave nature of electrons [BLO 29, WIL 31]. Hence, a good knowledge of solid-state physics, including the quantum theory of solids, is needed to address the theory, simulation and modeling of electronic devices.

The mechanical and electronics properties of solids can be modeled at various degrees of refinement from the atomic scale to continuous medium models (see Chapters 2–4). Electronic transport can be described by an even broader variety of formalisms (Chapters 5–8). The semiclassical theory of electronic transport, which essentially consists of describing electron "wavepackets" as point particles (see Chapters 5 and 8), has successfully accompanied CMOS technology up to gate lengths well below 100 nm. Formalisms including the relevant quantum phenomena must be used for simulating smaller CMOS transistors and nanoelectronics devices.

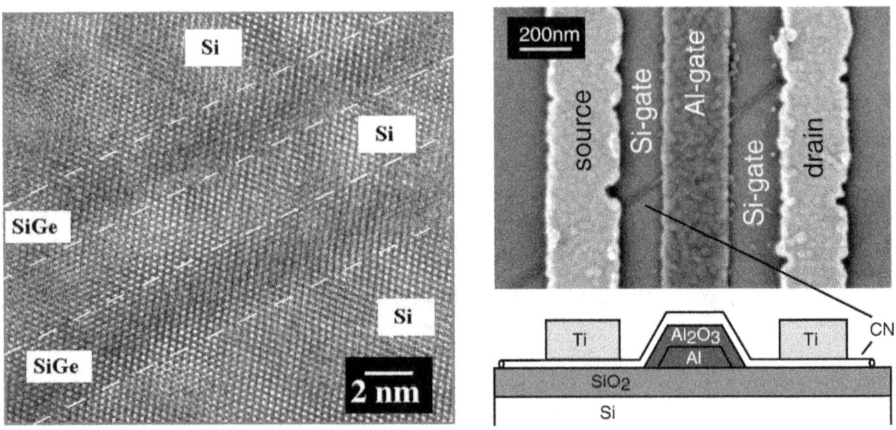

Figure 1.3. *Devices exploiting the wave nature of the electron. Left panel: resonant tunneling diode made up of a thin layer of silicon between two SiGe barriers [PAU 00]. Right panel: tunnel FET made up of a carbon nanotube channel controlled by an Al gate and a doped Si back gate [APP 04]*

To study a given type of device, we have to choose a good approximation for modeling the electronic properties of the materials, and an appropriate formalism for simulating electronic transport. This requires a good knowledge of the available formalisms and how they capture the quantum phenomena involved in the device operation. The main purpose of this book is to give an overview of some commonly used formalisms for electronic transport.

1.2. Basic notions of solid-state physics

1.2.1. *Simplifications of the many-body problem*

As will be detailed in Chapter 2, the quantum theory of atoms, molecules and solids is based on a complex-valued wavefunction $\Psi(\mathbf{r}_1, ..., \mathbf{r}_N; \mathbf{R}_1, ..., \mathbf{R}_{\mathcal{N}}; t)$ whose square modulus gives the probability to find, at time t, the N electrons of the system at positions $\mathbf{r}_1, ..., \mathbf{r}_N$ and its $\mathcal{N}$ nuclei at positions $\mathbf{R}_1, ..., \mathbf{R}_{\mathcal{N}}$. Ψ satisfies the many-body Schrödinger equation that includes the kinetic energies of all particles, the Coulomb electron-electron, electron–ion and ion–ion interactions, and some relativistic corrections. This equation is essentially exact at the energy scales considered in electronics and optoelectronics (a few electron volts). However, its resolution for systems containing more than a few particles is far beyond the capabilities of modern computers, and the numerical cost grows exponentially with N and $\mathcal{N}$. Approximations are thus required.

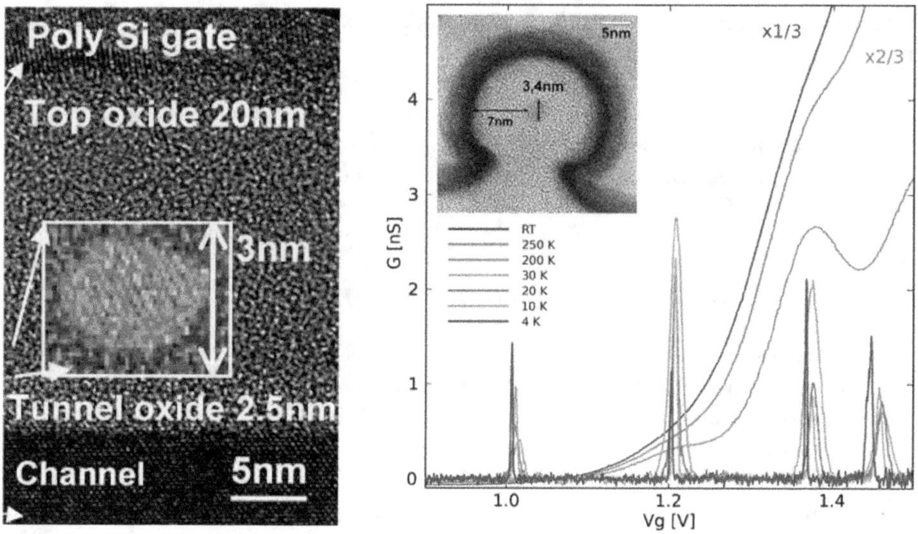

Figure 1.4. *Devices exploiting the granularity of the charge. Left panel: "flash" memory based on silicon nanocrystals embedded into the oxide between the channel and the gate. The memory effect is obtained by trapping/detrapping electrons into the nanocrystals [MOL 06]. Right panel: single-electron transistor made up of a very thin silicon nanowire of diameter $\simeq 3$ nm surrounded by a thick oxide and an Ω-gate. For each electron added into the channel, a conductance peak is observed, which is the signature of Coulomb blockade. In this particular sample, single electron effects remain visible at room temperature [LAV 15]*

The first approximation that is generally made is the Born–Oppenheimer approximation, described in Chapter 2. Since the motion of nuclei is much slower than the motion of electrons, we can consider that the electrons see a static Coulomb potential created by the nuclei. Conversely, the nuclei see an electronic cloud that rearranges quickly with respect to their motion, giving a contribution to their potential energy, which depends only on their positions and not on their velocities. The motion of the nuclei is discussed in Chapter 4. In this chapter, we focus on the electronic part. The wavefunction reduces to $\Psi(\mathbf{r}_1, ..., \mathbf{r}_N; t)$ and the Schrödinger equation contains a potential energy term created by nuclei at fixed positions. However, the problem remains unsolvable for more than a few electrons.

A second approximation is thus needed to treat the electronic and transport properties of large systems. One generally starts from an independent electron approximation, also called a mean-field approximation. Each electron sees an average potential $V(\mathbf{r}, t)$ created by the nuclei and by other electrons. The independent electron approximation seems crude but it actually works very well in many situations, provided that the average potential is well chosen (see Chapter 2). Many-body effects can then be reintroduced using approximate perturbation methods.

Hence, the one-electron Schrödinger equation is a good starting point for studying the quantum mechanics of electrons in solids together with the related transport properties. Before going further, it is necessary to introduce some basic notions of quantum mechanics.

1.2.2. *Basic notions of quantum mechanics*

This section summarizes some basic concepts of quantum mechanics. They are illustrated on a simple case: the one-electron Schrödinger equation. Much more details can be found in quantum mechanics textbooks [MES 99, COH 77].

1.2.2.1. *One-electron Schrödinger equation*

To a free electron of energy E and of momentum $\mathbf{p}$ is associated a wave of angular frequency $\omega = E/\hbar$ and wavevector $\mathbf{k} = \mathbf{p}/\hbar$. The corresponding wavelength $\lambda = 2\pi/k$ is called the de Broglie wavelength. More generally, with an electron moving in a potential $V(\mathbf{r}, t)$ is associated a complex-valued wavefunction $\psi(\mathbf{r}, t)$, which completely characterizes the state of the electron. In particular, $|\psi(\mathbf{r}, t)|^2 \, d^3r$ is the probability to find, at time t, the electron in a small volume d^3r around point $\mathbf{r}$. The wavefunction evolution is given by the one-electron Schrödinger equation:

$$i\hbar \frac{\partial \psi(\mathbf{r}, t)}{\partial t} = -\frac{\hbar^2}{2m_0} \Delta \psi(\mathbf{r}, t) + V(\mathbf{r}, t)\psi(\mathbf{r}, t) \qquad [1.1]$$

where m_0 is the electron mass and Δ is the Laplacian operator with respect to the position $\mathbf{r}$.

For a free electron ($V(\mathbf{r}, t) = 0$), the plane waves

$$\psi(\mathbf{r}, t) = e^{i\mathbf{k}.\mathbf{r}} e^{-i\omega t} \qquad [1.2]$$

are solutions of the Schrödinger equation if $\omega = \hbar k^2/2m_0$. We recover the classical mechanics relation between the energy and the momentum: $E = p^2/2m_0$. The

Laplacian term in the Schrödinger equation is hence associated with the kinetic energy. The link between wave propagation and classical mechanics will be further analyzed below and in Chapter 5.

1.2.2.2. *Static potential: stationary states*

For a static potential $V(\mathbf{r},t) = V(\mathbf{r})$, we can define particular solutions of the Schrödinger equation by decoupling the space and time variables:

$$\psi(\mathbf{r},t) = \phi(\mathbf{r})e^{-i\omega t} \qquad [1.3]$$

where $\phi(\mathbf{r})$ satisfies:

$$\left[-\frac{\hbar^2}{2m_0}\Delta + V(\mathbf{r})\right]\phi(\mathbf{r}) = \hbar\omega\phi(\mathbf{r}) \qquad [1.4]$$

This equation is called the time-independent Schrödinger equation. Wavefunctions of the form [1.3] are called "stationary" since their time evolution is given by a mere phase factor that does not change any physical property of the particle. For instance, the probability density $|\psi(\mathbf{r},t)|^2 = |\phi(\mathbf{r})|^2$ is time independent. Such a state evolves with a single angular frequency ω. Hence, it has a well-defined energy $E = \hbar\omega$.

The linear differential operator acting on $\phi(\mathbf{r})$ in equation [1.4] is called the Hamiltonian and denoted as $\hat{H}$:

$$\hat{H} = -\frac{\hbar^2}{2m_0}\Delta + V(\mathbf{r}) \qquad [1.5]$$

With this compact notation, the time-independent Schrödinger equation reads:

$$\hat{H}\phi(\mathbf{r}) = E\phi(\mathbf{r}) \qquad [1.6]$$

and the time-dependent Schrödinger equation reads:

$$i\hbar\frac{\partial\psi(\mathbf{r},t)}{\partial t} = \hat{H}\psi(\mathbf{r},t) \qquad [1.7]$$

$\hat{H}$ is a Hermitian linear operator in the complex vector space of wavefunctions. Its eigenvalues are the allowed total energies (kinetic + potential) for the electron. These energies can form a continuum (e.g. $E = \hbar^2k^2/2m_0$ for a free electron) or take discrete values (e.g. the energy spectrum of bound electrons in an atom).

Any solution of the time-dependent Schrödinger equation [1.1] can be expressed as a linear combination of stationary states with different energies. Each term of the sum evolves according to [1.3], with its own angular frequency $\omega = E/\hbar$. This will be illustrated in section 1.2.2.6 by studying the evolution of wavepackets.

1.2.2.3. *Dirac notation and basis sets for quantum states*

In Dirac notation, a wavefunction $\psi(\mathbf{r}, t)$ is denoted $|\psi(t)\rangle$, or $|\psi\rangle$ if we do not consider time evolution. It is an abstract vector of the complex vector space of wavefunctions, without specifying any basis set to expand it. The action of a linear operator $\hat{A}$ on this state is denoted $\hat{A}\,|\psi\rangle$, and the adjoint row vector of $|\psi\rangle$ is denoted $\langle\psi|$. Hence, the scalar product between two states $|\psi\rangle$ and $|\psi'\rangle$ reads $\langle\psi|\psi'\rangle$, and the matrix element of an operator $\hat{A}$ between such two states reads $\langle\psi|\,\hat{A}\,|\psi'\rangle$. The time-dependent Schrödinger equation reads:

$$i\hbar\frac{\mathrm{d}}{\mathrm{d}t}\,|\psi(t)\rangle = \hat{H}\,|\psi(t)\rangle \qquad [1.8]$$

and the stationary Schrödinger equation reads:

$$\hat{H}\,|\phi\rangle = E\,|\phi\rangle \qquad [1.9]$$

Depending on the studied quantum system, different choices can be made for the orthonormal basis set spanning the space of quantum states. The spatial basis set $\{|\mathbf{r}\rangle\}$ consists of 3D Dirac delta functions at each point $\mathbf{r}$. They satisfy the orthonormality relation:

$$\langle\mathbf{r}'|\mathbf{r}\rangle = \delta^{(3)}(\mathbf{r}' - \mathbf{r}) \qquad [1.10]$$

The notation $\psi(\mathbf{r})$ used in the previous sections is simply the representation of $|\psi\rangle$ on the spatial basis set:

$$\psi(\mathbf{r}) \equiv \langle\mathbf{r}|\psi\rangle \qquad [1.11]$$

$$|\psi\rangle = \int d^3r\,\langle\mathbf{r}|\psi\rangle\,|\mathbf{r}\rangle = \int d^3r\psi(\mathbf{r})\,|\mathbf{r}\rangle \qquad [1.12]$$

A useful basis set is the basis of plane waves $|\mathbf{k}\rangle$ defined as:

$$\langle\mathbf{r}|\mathbf{k}\rangle = \frac{1}{(2\pi)^{3/2}}e^{i\mathbf{k}.\mathbf{r}} \qquad [1.13]$$

and satisfying the orthonormality relation:

$$\langle \mathbf{k}' | \mathbf{k} \rangle = \delta^{(3)}(\mathbf{k}' - \mathbf{k}) \tag{1.14}$$

Note that these orthonormality relations are peculiar, since the basis vectors $|\mathbf{r}\rangle$ and $|\mathbf{k}\rangle$ are not normalized in the usual sense:

$$\langle \psi | \psi \rangle = \int d^3r \, |\psi(\mathbf{r})|^2 = 1 \tag{1.15}$$

However, the wavefunction of an electron must be normalized to unity, since the probability of finding the electron somewhere in space equals 1. As will be illustrated in section 1.2.2.5, such wavefunctions can be built from normalized linear combinations of plane waves $|\mathbf{k}\rangle$.

Another useful basis set is formed by the stationary states. For an electron confined in a potential well $V(\mathbf{r})$ (e.g. bound electron in an atom), we obtain a discrete set of normalized eigenvectors $|\phi_n\rangle$ of the Hamiltonian, with energies E_n indexed by an integer n:

$$\hat{H} \, |\phi_n\rangle = E_n \, |\phi_n\rangle \tag{1.16}$$

$$\langle \phi_n | \phi_m \rangle = \delta_{n,m} \tag{1.17}$$

where δ is the Kronecker delta. Alternatively, for an unbound electron, the eigenvalues of $\hat{H}$ form a continuum of energies and the eigenvectors are not normalized in the usual sense [1.15], but can satisfy a continuous orthonormalization relation such as [1.14]. For instance, the stationary states of a free electron are precisely the plane waves [1.13].

Many other basis sets can be used to describe electron wavefunctions in solids. An example is the "tight-binding" approach based on atomic orbitals (stationary states of bound electrons in isolated atoms). It is presented in Chapter 3.

1.2.2.4. *Observables*

Every measurable physical quantity is described by a Hermitian linear operator acting in the space of quantum states. This operator is called the "observable" associated with the physical quantity. Any measurement of this quantity returns one of the eigenvalues of the observable. For an observable $\hat{A}$ with an orthonormal basis of eigenvectors $|u_n\rangle$ and eigenvalues a_n, the probability for a measurement performed on a quantum state $|\psi\rangle$ to return the value a_n is:

$$P(a_n) = |\langle u_n|\psi\rangle|^2 \qquad [1.18]$$

Important physical quantities are the energy, the position and the momentum. In Chapter 5, we will also introduce the current operator, which is necessary to study electronic transport. The observable associated with the total energy is the Hamiltonian $\hat{H}$. The three observables associated with the position are the operators $\hat{x}$, $\hat{y}$ and $\hat{z}$. They are defined by their eigenvectors, which are the localized wavefunctions $|\mathbf{r}\rangle$, and by their eigenvalues, which are the Cartesian coordinates of $\mathbf{r}$. For instance:

$$\hat{x}\,|\mathbf{r}\rangle = x\,|\mathbf{r}\rangle \qquad [1.19]$$

From this definition we can deduce, by linearity, the action of the position operators on any state $|\psi\rangle$ expressed in real-space representation [1.11]:

$$\hat{x}\,|\psi\rangle = \int d^3r\psi(\mathbf{r})\hat{x}\,|\mathbf{r}\rangle = \int d^3r\psi(\mathbf{r})x\,|\mathbf{r}\rangle \qquad [1.20]$$

Hence, in real-space representation, the action of $\hat{x}$ reads:

$$\hat{x}\psi(\mathbf{r}) = x\psi(\mathbf{r}) \qquad [1.21]$$

The three observables associated with the momentum are the operators $\hat{p}_x$, $\hat{p}_y$ and $\hat{p}_z$. They are most easily defined on the basis of plane waves [1.13], since a plane wave of wavevector $\mathbf{k}$ corresponds to a free electron of momentum $\mathbf{p} = \hbar\mathbf{k}$. For instance:

$$\hat{p}_x\,|\mathbf{k}\rangle = \hbar k_x\,|\mathbf{k}\rangle \qquad [1.22]$$

It can be useful to express the action of the momentum operator in real-space representation. The decomposition of a wavefunction $\psi(\mathbf{r})$ in plane waves is given by its 3D Fourier transform:

$$\tilde{\psi}(\mathbf{k}) \equiv \frac{1}{(2\pi)^{3/2}}\int d^3re^{-i\mathbf{k}.\mathbf{r}}\psi(\mathbf{r}) \qquad [1.23]$$

The inverse Fourier transform reads:

$$\psi(\mathbf{r}) = \frac{1}{(2\pi)^{3/2}} \int d^3k \tilde{\psi}(\mathbf{k}) e^{i\mathbf{k}.\mathbf{r}} \quad [1.24]$$

By identification with [1.13], this is equivalent to:

$$|\psi\rangle = \int d^3k \tilde{\psi}(\mathbf{k}) |\mathbf{k}\rangle \quad [1.25]$$

Applying $\hat{p}_x$ on both sides of [1.25] and using linearity yields:

$$\hat{p}_x |\psi\rangle = \int d^3k \tilde{\psi}(\mathbf{k}) \hbar k_x |\mathbf{k}\rangle \quad [1.26]$$

Going back to real-space representation,

$$\begin{aligned} \langle \mathbf{r}| \hat{p}_x |\psi\rangle &= \frac{1}{(2\pi)^{3/2}} \int d^3k \tilde{\psi}(\mathbf{k}) \hbar k_x e^{i\mathbf{k}.\mathbf{r}} \\ &= \frac{1}{(2\pi)^{3/2}} \int d^3k \tilde{\psi}(\mathbf{k}) \frac{\hbar}{i} \frac{\partial}{\partial x} e^{i\mathbf{k}.\mathbf{r}} \\ &= \frac{\hbar}{i} \frac{\partial}{\partial x} \left[\frac{1}{(2\pi)^{3/2}} \int d^3k \tilde{\psi}(\mathbf{k}) e^{i\mathbf{k}.\mathbf{r}} \right] \\ &= \frac{\hbar}{i} \frac{\partial}{\partial x} \psi(\mathbf{r}) \end{aligned} \quad [1.27]$$

Hence, in real-space representation, the action of $\hat{p}_x$ reads:

$$\hat{p}_x \psi(\mathbf{r}) = \frac{\hbar}{i} \frac{\partial}{\partial x} \psi(\mathbf{r}) \quad [1.28]$$

As expected, we recover the kinetic energy operator of the Schrödinger equation, along the x direction:

$$\frac{\hat{p}_x^2}{2m_0} = -\frac{\hbar^2}{2m_0} \frac{\partial^2}{\partial x^2} \quad [1.29]$$

Equations [1.21] and [1.28] lead to the commutation relation between the position and momentum operators:

$$[\hat{x}, \hat{p}_x] = i\hbar \qquad [1.30]$$

where the square brackets stand for the commutator between two operators: $[\hat{A}, \hat{B}] \equiv \hat{A}\hat{B} - \hat{B}\hat{A}$. Since $\hat{x}$ and $\hat{p_x}$ do not commute, an eigenstate of p_x is not an eigenstate of $\hat{x}$ and vice versa. This leads to Heisenberg's uncertainty principle detailed below.

1.2.2.5. *Wavepackets: Heisenberg's inequality*

The momentum of a free electron ($V(\mathbf{r}, t) = 0$) is precisely known only if the electron is in a plane wave state $|\mathbf{k}\rangle$. Indeed, the measurement of the momentum on such state will always return the corresponding eigenvalue $\hbar\mathbf{k}$ of the momentum operator. The energy is also exactly known: $E = \hbar^2 k^2/2m_0$. However, plane waves are delocalized in the whole space, hence the position of the electron is completely undetermined. Moreover, plane waves are unphysical since they are not normalized. Physical states can be built by restricting plane waves in some region of space. Let us consider a plane wave of wavevector k_0 along the x direction and located in a region around x_0:

$$\psi(x) = f(x - x_0)e^{ik_0 x} \qquad [1.31]$$

where $f(x)$ is a real bell-shaped function centered around 0 with a certain width, and satisfying the normalization of ψ:

$$\int dx\, |f(x)|^2 = 1 \implies \int dx\, |\psi(x)|^2 = 1 \qquad [1.32]$$

To simplify the discussion, the y and z dependences are omitted, but the wavefunction can be localized in these directions using similar factors $f(y - y_0)$ and $f(z - z_0)$. A common choice for $f(x)$ is a Gaussian:

$$f(x) = \frac{1}{(2\pi)^{1/4}\sqrt{\Delta}} e^{-x^2/4\Delta^2} \qquad [1.33]$$

A wavepacket with a Gaussian envelope is shown in Figure 1.5. However, the following discussion is general to any smooth and localized envelope function. The uncertainty in the position x is defined by the standard deviation Δ_x of the values returned by many measurements performed on the same state $|\psi\rangle$. The probability of finding the electron in an interval x around dx is $|\psi(x)|^2\, dx$. Hence Δ_x is written:

$$\Delta_x^2 = \int dx\, (x - \bar{x})^2\, |\psi(x)|^2 \qquad [1.34]$$

where $\bar{x}$ is the mean value of x:

$$\bar{x} = \int dx x\, |\psi(x)|^2 \qquad [1.35]$$

With the Gaussian envelope [1.33], we obtain $\bar{x} = x_0$ and $\Delta_x = \Delta$.

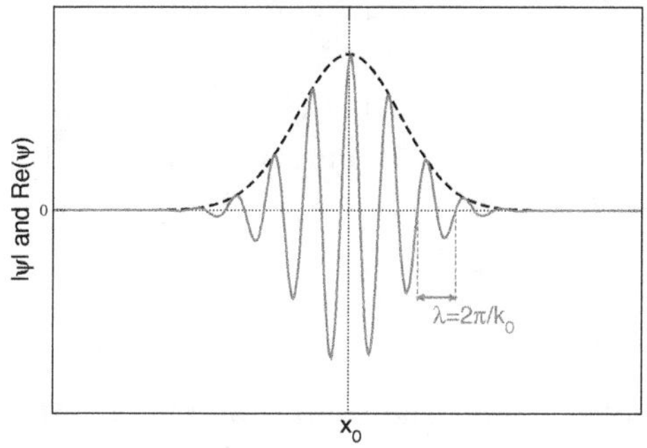

Figure 1.5. *Wavepacket with wavevector k_0 and Gaussian envelope centered around $x = x_0$. Dashed line: modulus of ψ. Full line: real part of ψ*

The uncertainty in the momentum is analyzed by expanding ψ in plane waves:

$$\psi(x) = \frac{1}{\sqrt{2\pi}} \int dk \tilde{\psi}(k) e^{ikx} \qquad [1.36]$$

where $\tilde{\psi}$ is the Fourier transform of ψ:

$$\tilde{\psi}(k) = \frac{1}{\sqrt{2\pi}} \int dx \psi(x) e^{-ikx} \qquad [1.37]$$

Inserting [1.31] into this equation yields:

$$\tilde{\psi}(k) = e^{-i(k-k_0)x_0} \tilde{f}(k - k_0) \qquad [1.38]$$

where $\tilde{f}$ is the Fourier transform of f:

$$\tilde{f}(k) = \frac{1}{\sqrt{2\pi}} \int dx f(x) e^{-ikx} \qquad [1.39]$$

For $|k| > 1/\Delta_x$, the integrand has several oscillations within the region of a few Δ_x where $f(x)$ is non-negligible. These oscillations tend to cancel each other out and the integral becomes small. Hence, we can infer that $\tilde{f}(k)$ is maximum at $k = 0$ and has a decay width of order $1/\Delta_x$. More precisely, from Fourier analysis, we can prove that:

$$\Delta_k \equiv \left(\int dk k^2 \left| \tilde{f}(k) \right|^2 \right)^{1/2} \geq \frac{1}{2\Delta_x} \qquad [1.40]$$

Going back to [1.38], we conclude that $\left| \tilde{\psi}(k) \right|^2$ is centered around $k = k_0$ with a standard deviation Δ_k given by [1.40]. Finally, we obtain Heinsenberg's inequality for position and momentum $p_x = \hbar k$:

$$\Delta_x \Delta_{p_x} \geq \hbar/2 \qquad [1.41]$$

The same inequality holds in the y and z directions. This is Heisenberg's uncertainty principle: the position and momentum of a particle cannot be both defined accurately at the same time. Note that Heisenberg's inequality becomes an equality in the special case of Gaussian wavepackets, since the Fourier transform of [1.33] is:

$$\tilde{f}(k) = \frac{1}{(2\pi)^{1/4}\sqrt{\Delta_k}} e^{-k^2/4\Delta_k^2} \qquad [1.42]$$

with $\Delta_k = 1/2\Delta$.

1.2.2.6. *Wavepacket evolution: group velocity*

We consider an electron propagating in free space ($V(\mathbf{r}) = 0$) along the x direction, from an initial state defined by [1.31]:

$$\psi(x, t = 0) = f(x - x_0)e^{ik_0 x} \qquad [1.43]$$

The time evolution of $\psi(x, t)$ is easily expressed by expanding it in plane waves. Each plane wave of wavevector k is a stationary state evolving with the angular frequency $\omega(k) = \hbar k^2/2m_0$. By linearity of the time-dependent Schrödinger equation, we obtain:

$$\psi(x,t) = \frac{1}{\sqrt{2\pi}} \int dk\tilde{\psi}(k)e^{i[kx-\omega(k)t]} \quad [1.44]$$

Each plane wave propagates with a different phase velocity $v_\phi(k) = \omega(k)/k = \hbar k/2m_0$. This leads to quantum interferences that change the shape and position of the wavepacket. Replacing $\tilde{\psi}(k)$ by its expression [1.38] yields:

$$\psi(x,t) = \frac{1}{\sqrt{2\pi}} \int dke^{-i(k-k_0)x_0}\tilde{f}(k-k_0)e^{i[kx-\omega(k)t]} \quad [1.45]$$

Assuming that the wavepacket has a well-defined wavevector ($\Delta_k \ll k_0$), we can linearize $\omega(k)$ around k_0:

$$\omega(k) \simeq \omega(k_0) + \omega'(k_0)(k-k_0) \text{ with } \omega'(k_0) = \left.\frac{\mathrm{d}\omega}{\mathrm{d}k}\right|_{k=k_0} \quad [1.46]$$

Insertion into [1.45] yields:

$$\begin{aligned}\psi(x,t) &\simeq \frac{1}{\sqrt{2\pi}} \int dke^{-i(k-k_0)x_0}\tilde{f}(k-k_0)e^{i\left[kx-\omega(k_0)t-\omega'(k_0)(k-k_0)t\right]} \\ &= e^{i\left[-\omega(k_0)t+\omega'(k_0)k_0t\right]}\frac{1}{\sqrt{2\pi}} \int dke^{-i(k-k_0)x_0}\tilde{f}(k-k_0)e^{i\left[k(x-\omega'(k_0)t)\right]} \\ &= e^{i\left[-\omega(k_0)t+\omega'(k_0)k_0t\right]}\psi(x-\omega'(k_0)t,0)\end{aligned} \quad [1.47]$$

Hence from time 0 to t, the envelope of the wavepacket is merely shifted along x by $\omega'(k_0)t$. This defines the velocity of the wavepacket, called the "group velocity":

$$v(k_0) = \left.\frac{\mathrm{d}\omega}{\mathrm{d}k}\right|_{k=k_0} = \frac{1}{\hbar}\left.\frac{\mathrm{d}E}{\mathrm{d}k}\right|_{k=k_0} \quad [1.48]$$

This concept of group velocity is common to all kinds of waves in physics (electromagnetic, acoustic, etc.). It is the velocity at which energy and/or matter are conveyed. The variation of the phase velocity v_ϕ with k is called "dispersion". In the presence of dispersion, the group velocity generally differs from the phase velocity. For the free electron wavepacket, we have:

$$v(k) = \frac{\hbar k}{m_0} = 2v_\phi(k) \qquad [1.49]$$

As expected, we recover the classical mechanics relations $v = p/m_0$ and $E = m_0 v^2/2$. A classical particle can be viewed as a quantum mechanical wavepacket whose spatial extension is much smaller than the other characteristic lengths of the system.

The quadratic term that has been neglected in [1.46] mainly leads to a spreading of the wavepacket during its evolution. This is shown by numerical simulations in Chapter 5. However, this does not change the group velocity.

Besides this, the evolution of wavepackets provides an example of another Heisenberg inequality, which relates the uncertainty in the energy to the characteristic evolution time of a quantum state. For a free wavepacket, the uncertainty Δ_{p_x} in the momentum corresponds to an uncertainty Δ_E in the energy through the relation $E = p_x^2/2m_0$:

$$\Delta_E = \frac{p_x \Delta_{p_x}}{m_0} \qquad [1.50]$$

The characteristic time of variation of a wavepacket within its region of width Δ_x can be defined as the time it takes to leave this region:

$$\Delta_t \simeq \frac{\Delta_x}{v(k)} = \frac{\Delta_x m_0}{p_x} \qquad [1.51]$$

From [1.41], we then obtain Heisenberg's inequality between time and energy:

$$\Delta_E \Delta_t \gtrsim \hbar \qquad [1.52]$$

This inequality can be proven more generally by developing any quantum state $|\psi(t)\rangle$ on the basis of stationary states (section 1.2.2.2). Each stationary state evolves with a phase factor $\exp(-iEt/\hbar)$ that depends on its energy E. The time evolution of $|\psi(t)\rangle$ is related to the dephasing of these different components. For a time evolution $\Delta_t \gtrsim 1/\Delta_E$, the dephasing, hence the modification of the state, is large. This justifies the generality of [1.52], E being the total energy (kinetic + potential). As discussed in section 1.2.2.2, if $|\psi(t)\rangle$ is a pure stationary state, it does not evolve in time ($\Delta_t = \infty$)

except for a phase factor, and its energy is known exactly. The time–energy Heisenberg inequality is important in nanoelectronics for the study of unstable quantum systems and their characteristic evolution times (e.g. travel time through a tunnel barrier).

1.2.3. *Bloch waves in crystals*

These concepts of quantum mechanics, presented above for the free electron case, remain fully relevant for the study of electrons in solids. We can now pursue the discussion of section 1.2.1. In the one-electron approximation, the stationary states satisfy the Schrödinger equation [1.4]:

$$\left[-\frac{\hbar^2}{2m_0}\Delta + V_c(\mathbf{r}) + V_{ext}(\mathbf{r})\right]\phi(\mathbf{r}) = E\phi(\mathbf{r}) \quad [1.53]$$

where the potential has been decomposed into $V_c(\mathbf{r})$, the average potential created by the nuclei and other electrons of the solid, and $V_{ext}(\mathbf{r})$, the potential induced by external electric fields. In perfect crystals, $V_c(\mathbf{r})$ has the periodicity of the Bravais lattice, which defines the 3D repetition of the elementary motive (unit cell). Due to this periodicity, and in the absence of external fields ($V_{ext} = 0$), the solutions of the stationary Schrödinger equation [1.53] take the form of Bloch waves [ASH 76]:

$$\phi_{n,\mathbf{k}}(\mathbf{r}) = u_{n,\mathbf{k}}(\mathbf{r})e^{i\mathbf{k}.\mathbf{r}} \quad [1.54]$$

where $\mathbf{k}$ is a wavevector, $u_{n,\mathbf{k}}(\mathbf{r})$ is a function with the periodicity of the Bravais lattice, and n is a "band index". The energy of $\phi_{n,\mathbf{k}}$ is denoted as $E_n(\mathbf{k})$. Bloch waves are plane waves modulated by periodic functions $u_{n,\mathbf{k}}(\mathbf{r})$ describing the variation of the wavefunctions at the interatomic scale. The set of functions $E_n(\mathbf{k})$ is called the electronic band structure of the material. They strongly differ from the free electron relation $E(\mathbf{k}) = \hbar^2k^2/2m_0$ due to the influence of the crystal potential $V_c(\mathbf{r})$. The band structure of bulk silicon is plotted in Figure 1.6. It is not possible to represent $E_n(\mathbf{k})$ graphically since it depends on a wavevector with three components. Hence, the curves $E_n(\mathbf{k})$ are plotted along particular segments in the 3D space of $\mathbf{k}$ vectors.

The Bloch states $\phi_{n,\mathbf{k}}$ are occupied by electrons according to Fermi–Dirac statistics, which arises from the combination of thermodynamics and quantum mechanics. The probability of occupation of a state of energy E is given by the Fermi–Dirac distribution:

$$f(E) = \frac{1}{1 + e^{(E-E_F)/k_BT}} \quad [1.55]$$

where E_F is the Fermi energy, k_B is the Boltzmann constant and T is the temperature. The value of E_F in a bulk solid is determined from the charge neutrality condition: the total number of electrons must be equal to the total number of protons in nuclei. At low temperature, the Fermi–Dirac distribution tends to a step function: the states with energies below E_F are fully occupied ($f = 1$), and the states with energies above E_F are empty ($f = 0$). Note that the maximum occupation is 1 due to Pauli's exclusion principle. In Figure 1.6, the bands with energies below 0 eV are the highest "valence bands" of silicon. At very low temperature, the valence bands are fully occupied by the valence electrons, and the bands with higher energies are unoccupied. In silicon, these higher energy bands, called the "conduction bands", are separated from the valence bands by an energy bandgap $E_g = 1.1$ eV. This is responsible for the semiconducting character of silicon, as discussed below. At room temperature, $k_BT \simeq 25$ meV is still much smaller than E_g: there are few electrons in the conduction bands, and few missing electrons (called "holes") in the valence band.

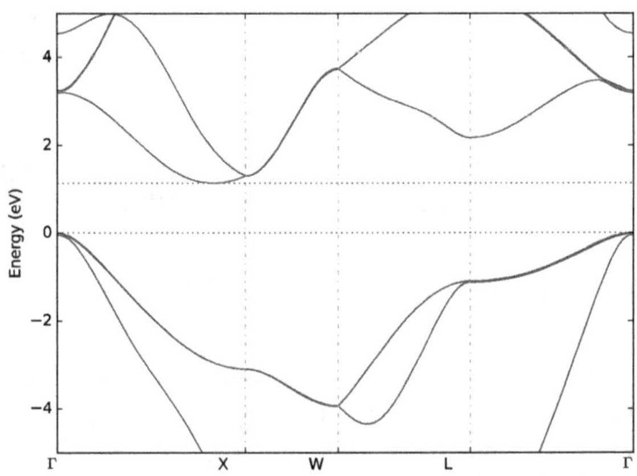

Figure 1.6. *Band structure of bulk silicon calculated from a tight-binding model (see Chapter 3) [NIQ 00]. The abscissa is measured along segments in* $\mathbf{k}$*-space that connect particular* $\mathbf{k}$*-points conventionally labeled* Γ*, X, W and L. For instance,* Γ *labels* $\mathbf{k} = \mathbf{0}$*. In pure silicon at room temperature, the valence bands (below* 0 *eV on this plot) are nearly completely occupied by electrons. The conduction bands are separated by an energy bandgap of* 1.1 *eV and are nearly unoccupied*

The band structure is the most important information needed to determine the electrical properties of materials. First, let us analyze the velocity of electrons. The equations of wavepacket propagation (section 1.2.2.6) remain perfectly valid for Bloch waves. Equation [1.48] generalizes to:

$$\mathbf{v}_n(\mathbf{k}) = \frac{1}{\hbar}\nabla_{\mathbf{k}} E_n(\mathbf{k}) \qquad [1.56]$$

where $\mathbf{v}_n(\mathbf{k})$ is the group velocity of a state in band n with wavevector $\mathbf{k}$, and $\nabla_{\mathbf{k}}$ is the gradient with respect to $\mathbf{k}$. In an ideal crystal, wavepackets would propagate at constant velocity without being scattered, hence behaving as free electrons but with the group velocity [1.56]. In real crystals, defects and lattice vibrations (phonons) yield scattering of Bloch waves toward other Bloch waves at different wavevectors (hence different velocities) and possibly other band indexes. This is one major breakthrough from the quantum theory of solids [BLO 29]: electrons are not scattered by each individual ion of the crystal, but by the deviations from ideal crystal periodicity due to defects and lattice vibrations. Hence, the mean free paths of electrons can be much larger than the interatomic distance, contrary to the hypothesis made by Drude in 1900 [ASH 76].

Second, let us analyze why the presence of an energy bandgap E_g makes the material semiconducting or even insulating [WIL 31]. At each Bloch state $\phi_{n,\mathbf{k}}$, corresponds a Bloch state $\phi_{n,-\mathbf{k}}$ with the same energy $E_n(-\mathbf{k}) = E_n(\mathbf{k})$. Hence, their group velocities are opposite: $\mathbf{v}_n(-\mathbf{k}) = \mathbf{v}_n(\mathbf{k})$. This property related to time-reversal symmetry is completely general. At thermodynamic equilibrium, states $\phi_{n,\mathbf{k}}$ and $\phi_{n,-\mathbf{k}}$ have the same occupation [1.55], hence their density$\times$velocity products cancel each other out and, as expected, the net electrical current is zero. In the presence of a bandgap and at low temperature ($k_B T \ll E_g$), the valence bands are fully occupied and the conduction bands are unoccupied. The only way to establish a current, in other words to break the cancellation of velocities, is to move some electrons from the valence bands to the conduction bands. This can only be done by excitations of energies larger than E_g (e.g. high electric field or photon absorption). However, if E_g is not too large like in silicon, the material can be made conductive by applying an external electric field or by chemical doping, which slightly depopulates the valence band (p-type doping) or populates the conduction band (n-type doping). Such materials are called semiconductors since their conductive character can be easily modulated, which is useful for electronics applications.

Similary to the electron wavefunctions, the vibrations of the crystal lattice are also described by Bloch waves and by relations $\omega_n(\mathbf{k})$ between their angular frequency and their wavevector. n indexes the acoustic and optical branches of the lattice vibrations. These relations constitute the phonon band structure. Chapters 2–4 show how the electron and phonon band structures can be calculated from first principles (quantum mechanics of electrons and nuclei in solids) and then described using simplified semiempirical models.

1.2.4. *Effective mass approximation*

The dispersion relations $E_n(\mathbf{k})$ of a crystal can often be approximated by quadratic functions in the vicinity of their extrema. Isoenergy surfaces in $\mathbf{k}$-space become ellipsoids close to each energy extremum $\mathbf{k}_0$. For simplicity, we limit here to $\mathbf{k}_0 = \mathbf{0}$ and to the isotropic case:

$$E_n(\mathbf{k}) \simeq E_n(\mathbf{0}) + \frac{\hbar^2 \mathbf{k}^2}{2m} \qquad [1.57]$$

where m is called the "effective mass". In addition, the stationary Schrödinger equation can be approximated by:

$$\left[-\frac{\hbar^2}{2m}\Delta + V_{ext}(\mathbf{r})\right]\phi(\mathbf{r}) = E\phi(\mathbf{r}) \qquad [1.58]$$

where $\phi(\mathbf{r})$ is now the "envelope" wavefunction, which does not include the variations of the "true" wavefunction, i.e. the solution of [1.53], at the atomic scale. Correspondingly, one can define a time-dependent envelope wavefunction whose evolution is governed by the following time-dependent Schrödinger equation:

$$i\hbar\frac{\partial\psi(\mathbf{r},t)}{\partial t} = -\frac{\hbar^2}{2m}\Delta\psi(\mathbf{r},t) + V_{ext}(\mathbf{r},t)\psi(\mathbf{r},t) \qquad [1.59]$$

Again, the variations of the true wavefunction at the atomic scale are not included in $\psi(\mathbf{r},t)$, but its propagation at larger scale is well described. The effect of the periodic crystal potential $V_c(\mathbf{r})$ is included into the effective mass. Only the external potential $V_{ext}(\mathbf{r})$ remains. From now on, it will be simply denoted as $V(\mathbf{r})$. We recover the Schrödinger equation [1.4] of an electron in vacuum, with the electron mass m_0 replaced by the effective mass m.

Throughout this chapter and Chapter 5, we will limit the discussion to this simple effective mass Schrödinger equation. It is sufficient to illustrate many important aspects of electronic transport in crystalline materials.

1.3. Quantum mechanics and electronic transport

1.3.1. *Wavepacket in a slowly varying potential: the semiclassical equations of motion*

We consider the propagation of an electron wavepacket $\psi(x,t)$ in a "slowly varying" static potential $V(x)$, i.e. a potential whose characteristic length of variation is much larger than the characteristic wavelength of the wavepacket. In this limit, the average position $\bar{x}$ and wavevector $\bar{k}$ of the wavepacket satisfy the semiclassical equations of motion:

$$\frac{\mathrm{d}\bar{x}}{\mathrm{d}t} \simeq v(\bar{k}) = \frac{1}{\hbar}\left.\frac{\mathrm{d}E(k)}{\mathrm{d}k}\right|_{k=\bar{k}} \qquad [1.60]$$

$$\hbar\frac{\mathrm{d}\bar{k}}{\mathrm{d}t} \simeq -\left.\frac{\mathrm{d}V}{\mathrm{d}x}\right|_{x=\bar{x}} \qquad [1.61]$$

where $E(k)$ is the energy of a state of wavevector k in the absence of external potential ($V = 0$). We call it "kinetic" energy, but it also contains the potential energy due to the periodic crystal potential $V_c(\mathbf{r})$, through the effective mass. The first equation simply expresses the group velocity of the wavepacket. The second equation looks like Newton's law if we interpret $\hbar\bar{k}$ as the electron momentum. A rigorous justification would rely on Ehrenfest's theorem [MES 99, COH 77]. Here, we just give a simple argument [ASH 76] that justifies equation [1.61] from energy conservation. During an infinitesimal time interval dt, the variation of the potential energy is:

$$dV = \frac{\mathrm{d}V}{\mathrm{d}x}v dt = \frac{\mathrm{d}V}{\mathrm{d}x}\frac{1}{\hbar}\frac{\mathrm{d}E}{\mathrm{d}k}dt \qquad [1.62]$$

and the variation of the "kinetic" energy is:

$$\frac{\mathrm{d}E}{\mathrm{d}k}d\bar{k} \qquad [1.63]$$

The conservation of the total energy then leads to equation [1.61].

These equations will be illustrated by numerical simulations in Chapter 5 and discussed in more detail in Chapter 8. Their generalization to Bloch waves reads:

$$\frac{\mathrm{d}\bar{\mathbf{r}}}{\mathrm{d}t} \simeq \mathbf{v}_n(\bar{\mathbf{k}}) = \frac{1}{\hbar}\left.\nabla_{\mathbf{k}}E_n(\mathbf{k})\right|_{\mathbf{k}=\bar{\mathbf{k}}} \qquad [1.64]$$

$$\hbar\frac{\mathrm{d}\bar{\mathbf{k}}}{\mathrm{d}t} \simeq -\left.\nabla_{\mathbf{r}}V_{ext}(\mathbf{r})\right|_{\mathbf{r}=\bar{\mathbf{r}}} \qquad [1.65]$$

Here, we see again that the knowledge of the band structure (section 1.2.3) is crucial for understanding the transport properties of semiconductors. It determines the carrier velocities and their evolution under the influence of an external electric field.

In silicon, the semiclassical theory remains generally valid at length scales larger than 20 nm. Hence, it is highly relevant for the simulation of current electronic devices. However, in the following, we discuss phenomena that are not described by the semiclassical theory.

1.3.2. *Square potential barrier: tunneling and quantum reflection*

We consider here the simple but relevant case of a square potential barrier of width a and height V_0, as schematized in Figure 1.7. We will study the propagation of an incident plane wave coming from the left side, in both cases of an energy E greater and smaller than V_0, respectively. It will allow us to illustrate two important quantum phenomena: the resonance and the tunneling effect.

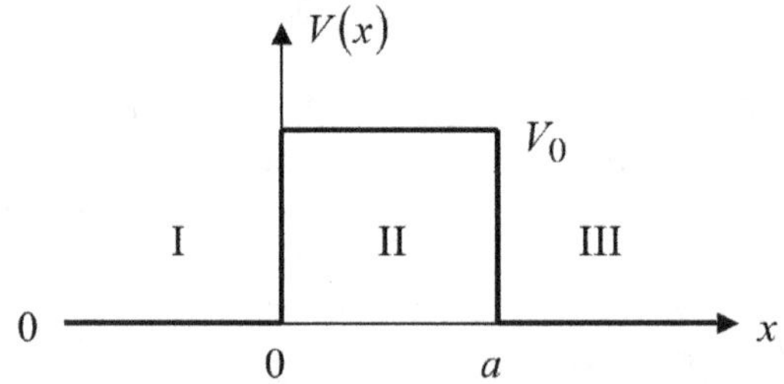

Figure 1.7. *Square potential barrier of width a and height V_0, defining three different regions labeled I, II and III*

As in all problems of wave function propagation, we will have to consider the relevant continuity conditions of the wave function. Even in the presence of a discontinuity of the potential $V(x)$, the wavefunction $\phi(x)$ must be regular enough to guarantee the continuity of the probability of presence and of the probability current [COH 77]. If the discontinuity of potential separates two media where the effective mass of the particle is different, these two conditions lead to:

$$\phi(x) \text{ and } \frac{1}{m(x)}\frac{\mathrm{d}\phi(x)}{\mathrm{d}x} \text{ are continuous functions.} \qquad [1.66]$$

Of course, if the effective mass does not change on either side of the discontinuity, we just have to consider that the wave function and its first derivative are continuous.

1.3.2.1. *Resonance ($E > V_0$)*

In the three regions (labeled I, II and III in Figure 1.7), the energy of the incident wave is higher than the potential energy. Hence, the solutions of the Schrödinger equation take the form:

$$\begin{cases} \phi_1(x) = A_1 e^{ik_1 x} + A_1' e^{-ik_1 x} \\ \phi_2(x) = A_2 e^{ik_2 x} + A_2' e^{-ik_2 x} \\ \phi_3(x) = A_3 e^{ik_1 x} + A_3' e^{-ik_1 x} \end{cases} \quad [1.67]$$

where the wavevectors k_1 and k_2 are written:

$$k_1 = \sqrt{\frac{2mE}{\hbar^2}} \text{ and } k_2 = \sqrt{\frac{2m(E - V_0)}{\hbar^2}} \quad [1.68]$$

In the case of an incident wave coming from the left, we have $A_3' = 0$. Applying the continuity conditions [1.66] at positions $x = 0$ and $x = a$, we obtain the reflection and transmission coefficients, R and T, respectively, of the probability current as:

$$R = \left|\frac{A_1'}{A_1}\right|^2 = \frac{\left(k_1^2 - k_2^2\right)^2 \sin^2(k_2 a)}{4k_1^2 k_2^2 + \left(k_1^2 - k_2^2\right)^2 \sin^2(k_2 a)} \quad [1.69]$$

$$T = \left|\frac{A_3}{A_1}\right|^2 = \frac{4k_1^2 k_2^2}{4k_1^2 k_2^2 + \left(k_1^2 - k_2^2\right)^2 \sin^2(k_2 a)}$$
$$= \frac{4E(E - V_0)}{4E(E - V_0) + V_0^2 \sin^2\left(\sqrt{2m(E - V_0)}a/\hbar\right)} \quad [1.70]$$

It is easy to check that $R + T = 1$. As shown in Figure 1.8, for given values of E and V_0, the transmission coefficient oscillates periodically as a function of the barrier width a between 1 and a minimum value

$$T_{min} = \left[1 + \frac{V_0^2}{4E(E - V_0)}\right]^{-1} \quad [1.71]$$

The resonance condition $T = 1$ is satisfied each time that a is a multiple of half wavelength in region II, which corresponds physically to the condition for forming standing waves in this region. The resonances are of course also observed when the transmission coefficient is displayed as a function of energy for a given barrier width (Figure 1.8). It should be noted that these results are qualitatively independent of the sign of V_0.

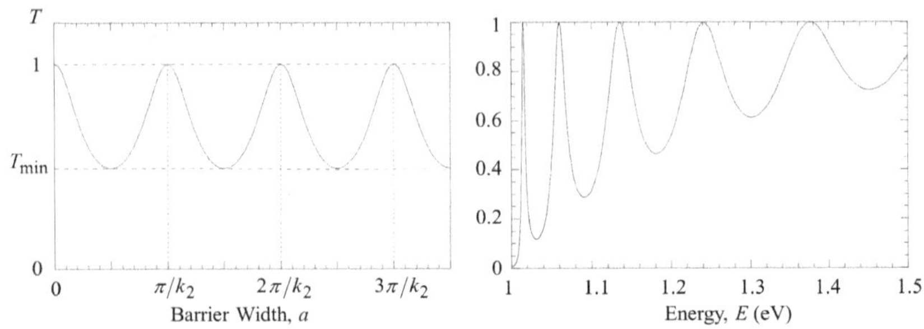

Figure 1.8. *Left: transmission coefficient as a function of the barrier width for $E = 1.2$ eV and $V_0 = 1$ eV. Right: transmission coefficient as a function of energy for $V_0 = 1$ eV and $a = 5$ nm*

1.3.2.2. *Tunneling effect ($E < V_0$)*

In this case $E < V_0$, the wavefunction in region II takes the form:

$$\phi_2(x) = B_2 e^{\rho_2 x} + B_2' e^{-\rho_2 x} \text{ with } \rho_2 = \sqrt{\frac{2m(V_0 - E)}{\hbar^2}} \quad [1.72]$$

A derivation similar to the previous case leads to the following transmission coefficient:

$$\begin{aligned} T = \left|\frac{A_3}{A_1}\right|^2 &= \frac{4k_1^2\rho_2^2}{4k_1^2\rho_2^2 + \left(k_1^2 + \rho_2^2\right)^2 \sinh^2(\rho_2 a)} \\ &= \frac{4E(V_0 - E)}{4E(V_0 - E) + V_0^2 \sinh^2\left(\sqrt{2m(V_0 - E)}a/\hbar\right)} \end{aligned} \quad [1.73]$$

In the barrier, the wave function has an evanescent shape of decaying length $1/\rho_2$ that depends on the energy E. When the barrier width is smaller than this length, the transmission probability by tunneling effect is non-negligible, as shown in Figure 1.9. These examples illustrate two important results of quantum mechanics with respect to classical mechanics:

– the transmission of a particle above a potential barrier can give rise to a partial reflection, while in classical mechanics there is a full transmission with a change in kinetic energy;

– there is a finite probability of a particle crossing (tunnel through) a potential barrier while in classical mechanics the particle is just fully reflected.

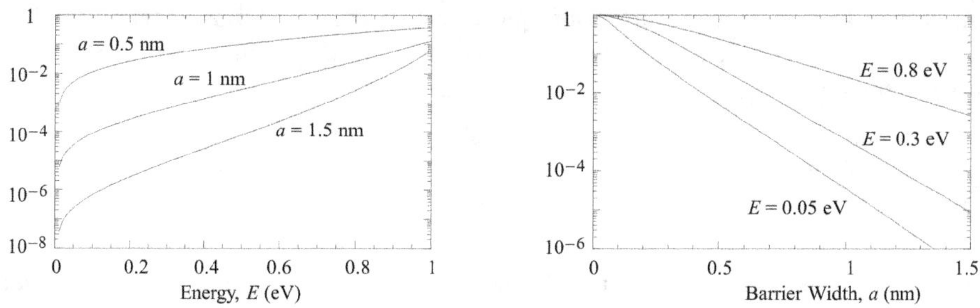

Figure 1.9. *Right: tunneling transmission coefficient as a function of the barrier width for $V_0 = 1$ eV and different values of E. Left: tunneling transmission coefficient as a function of energy for $V_0 = 1$ eV and different values of width a*

The transmission method used here for a square potential barrier will be generalized in Chapter 5 to study the propagation of waves in more complex systems. The general method is based on the Green's functions formalism. Its application to the conduction of molecular systems is also presented in Chapter 2.

1.3.3. *Quantum confinement*

Besides the tunneling effect, another quantum effect is of great importance in electronic devices: quantum confinement. In the semiconducting channel of field-effect transistors, electrons are attracted by the gate electric field close to the semiconductor/oxide interface. The electric field creates a potential well with a width of a few nanometers, smaller than the typical electron wavelengths. Besides, some transistor technologies use ultrathin semiconducting films [LIU 13] or nanowires [NGU 14] as transistor channels. The electrons are confined in these thin structures due to the high potential barrier at the semiconductor/oxide interface. This potential barrier is not related to an external electric field. It is the intrinsic energy difference between the conduction band of the oxide and the conduction band of the semiconductor related to the different crystal potentials $V_c(\mathbf{r})$ in each material. This barrier is generally large, since the oxides used in electronics are good insulators, with a much larger bandgap than semiconductors. For instance, the potential barrier at a Si/SiO_2 interface is 3.15 eV.

Let us consider, for instance, a semiconductor film of thickness a sandwiched between two oxide layers. z is the direction normal to the layers. Electrons are assumed to be described by the effective mass Schrödinger equation:

$$\left[-\frac{\hbar^2}{2m_x}\frac{\partial^2}{\partial x^2} - \frac{\hbar^2}{2m_y}\frac{\partial^2}{\partial y^2} - \frac{\hbar^2}{2m_z}\frac{\partial^2}{\partial z^2} + V(\mathbf{r})\right]\phi(\mathbf{r}) = E\phi(\mathbf{r}) \qquad [1.74]$$

where we have considered an anisotropic effective mass for more generality. The potential $V(\mathbf{r})$ is assumed to be 0 in the semiconductor and V_0 in the oxide. Hence, it only depends on z:

$$\begin{array}{l} V(z) = 0 \qquad \text{for} \qquad 0 < z < a \\ V(z) = V_0 \text{ elsewhere} \end{array} \qquad [1.75]$$

We can then separate the z direction from the x and y directions in the Schrödinger equation, and find solutions of the form:

$$\phi(x, y, z) = \chi(z) e^{i(k_x x + k_y y)} \qquad [1.76]$$

which are solutions of [1.74] if $\chi(z)$ satisfies the 1D Schrödinger equation

$$\left[-\frac{\hbar^2}{2m_z} \frac{\mathrm{d}^2}{\mathrm{d}z^2} + V(z) \right] \phi(z) = E_\perp \phi(z) \qquad [1.77]$$

The total energy is:

$$E = E_\perp + \frac{\hbar^2 k_x^2}{2m_x} + \frac{\hbar^2 k_y^2}{2m_y} \qquad [1.78]$$

$E_\perp$ is the kinetic + confinement energy along the z direction. The two other terms give the in-plane kinetic energy. Equation [1.77] can be solved in the same way than for the square barrier in the previous section. We obtain a continuum of solutions with $E_\perp > V_0$ (plane wave solutions in the oxides) and a finite number of solutions with $E_\perp < V_0$. The latter solutions are confined in the semiconductor, with evanescent tails in the oxide layers. Here, we only solve the simplified case of an infinite barrier $V_0 = \infty$, which imposes $\chi(z) = 0$ at both interfaces. The solutions are:

$$\chi_p(z) = \sin\left(\frac{p\pi z}{a}\right) \qquad [1.79]$$

They are indexed by an integer $p > 0$. The energy is:

$$E_{\perp,p} = \frac{p^2 \pi^2 \hbar^2}{2m_z a^2} \qquad [1.80]$$

This is the kinetic energy associated with the wavevector $k_z = p\pi/a$, which is restricted to quantized values by boundary conditions. Finally, the solutions of the 3D

Schrödinger equation [1.74] are indexed by the integer $p > 0$ labeling the confined wavefunction along z, and by the in-plane wavevector (k_x, k_y):

$$\phi_{n,k_x,k_y}(x, y, z) = \chi_p(z)e^{i(k_x x + k_y y)} \qquad [1.81]$$

The total energy is:

$$E_p(k_x, k_y) = \frac{p^2\pi^2\hbar^2}{2m_z a^2} + \frac{\hbar^2 k_x^2}{2m_x} + \frac{\hbar^2 k_y^2}{2m_y} \qquad [1.82]$$

This can be viewed as a band structure with a 2D wavevector and bands indexed by p. Here, the different bands are simply paraboloids shifted by the confinement energy $E_{\perp,p}$. In a silicon film of thickness $a = 5$ nm and confinement mass $m_z = 0.92 \times m_0$, the confinement energies are:

$$E_{\perp,p} = p^2 \times 16.3\,\text{meV} \qquad [1.83]$$

The energy difference between the first levels is comparable to the thermal energy at room temperature, $k_B T = 25$ meV. Hence, quantum confinement plays an important role in field effect transistors. For instance, if the Fermi level E_F is between $E_{\perp,1}$ and $E_{\perp,2}$, then only the states ϕ_{1,k_x,k_y} with energies $E_1(k_x, k_y) \lesssim E_F$ are occupied. The electron density is then proportional to:

$$n(z) \propto |\chi_1(z)|^2 = \sin^2\left(\frac{\pi z}{a}\right) \qquad [1.84]$$

The electron density is small close to the interfaces and maximum in the center of the film, while the semiclassical theory would predict a uniform density. This behavior is general. Electrons attracted in a triangular potential well created by a gate electric field have a region of small density at the semiconductor/oxide interface, called the "dark space" region. This is due to the fact that the few lowest energy states that contribute to the density have a small kinetic energy along z, hence a smooth variation, and are thus spread out from the interface (with a distance of a few nm for Si/SiO_2 interfaces). This can be observed by measuring the electrical capacitance of the gate–oxide–semiconductor junction. The dark space region is equivalent to an effective increase in the oxide thickness, hence it decreases the gate capacitance.

1.3.4. *Scattering*

To calculate the eigenstates (band structure) of an electron in a perfect crystal (see Chapters 2 and 3), it is necessary to make some approximations and in particular to neglect some interaction terms in the Hamiltonian, assuming that these terms are weak perturbations. These neglected terms are, for instance, the effect of impurities and of the atomic vibrations on the electrons, i.e. the electron–impurity and the electron–phonon interactions. However, these interactions play a very important role in transport properties and in the relaxation of electrons subjected to an external excitation. Actually, these terms are treated as time-dependent perturbations that do not modify the eigenstates of the electrons but give rise to transitions between these states [COH 77]. These transitions, usually considered to be instantaneous, are characterized by a scattering rate. A perturbing process is described by an additional term $\hat{H}_p(t)$ in the Hamiltonian, with a sinusoidal time dependence of angular frequency ω_p. For instance, when considering electron–phonon interaction, ω_p is the angular frequency of a phonon mode (see Chapter 4). For a static perturbation (e.g. the potential created by an impurity), we have $\omega_p = 0$. The first-order perturbation theory tells us that the transition probability per unit of time from Bloch state $\phi_{n,\mathbf{k}}$ to Bloch state $\phi_{n',\mathbf{k}'}$ is given by the Fermi's golden rule:

$$s(n,\mathbf{k};n',\mathbf{k}') = \frac{2\pi}{\hbar}\left|\langle\phi_{n',\mathbf{k}'}|\,\hat{H}_p\,|\phi_{n,\mathbf{k}}\rangle\right|^2 D_s(n',\mathbf{k}') \\ \times\delta\left(E_{n'}(\mathbf{k}') - E_n(\mathbf{k}) \pm \hbar\omega_p\right) \quad [1.85]$$

where $D_s(n',\mathbf{k}')$ is the density of states around the final state, i.e. the number of Bloch states in subband n' around wavevector $\mathbf{k}'$ per unit of energy, δ is the Dirac delta function and the quantity

$$M = \langle\phi_{n',\mathbf{k}'}|\,\hat{H}_p\,|\phi_{n,\mathbf{k}}\rangle = \int d^3r\,\phi^*_{n',\mathbf{k}'}(\mathbf{r})\hat{H}_p\phi_{n,\mathbf{k}}(\mathbf{r}) \quad [1.86]$$

is the transition matrix element of the scattering process between the initial and final Bloch states. Static perturbations ($\omega_p = 0$) lead to elastic scattering: the Dirac delta factor selects transitions for which the energies of the initial and final states are equal. Time-dependent perturbations ($\omega_p \neq 0$) lead to inelastic scattering, with a variation $\pm\hbar\omega_p$ of the energy. For instance, the absorption of a phonon leads to:

$$E_{n'}(\mathbf{k}') = E_n(\mathbf{k}) + \hbar\omega_p \quad [1.87]$$

Details on the calculation of scattering rates and their treatment in transport problems are given in Chapters 2 and 8. In Chapter 5, a method to calculate exactly elastic scattering by defects is presented, and it is shown that the first-order perturbation theory can be inaccurate in some cases.

1.4. Conclusion

In this chapter, we have highlighted the importance of quantum mechanics for understanding transport in electronic devices. The notion of electronic band structure has been introduced. The main implications of quantum mechanics on transport have been illustrated using the effective mass approximation, which is the simplest band structure model. Any reader who needs more details about wavepacket propagation may skip to section 5.1 of Chapter 5 before going through Chapter 2.

With the introduction of new materials and interfaces in nanoelectronics, there is an increasing need for accurate calculations of electron and phonon band structures. Chapter 2 presents first principle (also called *ab initio*) methods based on the density functional theory for computing electron and phonon properties. The coupling of these methods to electronic transport simulations is also addressed. Chapters 3 and 4 present semiempirical methods to describe electron band structures and phonon band structures, respectively. These descriptions, which are simplified with respect to *ab initio* methods, are necessary to study transport in systems of realistic size (more than 10 nm).

Chapters 5–8 are devoted to various simulation methods for electronic transport. We have chosen to start from the quantum mechanical methods, which are the most fundamental ones and allow to understand better the domain of validity of the semiclassical transport theory, presented in Chapter 8. Chapter 5 deals with quantum transport in the coherent regime, i.e. without inelastic interactions such as electron–phonon coupling. It presents the Green's function formalism in the one-electron approximation. Chapter 6 generalizes this approach to quantum transport with electron–phonon coupling, based on the non-equilibrium Green's functions formalism. Chapter 7 presents an alternative quantum transport formalism based on Bohmian trajectories, which offers promising perspectives for the simulation of transport at high frequency and in transient regimes, including current noise. Chapter 8 presents the semiclassical theory of transport based on the Boltzmann transport equation and the Monte Carlo approach to solve this equation. It details the various scattering mechanisms. It also presents the multi-subband Boltzmann method, which combines quantum confinement in the transverse direction and semiclassical transport in the channel direction. Finally, it presents the Wigner formalism, which is an alternative formulation of quantum transport that resembles the Boltzmann equation.

1.5. Bibliography

[APP 04] APPENZELLER J., LIN Y.-M., KNOCH J. *et al.*, "Band-to-band tunneling in carbon nanotube field-effect transistors", *Physical Review Letters*, vol. 93, p. 196805, 2004.

[ASH 76] ASHCROFT N.W., MERMIN N.D., *Solid State Physics*, Brooks/Cole, Salt Lake City, 1976.

[BLO 29] BLOCH F., "Über die Quantenmechanik der Elektronen in Kristallgittern", *Zeitschrift für Physik*, vol. 52, no. 7, pp. 555–600, 1929.

[COH 77] COHEN TANNOUDJI C., DIU B., LALOË F., *Quantum Mechanics*, Wiley, New York, 1977.

[LAV 15] LAVIEVILLE R., TRIOZON F., BARRAUD S. *et al.*, "Quantum dot made in metal oxide silicon-nanowire field effect transistor working at room temperature", *Nano Letters*, vol. 15, no. 5, pp. 2958–2964, 2015.

[LIU 13] LIU Q., VINET M., GIMBERT J. *et al.*, "High performance UTBB FDSOI devices featuring 20 nm gate length for 14 nm node and beyond", *IEEE International Electron Devices Meeting (IEDM)*, pp. 9.2.1–9.2.4, 2013.

[MES 99] MESSIAH A., *Quantum Mechanics*, Dover, New York, 1999.

[MOL 06] MOLAS G., DELERUYELLE D., DE SALVO B. *et al.*, "Degradation of floating-gate memory reliability by few electron phenomena", *IEEE Transactions on Electron Devices*, vol. 53, no. 10, pp. 2610–2619, 2006.

[NGU 14] NGUYEN P., BARRAUD S., TABONE C. *et al.*, "Dual-channel CMOS co-integration with Si NFET and strained-SiGe PFET in nanowire device architecture featuring sub-15nm gate length", *IEEE International Electron Devices Meeting (IEDM)*, pp. 16.2.1–16.2.4, 2014.

[NIQ 00] NIQUET Y.M., DELERUE C., ALLAN G. *et al.*, "Method for tight-binding parametrization: application to silicon nanostructures", *Physical Review B*, vol. 62, pp. 5109–5116, 2000.

[PAU 00] PAUL D.J., SEE P., ZOZOULENKO I. V. *et al.*, "Si/SiGe electron resonant tunneling diodes", *Applied Physics Letters*, vol. 77, no. 11, 2000.

[WIL 31] WILSON A.H., "The theory of electronic semi-conductors", *Proceedings of the Royal Society of London A: Mathematical, Physical and Engineering Sciences*, vol. 133, no. 822, pp. 458–491, 1931.

2

Electronic Transport: Electrons, Phonons and Their Coupling within the Density Functional Theory

2.1. Introduction

The design of electronic circuits has heavily relied on the simulation of their DC and AC electric responses, for instance with the use of the SPICE program, which contains built-in models for semiconductor devices [ANT 88]. A more accurate description of the electronic response requires the simultaneous solution of the transport equations governing the charge flow and of Maxwell's equations for the time-dependent electric and magnetic fields [VAS 11].

In this chapter, we consider the coupling between the *ab initio* description of the electronic and vibrational properties of materials (sections 2.1, 2.2 and 2.3), and the solution of transport equations. The purpose is to provide a more accurate description of transport with respect to usual models.

Transport properties can be tackled from at least two viewpoints [VEN 08]. In the first approach, an electrical current is seen as the consequence of an applied electric field. This led to the Drude model [ASH 76], Kubo formalism [KUB 57, KUB 66] and the Boltzmann transport equation (BTE). Only the latter formalism, and its coupling with the density functional theory (DFT), are going to be discussed in this chapter (section 2.5). Following the second viewpoint, the current flux is determined by the boundary conditions at the surfaces of the sample, and the electric field is a consequence of the carrier flow [FRE 30]. This viewpoint is adopted in the Landauer approach, which is also going to be detailed in present chapter to calculate quantum transport from first principles (section 2.6).

Chapter written by Nathalie VAST, Jelena SJAKSTE, Gaston KANÉ and Virginie TRINITÉ.

BTE has accompanied the development of the industry of transistors. In BTE, upon application of an electric field $\mathbf{E}$, electrons are accelerated according to Bloch's acceleration theorem,

$$\frac{d\mathbf{k}}{dt} = -\frac{e}{\hbar}(\mathbf{E} + \mathbf{v} \wedge \frac{\mathbf{B}}{c}), \qquad [2.1]$$

where $\mathbf{k}, -e\, \mathbf{v}$ are, respectively, the electronic crystal momentum, charge and velocity, $\mathbf{B}$ is the magnetic field and c is the speed of light (see also equation [2.69] below). A balance comes from the electron backscattering due to their interaction with phonons and lattice defects [IMR 99]. The system is thus open [VEN 08] and, in general, phonons are considered as a bath at thermodynamical equilibrium. BTE for charge transport has been solved mainly with Monte Carlo simulations in which the scattering was modeled with a few parameters fitted on the experiment [JAC 83, VAS 11].

Recent progress toward a parameter-free description of the electronic transport has been achieved, in which algorithms to solve BTE and *ab initio* calculation are coupled [MUR 06, MUR 08, RES 09, WAN 11]. In this chapter, we are going to introduce the notions that are necessary to the *ab initio* calculation of the ground state of the electronic system and of the scattering cross-sections, with a particular emphasis on the electronic scattering by phonons. The theoretical framework is the time-independent DFT presented in section 2.2. Phonons are treated as static perturbations in DFT. The theoretical framework that is applied, known as the density functional perturbation theory (DFPT), is presented in section 2.3. The formalism used to calculate scattering cross-sections for the electron–phonon interaction is detailed for DFPT in section 2.4. Finally, BTE and the various scattering mechanisms included in BTE, among which the electronic scattering by phonons via the electron–phonon coupling, are described in section 2.5, together with the limitations of BTE. Last but not least, in section 2.6, we address the question of quantum approaches to the electronic transport.

Indeed, modern technology has enabled the fabrication of samples of very small size, in which the BTE has become pointless: it is formulated in terms of wavepackets whose dimensions in space are of the order of a few tens of nanometers, much larger than, for instance, the size of junctions (Figure 2.1). Advances in the calculation of the electronic transport of such systems have made the methods evolve from the semiclassical level (BTE) to the quantum one, with the master equation, non-equilibrium Green's function (NEGF) and Landauer–Büttiker approaches (Table 2.1). The main difference between these quantum approaches lies in the fact that the Landauer–Büttiker approach describes the ballistic transport only [IMR 99]. This so-called coherent transport occurs in the region of the junction drawn in black color in Figure 2.1. The treatment of electronic transport in the Landauer–Büttiker approach can be contrasted with the master equation [ROS 11, GEB 06] and NEGF approaches [DAT 97, BUO 09, LIA 11], which are able to describe both coherent and incoherent transports on the quantum level. Examples of existing methods are presented in Table 2.2, together with their main limitations. Each of these methods is presented in this chapter and listed in Table 2.2. The description of the quantum master equation, section 2.6.2, requires new definitions that make this section appear as highly specialized. It can be skipped at the time of the first reading of the chapter.

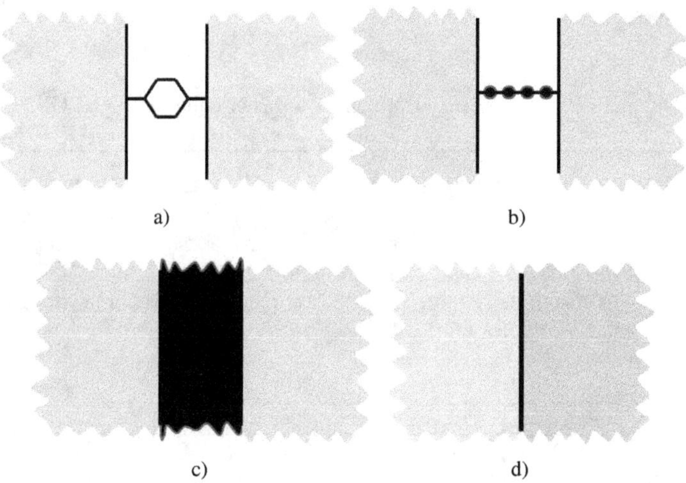

Figure 2.1. *Several examples of junctions (in black) on the nanometer scale, the electronic transport of which have been calculated with the help of density functional theory: a) 0D junction formed in a structure consisting of a single molecule sandwiched between two semi-infinite conducting leads; b) 1D junction formed by a monoatomic wire of metal atoms between two leads; c) 2D heterojunction formed by a layer of material acting as a barrier sandwiched between two leads; d) 2D heterojunction formed by the interface between two conducting materials. In all of the cases, the junction consists of the most resistive part of the structure and is the region where coherent transport may occur, drawn in black*

Two crucial ingredients necessary for the quantum methods are the electronic band structure and the electron–phonon interaction. The former is described in section 2.2 on the level of the DFT. Recent progress has allowed for the calculation of electron–phonon scattering without any fitting parameter, which is described in section 2.4. For each existing method, the approach used to compute the electron–phonon interaction is summarized in Table 2.2, as well as their main limitations.

This chapter is devoted to researchers and PhD students interested in the field of electronic transport properties computed from first principles, and requires an adequate knowledge in quantum mechanics. It is not intended to be an exhaustive or extensive review of today's state-of-the-art, and the authors apologize in advance to scientists whose contribution has been left out. Rather, the chapter has been designed to provide guidelines for the calculation of the electronic transport properties coupled with *ab initio* calculations. Whenever possible, we have chosen to cite review papers and text books related to the subject.

Level of Theory	Coherent	$\rightarrow$	Coherent + incoherent	
Semiclassical	-		Boltzmann transport equation (f)	
$\downarrow$			$\downarrow$	
Quantum level	Landauer-Büttiker (G)		Master equation (ρ_M)	NEGF ($G^<$)

Table 2.1. *Methods for the calculation of electronic transport: from semiclassical to quantum treatments (vertical axis) and from coherent transport to both coherent and incoherent transport (horizontal axis). In parentheses is indicated the main quantity that is obtained: f is the classical distribution function of the electrons, G is the quantum conductance, ρ_M is the density matrix and $G^<$ is the retarded Green function. Each of these methods has been combined with the density functional theory*

Method	Reference	Electron–phonon coupling	Section	Limits
BTE	[WAN 11]	Explicit and *ab initio*	2.5	No quantum effects
Landauer–Büttiker	[CHO 99]	Two chemical pot. in the leads	2.6.1	Ballistic and low bias
Master equation	[KAN 12]	Explicit and *ab initio*	2.6.2	No electron–electron scatt.
Master equation	[GEB 04a]	Model	2.6.2	Approximate scatt.
NEGF	[LIA 11]	Explicit	2.6.4	Approximate self-energy

Table 2.2. *Calculation of electronic transport with the use of density functional theory: name of the method that will be described in present chapter, one selected reference, treatment of the electron–phonon interaction used in the method, sections where the method is described, known present limitations of the method. In the Landauer–Büttiker approach, the scattering processes necessary to obtain the balance between the acceleration by the electric field is deported to the leads in contact with the molecule and modeled by two different chemical potentials in the two leads. BTE: Boltzmann transport equation; NEGF: non-equilibrium Green's function*

2.2. Electronic structure

In this section, we are going to introduce the notions that are necessary to calculate the ground state of the electronic system. In the absence of a time-dependent external field, the Schrödinger equation has stationary solutions: its solutions are wavefunctions that can form standing waves, and are well represented by the time-independent Schrödinger equation (section 2.2.1). The main hypothesis that underlies most of the calculations of these solutions is the adiabatic approximation, which allows for disentangling the ionic and electronic motions (section 2.2.2). The electronic part of the Hamiltonian can be treated with "exact" methods for systems with few electrons and in section 2.2.3, the principles underlying these treatments are described and the exponential scaling of these methods with the number of electrons is recalled, illustrating their limitation for large systems. The DFT gives up the exact treatment of all of the electronic positions, and instead, uses the electronic density as the fundamental variable. Its principles are developed in section 2.2.4. Finally, effects beyond the DFT will be presented in section 2.2.5, and conclusions will be drawn. In the following, we will treat ground and excited states as non-degenerate states.

2.2.1. *The time-independent Schrödinger equation*

A crystal or a molecule is made up of N ions and $\mathcal{N}$ electrons, and the system is characterized by $3N$ atomic and $3\mathcal{N}$ electronic degrees of freedom, labeled in the following, respectively, by $R_I = \{\mathbf{R}_1, ..., \mathbf{R}_N\}$ and $r_e = \{\mathbf{r}_1, ..., \mathbf{r}_{\mathcal{N}}\}$. Each atomic nucleus has a mass M_L and charge Z_L, and we define the atomic displacement as $\mathbf{u}_L = \mathbf{R}_L - \mathbf{R}_L^{eq}$, around the static equilibrium position $\mathbf{R}_L^{eq}$. The Hamiltonian of the system of nuclei and electrons is written in terms of the kinetic operator for the nuclei, T_I, for the electronic system, T_e, and of the potential energy operator $U(R_I, r_e)$, which consists of Coulombian interactions: the repulsive interatomic Coulombian potential (CP) U_I, the attractive ionic-electronic CP U_{Ie} and the repulsive interelectronic CP U_{ee}. In the following, the unit of energy is the Rydberg, $e^2 = 1, \hbar = 1, m = \frac{1}{2}$, spin degeneracy is supposed – indices for position and spin are condensed into r_e – and the Schrödinger equation of the many-body problem reads:

$$(H - E^m)\,\Psi^m(R_I, r_e) = 0, \qquad [2.2]$$

$$H = T_I + T_e + U(R_I, r_e), \qquad [2.3]$$

$$U(R_I, r_e) = U_I + U_{Ie} + U_{ee}$$

$$= \frac{1}{2}\sum_{LM}{}' \frac{Z_L Z_M}{|\mathbf{R}_L - \mathbf{R}_M|} - \sum_{iL} \frac{Z_L}{|\mathbf{r}_i - \mathbf{R}_L|} + \frac{1}{2}\sum_{ij}{}' \frac{1}{|\mathbf{r}_i - \mathbf{r}_j|}. \qquad [2.4]$$

In the sum, the prime ($'$) symbol stands for the exclusion of identical values of the indices; the symbol m is the index running over the number of electronic states.

2.2.2. *Born–Oppenheimer separation*

The first approximation consists in dissociating the quantum motion of atoms from the electronic one: $\Psi = \psi(R_I)\,\phi_e(R_I, r_e)$ [BOR 54]. The mass ratio between protons and electrons being 1,836, the electronic response to a change in the potential energy U_{Ie} is considered as instantaneous with respect to the slow ionic motion causing the variation of U_{Ie}. The electronic wavefunction ϕ_e^m of the mth electronic state is computed as if instantaneous positions of ions were fixed:

$$(H_e - E_e^m)\,\phi_e^m(r_e) = 0, \qquad [2.5]$$

$$H_e = T_e + U(R_I, r_e),\ fixed\,R_I. \qquad [2.6]$$

The second approximation is to treat the electronic response to a change in the potential U_{Ie} as adiabatic: the electronic wavefunction ϕ_e^m depends implicitly on the nuclear motion and is progressively distorted when the set R_I varies. The electron–ion coupling is supposed to be unable to provoke an electronic transition from the mth state to another state. In other words, during the ionic motion, the electronic system remains in its ground state, and explores the potential energy surface called the Born–Oppenheimer PES (BO-PES):

$$E_e(R_I) = E_e^{GS}(R_I) + E_I(R_I) = \langle\phi_e^{GS}|H_e|\phi_e^{GS}\rangle + E_I(R_I), \qquad [2.7]$$

where E_I reads

$$\frac{1}{2}\sum_{LM}{}' \frac{Z_L Z_M}{|\mathbf{R}_L - \mathbf{R}_M|}. \qquad [2.8]$$

From now on, the parametric dependence on R_I, of the quantities describing the electronic system, is going to be omitted in the notation until the end of the section. For a study of the effects beyond the adiabatic BO approximation, on the ion dynamics, the reader is referred to [ABE 10] and references therein.

2.2.3. *The many-electron problem*

2.2.3.1. *The variational method*

One is left with the Schrödinger equation [2.5] for the electronic system, in which r_e is the set of electronic positions, $r_e = \{\mathbf{r}_1, ..., \mathbf{r}_{\mathcal{N}}\}$. This forms an eigenvalue problem, in which the E_e^m are the eigenvalues and the ϕ_e^m are the eigenvectors.

If the Hamiltonian of equation [2.6] is bounded from below, a mathematical expression to say that its lowest eigenvalue cannot be infinite, which is the case for a Coulomb potential [KAT 51], then an upper bound to the lowest lying eigenvalue E_e of equation [2.7] can be found from the Rayleigh–Ritz variational method.

In this variational method, a guess of the form of the wavefunction ϕ_e is designed. The Ansatz wavefunction ϕ_e^a contains a number P of parameters p_i. The expectation value of the Hamiltonian in equation [2.7] is calculated with the Ansatz wavefunction and minimized with respect to all of the parameters p_i of the trial wavefunction:

$$E_e^0(R_I) = min_{\phi_e^a} \langle \phi_e^a | H_e | \phi_e^a \rangle. \qquad [2.9]$$

To solve equation [2.9] for small systems, the "gold standard" in quantum chemistry is the method called "coupled cluster method with triple excitations" (CCSD(T)) where the Ansatz wavefunction has the form [BAR 07]:

$$\phi_e^a = \mathrm{e}^{\hat{T}(p_i)} \Phi_0, \qquad [2.10]$$

where Φ_0 is typically a Slater determinant of Hartree–Fock wavefunctions [HED 69] and $\hat{T}$ is the cluster operator whose action on Φ_0 yields linear combinations of Slater determinants. Action of the operator $\hat{T}$ improves the description of electronic correlations with respect to the single-determinant approximation.

Alternatively, the energy of equation [2.7] can be calculated with the variational and diffusion quantum Monte Carlo methods. In variational Monte Carlo, the trial wavefunction has the form

$$\phi_e^a = \mathrm{e}^{J(p_i)} \Phi_0, \qquad [2.11]$$

where Φ_0 is a determinant of single-particle wavefunctions, either coming from Hartree–Fock [HED 69], or from the (below-described) DFT, and J is the Jastrow factor containing isotropic electron–electron, electron–nucleus and electron–electron–nucleus terms [NEE 10]. In general, the use of the variational Monte Carlo method requires a preliminary knowledge about the system. When this is not the case, it is supplemented by diffusion Monte Carlo (DMC) calculations [NEE 10]. For more details the reader is referred to the textbook [MAR 16a].

2.2.3.2. *Use of symmetry*

In the search for the minimum of the expectation value of the Hamiltonian, the number of parameters of the trial wavefunction is reduced by the use of symmetry. The electronic Hamiltonian of equation [2.6] has the symmetry of the system under consideration: discrete rotations and reflections defining the point group of a molecule, discrete translation symmetry in the n directions of a n-dimensional crystal, which are eventually combined to the symmetry operations of the crystal point group to form the space group [VIT 96]. Both the solution ϕ_e and the designed Ansatz wavefunction ϕ_e^a are invariant under all of the symmetry operations of the molecular point group or crystal space group.

In addition, the Pauli principle requires that

$$P_{jj'}\phi_e^m(r_e) = -\phi_e^m(r_e) \qquad [2.12]$$

where the operator $P_{jj'}$ permutes the space and spin coordinates of electrons j and j' [KOH 99]. In simpler words, indiscernible fermions cannot occupy the same quantum state and must, for instance, have opposite spins.

2.2.3.3. *Many electrons: "encountering an exponential wall"*

Notwithstanding the reduction in size of our problem with the help of symmetry, the number of parameters depends exponentially on the number of electrons in our systems, making the minimization of the energy with respect to all of the parameters rapidly intractable as the number of particles increases. An estimate of the number of parameters is given in [KOH 99]. When the wavefunction depends on a set of $3\mathcal{N}$ continuous variables r_i, and if p is the number of parameters needed to describe the dependence of ϕ_e^a on one single r_i within a given accuracy $\mathcal{O}\ (j)$, a rough estimate of the total number P of parameters is:

$$P = p^{3\mathcal{N}}, \qquad [2.13]$$

or

$$\mathcal{N} = \frac{1}{3}\frac{ln(P)}{ln(p)}. \qquad [2.14]$$

Let us suppose that the energy minimization can be performed for a space of one billion parameters, P = 10^9, a formidable task even for today's computers. If we suppose that we need a minimum p = 3 parameters per r_i, as we have to minimize with respect of all of the parameters, then the exponential of equation [2.13] yields a maximum number of electrons that can be treated with the variational method that turns out to be small: $\mathcal{N}$ = 6 only [KOH 99]. This is the reason why alternative methods to treat large systems are required: the DFT [KOH 99] has proved to be the most successful one.

2.2.4. *The DFT*

2.2.4.1. *Introduction*

DFT is a very powerful mean-field theory to tackle the many-body problem raised by equation [2.5] without introducing empiricism [KOH 99, JON 89, PAR 89, DRE 90, GRO 95, SEM 96, DOB 98]. When the large number of degrees of freedom prevents the calculation with "exact" quantum methods like CCSD(T) or DMC, the use of the ground-state electronic density $\rho_e(\mathbf{r})$ as the basic variable makes the problem depend on only three or four variables: the Cartesian components (r_x, r_y, r_z) of the $\mathbf{r}$ vector, plus eventually the spin (which we will neglect in the following). It is formally expressed from the many-body wavefunction by integrating out most of the electronic degrees of freedom:

$$\rho_e(\mathbf{r}) = N \int \phi_e^*(\mathbf{r}, \mathbf{r}_2..., \mathbf{r}_{\mathcal{N}}) \phi_e(\mathbf{r}, \mathbf{r}_2, ..., \mathbf{r}_{\mathcal{N}}) d\mathbf{r}_2 d\mathbf{r}_3...d\mathbf{r}_{\mathcal{N}}. \quad [2.15]$$

There are many possibilities to calculate the integral in equation [2.15] and the expression employed in the DFT is based on the approximation of non-interacting electrons, where the many-body wavefunction is reduced to: $\phi_e(\mathbf{r}, \mathbf{r}_2..., \mathbf{r}_{\mathcal{N}}) = \prod_i \varphi_i(\mathbf{r}_i)$ (see also equation [2.25] below).

DFT simplifies the search for the ground state $\phi_e(\mathbf{r}_1, ..., \mathbf{r}_{\mathcal{N}})$ of a system of $\mathcal{N}$ interacting electrons defined by equation [2.5]. Instead, the problem is reduced to non-interacting electrons moving in an effective potential v_{eff}, a kind of mean field, and many-body interactions are accounted for via the electronic density $\rho(\mathbf{r})$ and the functional dependence of the effective potential on the density, $v_{eff}[\rho(\mathbf{r})]$ [SCH 95].

The kinetic energy operator T_e is taken to be the expression for non-interacting (ni) electrons ($\frac{\hbar^2}{2m} = 1$),

$$T_e^{ni} = -\nabla^2, \quad [2.16]$$

which is also used in the Hartree–Fock approach [HED 69], yielding the equation:

$$[-\nabla^2 + v_{eff}(\mathbf{r}) - \varepsilon^m]\varphi^m(\mathbf{r}) = 0. \quad [2.17]$$

The Kohn–Sham equation [2.17] is analogous to the Schrödinger equation for one single particle. It also forms an eigenvalue problem, in which $H^{KS} = -\nabla^2 + v_{eff}[\rho]$ is the Kohn–Sham Hamiltonian to be diagonalized, and the ε^m and φ^m are the eigenvalues and eigenvectors. However, we are going to show that the eigenvalues are not the energy values of the particle (section 2.2.4.4), at variance with the E_e^m of the many-body Schrödinger equation [2.5], in which, for instance, E_e^0 is the total energy of the ground state and E_e^1 is the total energy of the first excited state.

2.2.4.2. *Total energy*

Equation [2.17] is based on two theorems formulated by Hohenberg and Kohn [HOH 64] and Kohn and Sham [KOH 65].

To express the total energy in DFT, let us first consider the ground state of $\mathcal{N}$ interacting electrons in an external potential v_{ext}, which represents the electron–ion Coulomb interaction for an electron at the $\mathbf{r}$ point (see U_{Ie} in equation [2.4]):

$$v_{ext}(\mathbf{r}) = \sum_L \frac{Z_L}{|\mathbf{r} - \mathbf{R}_L|}. \qquad [2.18]$$

A quantity F is defined as

$$F = \langle \phi_e | T_e + U_{ee} | \phi_e \rangle, \qquad [2.19]$$

in which T_e and U_{ee} are the many-body kinetic energy operator and electron–electron Coulomb interaction of equation [2.4].

The expression of the total energy of the electronic systems depends on the external potential and reads:

$$E_e[v_{ext}] = \int d\mathbf{r}\, v_{ext}(\mathbf{r})\, \rho(\mathbf{r}) + F, \qquad [2.20]$$

where the integral in equation [2.20] runs over the volume of the system.

2.2.4.3. *Hohenberg–Kohn theorem*

The first theorem yields a positive answer to the following question: does the knowledge, at each $\mathbf{r}$ point of space, of the electronic ground-state density $\rho_{e,GS}$ uniquely define the external potential v_{ext} from which $\rho_{e,GS}$ arises? Hohenberg and Kohn proved that this is exact. As moreover, the knowledge of v_{ext} completely defines the Hamiltonian of the many-body electronic system (equation [2.6]), one consequence of the theorem is that the knowledge of $\rho_{e,GS}$ determines all of the ground-state properties: the wavefunction $\phi_e(\mathbf{r}_1, ..., \mathbf{r}_\mathcal{N})$ and the electronic energy E_e are unique functionals of $\rho_{e,GS}$, $\phi_e[\rho_{e,GS}]$ and $E_e[\rho_{e,GS}]$. Moreover, the quantity F defined by equation [2.19] has also a functional dependence on the ground-state density $F[\rho_e]$, which is universal, i.e. independent from the system under study, and from the external potential.

The functional form of the total energy as a function of ρ_e then reads:

$$E_e^{DFT}[\rho_e] = \int d\mathbf{r}\, v_{ext}(\mathbf{r})\, \rho_e(\mathbf{r}) + F[\rho_e]. \qquad [2.21]$$

The application of the Rayleigh–Ritz principle can then be performed not with respect to a trial wavefunction ϕ_e^a as in section 2.2.3, but w.r.t. a trial density ρ [HOH 64]. When the density is equal to the ground-state density at each point $\mathbf{r}$, $E_e^{DFT}[\rho]$ has a minimum, which is unique, and $E_e^{DFT}[\rho_{GS}] = E_e[\rho_{e,GS}]$. During the search for the minimum, the constraint is that the total number of electrons is constant:

$$\int \rho(\mathbf{r})\, dr = \mathcal{N}. \qquad [2.22]$$

Referring to the calculus of variations [BYR 69, KÜB 92], an extended Lagrangian is constructed that takes into account the expression of the constraint of equation [2.22]:

$$\mathcal{L} = \int d\mathbf{r}\, v_{ext}(\mathbf{r})\, \rho(\mathbf{r}) + F[\rho] + \mu(\int \rho(\mathbf{r})\, dr - n), \qquad [2.23]$$

in which μ is the Lagrange multiplier associated with the constraint of conserving the number of electrons.

The Euler–Lagrange equation $\frac{\partial \mathcal{L}}{\partial \rho} = 0$ associated with the variational principle is then calculated [LEV 79, KÜB 92, KOH 99], yielding

$$v_{ext}(\mathbf{r}) + [\frac{\partial F(\rho)}{\partial \rho}]_{\rho=\rho_{GS}} = \mu. \qquad [2.24]$$

Physically, μ is a chemical potential, or, at zero temperature, the Fermi level energy. Equation [2.24] is formally exact, and together with the knowledge of the exact functional form of F, would enable us to compute the exact ground-state density. In practice, an additional approximation is needed, as we do not know F, to perform specific applications and calculate the total energy of atoms, molecules, crystals and disordered systems.

2.2.4.4. *Kohn–Sham theorem*

The electronic kinetic energy is not expressed satisfactorily with functionals of the density, like the Thomas–Fermi or von Weizsäcker expressions [PAR 89, WEI 35]. Instead, Kohn and Sham have used the orbital-dependent expression for non-interacting (ni) electrons (equation [2.16]) and have expressed the electronic density as a sum over those monoelectronic states:

$$\rho(\mathbf{r}) = \rho^{ni}(\mathbf{r}) = 2 \sum_{m=1}^{\frac{\mathcal{N}}{2}} \varphi^{m^*}(\mathbf{r}) \varphi^m(\mathbf{r}), \qquad [2.25]$$

where the sum runs over one-electron states with $\varepsilon^m < \varepsilon_F$, ε_F being the Fermi energy, and the factor of 2 stands for the spin degeneracy. The constraint of equation [2.22] is changed to a normalization condition of each single-particle wavefunction,

$$\int \varphi^{m^*}(\mathbf{r})\varphi^m(\mathbf{r})\, dr = 1, \quad \forall m, \qquad [2.26]$$

and each of them is associated with a Euler–Lagrange parameter ε^m, so that the extended Lagrangian reads

$$\mathcal{L} = \int d\mathbf{r}\, v_{ext}(\mathbf{r})\, \rho(\mathbf{r}) + F[\rho] + \sum_m [\varepsilon^m (\int \varphi^{m^*}(\mathbf{r})\varphi^m(\mathbf{r})\, d\mathbf{r} - 1)]. \qquad [2.27]$$

The variational principle comes in terms of the single-particle wavefunction, and corresponding Euler–Lagrange equations read

$$\frac{\partial \mathcal{L}}{\partial \varphi^{m^*}} = 0, \quad \forall m, \qquad [2.28]$$

yielding the Kohn–Sham equation [2.17]. We point out that in above formalism, ε^m are Euler–Lagrange parameters and not the values of the exact excitation energy. Nevertheless, it has been shown that the uppermost occupied DFT eigenvalue equals the ionization potential [ALM 85].

2.2.4.5. *Effective potential and self-consistent solution*

Kohn–Sham equation [2.17] has the structure of a single-particle Schrödinger equation, in which the external potential, which is the Coulomb interaction between the electronic system and the ions (equation [2.18]), is supplemented by a screening potential $\partial F[\rho]/\partial \varphi^{m^*}$:

$$v_{ext}(\mathbf{r}) + \frac{\partial F[\rho]}{\partial \varphi^{m^*}}. \qquad [2.29]$$

We have now to express $F[\rho]$, which is not a simple task. The classical electrostatic potential v_H coming from a charge distribution ρ is given by [JAC 99]

$$v_H(\mathbf{r}) = \int \frac{\rho(\mathbf{r}')}{|\mathbf{r} - \mathbf{r}'|}\, d\mathbf{r}', \qquad [2.30]$$

and we suppose that this is the main contribution – the Hartree term – in our electron–electron interaction. The corresponding contribution to the total energy,

$$E_H[\rho] = \frac{1}{2} \int \rho(\mathbf{r}) v_H(\mathbf{r})\, d\mathbf{r}, \qquad [2.31]$$

is a functional of ρ, and we can define the remaining unknown part as

$$E_{xc}[\rho(\mathbf{r})] = F[\rho] - E_H[\rho]. \qquad [2.32]$$

The exchange and correlation functional E_{xc} contains all of the quantum many-body effects, plus the interacting part of the kinetic energy $T_e - T_e^{ni}$, and requires an approximation that is a functional of ρ.

Finally, the functional form of F reads:

$$F[\rho] = -\sum_{m=1}^{\frac{\mathcal{N}}{2}} \int d\mathbf{r} \varphi^{m^*}(\mathbf{r}) \nabla_{\mathbf{r}}^2 \varphi^m(\mathbf{r}) + \int d\mathbf{r} \int d\mathbf{r}' \frac{\rho(\mathbf{r})\rho(\mathbf{r}')}{|\mathbf{r}-\mathbf{r}'|} + E_{xc}[\rho(\mathbf{r})], \qquad [2.33]$$

and the effective potential reads

$$v_{eff}(\mathbf{r}) = v_{ext}(\mathbf{r}) + v_H(\mathbf{r}) + v_{xc}[\rho(\mathbf{r})], \qquad [2.34]$$

where $v_{xc} = \partial E_{xc}/\partial \rho$ is the exchange and correlation potential.

As the effective potential depends on the density through the Hartree and exchange and correlation terms, and as the density is itself obtained from the solution of equation [2.17] substituted into equation [2.25]. [2.17], [2.34], and [2.25] form a set of equations that must be solved self-consistently. The ground-state density is obtained once the input density and the output density are equal with an *a priori* given tolerance threshold, which fixes the point at which self-consistency is reached. Chemical and physical properties can then be computed with ρ_{GS}.

An approximation of E_{xc} is mandatory to use the DFT in practical applications. The success of DFT has relied on the ability of the local density approximation (LDA) in reproducing the physical properties of atoms, molecules and solids [JON 89]. In the LDA, the exchange and correlation energy at the point $\mathbf{r}$, where the density is $\rho(\mathbf{r})$, is approximated by the exchange and correlation energy of a uniform electron gas of the same density [CEP 80, PER 81],

$$E_{xc}[\rho(\mathbf{r})] = \int \rho(\mathbf{r})\epsilon_{xc}[\rho(\mathbf{r})]\, d\mathbf{r}. \qquad [2.35]$$

Inclusion of gradient corrections in equation [2.35], $E_{xc}[\rho, \nabla\rho]$ as in the generalized gradient approximation (GGA) [LAN 83, LAN 84, PER 03] yields improvements particularly on the cohesive energy, and is nowadays favored in the literature because error bars with respect to experiments are reduced in many cases. Yet, LDA and GGA contain the same physical ingredients, and a result that can be explained by a calculation performed in GGA and not in LDA must be taken with extreme caution.

Finally, as the Coulomb interaction between the nuclei U_I is independent both from the electronic coordinates (equation [2.4]), and from the electronic density, its contribution to the electronic energy is the constant E_I defined in equation [2.8], and the total energy of the electronic systems reads:

$$E_{TOT}^{KS}[\rho, R_I] = \int d\mathbf{r}\, v_{ext}(\mathbf{r})\,\rho(\mathbf{r}) + F[\rho] + E_I. \qquad [2.36]$$

It can also be expressed in terms of Euler–Lagrange parameters ε_m, i.e. the Kohn–Sham eigenvalues of equation [2.17] [KÜB 92, IHM 79]

$$E_{TOT}^{KS}[\rho, R_I] = \sum_{m=1}^{\frac{\mathcal{N}}{2}} \varepsilon^m - \frac{1}{2}\int d\mathbf{r}\, v_H(\mathbf{r})\,\rho(\mathbf{r}) + \int d\mathbf{r}\,(\epsilon_{xc} - v_{xc})v_{ext}(\mathbf{r})\,\rho(\mathbf{r}) + E_I. \qquad [2.37]$$

2.2.5. *Conclusions and remarks about and beyond DFT*

As many physical or chemical quantities mainly depend on the electronic density, the DFT is a convenient way to estimate the magnitude of a wealth of physical properties, as well as their behavior under an external parameter such as pressure or temperature, yet at the price of a built-in inaccuracy: increasing the accuracy and convergence of all of the quantities involved in the calculation does not lead to improved estimations. One way to understand it is that, the density being the integrated quantity of equation [2.15], the reduction in the number of degrees of freedom is paid by renouncing to express the correlations between them that require non-locality. For instance, effects of correlation that depend on two space variables, **r** and **r**′ cannot be expressed through a mean-field term like the Hartree potential, nor through the exchange and correlation potential, as long as the latter depends on one single space variable **r**. Theoretical developments to express correlation functions as functions of density and of its gradients are in progress [LEE 98].

When LDA or GGA are used, calculations can be considered as parameter free. When those approximations fail in reproducing the ground-state density however, many "meta" exchange and correlation functionals must be tested, as well as methods beyond DFT like the DFT + U model, in which (briefly said) the U parameter is a parameter aimed at treating intraatomic correlations. In transition metal oxides for instance, where the above-mentioned electronic correlations are significant (section 2.2.4.1), a hierarchy of exchange and correlation functionals can be tested [YAN 13]. This reintroduces a large part of empiricism in the calculations, and confirms the need of theoretical approaches beyond the DFT [ONI 02, GEO 96, MAR 16a].

Another limitation of DFT-LDA and DFT-GGA is that, in semiconductors and insulators, when the bandgap is evaluated as the difference between the highest occupied eigenvalue and the lowest unoccupied one, it is underestimated by around 50% [AUL 99]. The reason is that Kohn–Sham eigenvalues are not excitation energies – a statement previously made in section 2.2.4.1 – and Koopman's theorem [HED 69] is not fulfilled. The difference between the bandgap measured in the experiments, and the DFT-KS one has two origins: (1) the exchange and correlation potential must be discontinuous when the number of particles in the electronic system is changed from $\mathcal{N}$ to $\mathcal{N} \pm 1$ during an excitation [GRU 06]; (2) the inaccuracy of LDA and GGA. While the latter can be corrected by including a part of exact exchange energy in the functional, the former problem can only be solved through a proper evaluation of long-range correlations, as performed for instance with the GW method [HED 69].

Last but not least, neither the DFT nor the GW method are able to describe the electron–hole interaction present in the absorption spectra of materials. An effective Bethe–Salpeter Hamiltonian must be solved to account for excitonic effects, and a review can be found in [ONI 02].

Coupled to the LDA or to the GGA for the exchange and correlation functional, DFT is able to predict ground-state properties such as geometries (lattice parameters, bond lengths and bond angles) within 1% [KOH 99], and phonon frequencies within a few percent [BAR 01]. Mechanical properties such as bulk modulus and elastic constants are predicted within 5% [GOL 13], and electron–phonon coupling constants [SJA 06] within 10% with respect to experiments. Cohesive energies are usually more accurately predicted in DFT-GGA than in LDA [KOH 99].

DFT has been implemented in a flavor of methods [KÜB 92, AND 75]. The all-electrons full-potential linearized plane wave method offers the best accuracy for band structure calculations, and represents the state-of-the-art reference to compare results obtained with other methods [SCH 02, BLU 06]. The pseudopotential approach, together with the expansion of Kohn–Sham orbitals onto a plane wave basis set, allows for treating only valence electrons, and offers a good compromise between accuracy and computational load [IHM 79, BAC 82, PIC 89]. It is implemented in many packages [GIA 09, SUN 03, HAF 08, GON 09, SOL 02, SHI 00] (to name but a few). To treat very large systems, methods that scale linearly with the number of atoms (order-$\mathcal{N}$ methods) are needed. Current implementations involve basis sets of atomic orbitals [SOL 02, BLU 09], or wavelets on adaptative grids [GEN 08].

2.3. Phonons

In this part, we are going to introduce the notions that are necessary to the calculation of the vibrational properties of materials [BOR 54, BAR 01], starting from the Schrödinger equation for the ions (section 2.3.1). The second hypothesis that comes after the adiabatic approximation is the harmonic approximation for the ion dynamics (section 2.3.2). The definition of the forces that the electronic system exerts on the ions – first-order derivatives of the total energy with respect to an atomic displacement – and the matrix of force constants (second-order derivatives) are going to be presented. The Hellmann–Feynman theorem will be used to calculate these derivatives (section 2.3.3), yielding expressions that will be evaluated within the linear response theory (section 2.3.4). Turning to practical phonon calculations, DFT will be used in conjunction with the frozen-phonon method (section 2.3.5), and then DFPT will be presented (section 2.3.6). Finally, anharmonic effects will be discussed in section 2.3.7.

2.3.1. *Schrödinger equation for the ions*

In the adiabatic approximation (section 2.2.2), the vibrational properties of the ions are determined by the eigenvalues $\mathcal{E}^\nu$ and eigenfunctions $\psi^\nu(R_I)$ of the Schrödinger equation for the ions:

$$(H_I - \mathcal{E}^\nu)\,\psi^\nu(R_I) = 0, \qquad [2.38]$$

$$H_I = T_I + E_e(R_I), \qquad [2.39]$$

where the potential felt by the ions is the Born-Oppenheimer potential energy surface (BO-PES) $E_e(R_I)$ given by equation [2.7].

In the adiabatic approximation, the displacements of the atoms characterizing a phonon act as a static perturbation on the electronic system. The first requirement is the fulfillment of the condition of mechanical equilibrium, following which the force acting on each ion vanishes. Equilibrium geometries, and in particular the set of equilibrium positions R_I^{eq} of the ions, are obtained from:

$$\mathbf{F_L} = -\nabla_{\mathbf{R_L}} E_e = \mathbf{0}, \qquad [2.40]$$

where the symbols L and M in the equation [2.41], condense the position of the unit cell and the position of the atom in the unit cell.

2.3.2. *The harmonic approximation*

The harmonic approximation consists in calculating the vibrational frequencies ω and the corresponding atomic displacement patterns ξ as, respectively, the eigenvalues and eigenvectors of the Hessian, or matrix of force constants $\mathcal{C}$, of the BO-PES, neglecting the anharmonic terms (generally given by third-order and quartic derivatives of the BO-PES), as well as the terms of higher orders:

$$\mathcal{C}_{LM\alpha\beta} = \frac{1}{\sqrt{M_L M_M}} \frac{\partial^2 E_e}{\partial R_{L\,\alpha} \partial R_{M\,\beta}}, \qquad [2.41]$$

$$(\mathcal{C} - \mathcal{I}\omega^2)|\xi\rangle = 0, \qquad [2.42]$$

where $\mathcal{I}$ is the identity matrix, ξ yields the atomic displacement patterns, andα and β are indices of Cartesian coordinates.

2.3.3. *The Hellmann–Feynman theorem*

The Hellmann–Feynman theorem enables us to calculate the first-order derivative of the BO-PES in equation [2.40] and to simplify the second-order derivative of equation [2.41]. Writing $E_e = \langle\phi_e^0|E_e|\phi_e^0\rangle$, and noting that the norm of the many-body wavefunction is constant, we have the expression $\langle\partial\phi_e^0|\phi_e^0\rangle = -\langle\phi_e^0|\partial\phi_e^0\rangle$, so that the first-order derivative of the BO-PES is the expectation value of the first-order derivative of the Hamiltonian [HEL 37, FEY 39].

$$F_{L\,\alpha} = -\langle\phi_e^0|\frac{\partial H_e}{\partial R_{L\,\alpha}}|\phi_e^0\rangle, \qquad [2.43]$$

$$\mathcal{C}_{LM\alpha\beta} = \frac{1}{\sqrt{M_L M_M}}\langle\phi_e^0|\frac{\partial^2 H_e}{\partial R_{L\,\alpha}\partial R_{M\,\beta}}|\phi_e^0\rangle + (\langle\frac{\partial\phi_e^0}{\partial R_{L\,\beta}}|\frac{\partial H_e}{\partial R_{L\,\alpha}}|\phi_e^0\rangle + c.c.), \qquad [2.44]$$

where *c.c.* stands for "complex conjugate expression".

2.3.4. *The linear response*

Thus, when nuclei are displaced from their equilibrium positions, the variation of the BO surface has several origins [BAL 83], which are obtained from equation [2.4]: the Coulombian interaction between the nuclei, whose second-order variation yields the contribution $\mathcal{C}^{\mathcal{I}}$ to the matrix of force constants; the electron–ion interaction, U_{Ie}, which, to first order, gives rise to the electron–phonon coupling constants described in section 2.4, and, to second order, yields a contribution $\mathcal{C}^{e}$ to the force-constant matrix.

Noting that the energy of the electron–ion interaction reads $E_{Ie} = \int U_{Ie}(\mathbf{r})\rho(\mathbf{r})d\mathbf{r}$, where the many-body electronic density ρ is given by equation [2.15], and depends on the Cartesian coordinate α of the position of the atom L (indexing the atom in the crystal) both explicitly through U_{Ie}, and implicitly through ρ, we find that [PIC 70, CIC 69]

$$F_{L\,\alpha} = -\int \rho(\mathbf{r}, \mathbf{R_I}) \frac{\partial U_{Ie}}{\partial R_{L\,\alpha}} d\mathbf{r}, \qquad [2.45]$$

$$\begin{aligned} \mathcal{C}^{e}_{LM\alpha\beta} &= -\frac{\partial F_{L\,\alpha}}{\partial R_{M\,\beta}} \\ &= \frac{1}{\sqrt{M_L M_M}} \int \Big(\frac{\partial \rho}{\partial R_{M\,\beta}} \frac{\partial U_{Ie}}{\partial R_{L\,\alpha}} + \rho \frac{\partial^2 U_{Ie}}{\partial R_{L\,\alpha} \partial R_{M\,\beta}} d\mathbf{r}\Big). \end{aligned} \qquad [2.46]$$

The second-order derivative of the many-body energy with respect to atomic displacements requires the knowledge of the ground-state density, and of its *linear response* to the same perturbation. When simplifying the many-body problem with the use of the DFT, then the electronic density is calculated in DFT (equation [2.25]), and the first-order derivative of the electronic density in DFPT (equation [2.49]).

2.3.5. *The frozen-phonon method*

Contrarily to the linear response, the frozen-phonon method does not require, in principle, perturbation theory, and only the adiabatic approximation is invoked (section 2.2.2). It consists of imposing finite and periodic displacements u_I of atoms with respect to their equilibrium position, "frozen phonons", in a crystal described by a supercell whose length is a multiple of the wavelength of the phonon under consideration, and in comparing the total energy $E_e(R_I)$ with and without atomic displacements.

DFT has been employed together with the frozen-phonon approach in early calculations [KUN 83]. The vibrational frequency is extracted by a fit of the total energy curve $E^{KS}_{TOT}[\rho, \mathbf{R}^{\mathrm{eq}}_{\mathbf{I}} + \mathbf{u_I}]$ as a function of the displacement $\mathbf{u_I}$

(equation [2.36]). Harmonic frequencies are obtained through a fit with a polynomial of second order, but the frozen method allows for estimating the magnitude of the anharmonic terms [KUN 83]. Its use is, however, limited by two factors: atomic displacement patterns of the phonon must be known *a priori*, and the phonon wavelength must be commensurate with the size of the supercell. These limitations are not present in the DFPT.

Finally, we note that in the early times, obtaining the self-consistency of Kohn–Sham equations for each displacement surpassed the computer capability, and total energy changes were approximated as the change in the one-electron eigenvalues $\Delta(\sum \varepsilon^m)$, with the help of the force theorem [CHR 84, MET 93].

2.3.6. *Density functional perturbation theory*

Applying the Hellmann–Feynman theorem and the linear response to the total energy of DFT equation [2.36], expressions for the force and for the second-order derivatives of the total energy read:

$$F_{L\,\alpha} = -\int \rho(\mathbf{r}) \frac{\partial v_{ext}(\mathbf{r})}{\partial R_{L\,\alpha}} d\mathbf{r}, \qquad [2.47]$$

$$C_{LM\alpha\beta} = \frac{1}{\sqrt{M_L M_M}} \int \rho(\mathbf{r}) \frac{\partial^2 v_{ext}(\mathbf{r})}{\partial R_{L\,\alpha} \partial R_{M\,\beta}} + \int \frac{\partial \rho(\mathbf{r})}{\partial R_{L\,\alpha}} \frac{\partial v_{ext}(\mathbf{r})}{\partial R_{M\,\beta}} d\mathbf{r}. \qquad [2.48]$$

In the pseudopotential + plane wave approach [IHM 79, BAC 82, PIC 89], analytical expressions exist for the derivatives of the electron–ion Coulomb potential $\frac{\partial v_{ext}}{\partial R_{L\,\alpha}}$ and $\frac{\partial^2 v_{ext}}{\partial R_{L\,\alpha} \partial R_{M\,\beta}}$ [BAR 01]. The calculation of the first-order derivative of the electronic density requires solving linearized Kohn–Sham equations.

2.3.6.1. *Linearization of Kohn–Sham equations*

Equations [2.17], [2.25] and [2.34] are linearized to first order in ΔR_L [BAR 01, GON 95]. The full variation of the electronic density reads:

$$\Delta\rho(\mathbf{r}) = \sum_{L\,\alpha} \frac{\partial \rho(\mathbf{r})}{\partial R_{L\,\alpha}} \Delta R_{L\,\alpha} = 4Re \sum_{m=1}^{\frac{\mathcal{N}}{2}} \varphi^{m^*}(\mathbf{r}) \Delta\varphi^m(\mathbf{r}), \qquad [2.49]$$

in which the electronic density response $\frac{\partial \rho(\mathbf{r})}{\partial R_{L\,\alpha}}$ is the quantity needed in equation [2.48].

The variation of Kohn–Sham orbitals $\Delta\varphi^m(\mathbf{r})$ is obtained by standard perturbation theory [BAR 01]:

$$[-\nabla^2 + v_{eff}(\mathbf{r}) - \varepsilon^m] \Delta\varphi^m(\mathbf{r}) = -(\Delta v_{eff}(\mathbf{r}) - \Delta\varepsilon^m)\varphi^m(\mathbf{r}), \qquad [2.50]$$

where

$$\Delta v_{eff}(\mathbf{r}) = \Delta v_{ext}(\mathbf{r}) + \Delta v_H(\mathbf{r}) + \Delta v_{xc}[\rho(\mathbf{r})], \qquad [2.51]$$

is the first-order correction to the effective potential, and $\Delta\varepsilon^m$ is the first-order variation of the Kohn–Sham eigenvalue,

$$\Delta\varepsilon^m = \int \varphi^{m^*}(\mathbf{r})\,\Delta v_{eff}(\mathbf{r})\,\varphi^m(\mathbf{r}). \qquad [2.52]$$

Applying the projector $\hat{P}_c = \hat{1} - \sum_v |\varphi_v\rangle\langle\varphi_v|$ on the left-hand side of each member of equation [2.50], and as $\langle\varphi_v|[-\nabla^2 + v_{eff}(\mathbf{r})] = \langle\varphi_v|\varepsilon_v$, we are left with:

$$[-\nabla^2 + v_{eff}(\mathbf{r}) - \varepsilon^m]\Delta\varphi^m(\mathbf{r}) = -\hat{P}_c\Delta v_{eff}(\mathbf{r})\varphi^m(\mathbf{r}), \qquad [2.53]$$

The structure of the set of equations [2.49], [2.51] and [2.53] – or, equivalently, equations [2.57], [2.59] and [2.60] – is analogous to the set of Kohn–Sham equations [2.17], [2.25] and [2.34], but for the presence of a right-hand side in equation [2.50]. It is no longer an eigenvalue problem, and requires algorithms for the solution of a linear system. These algorithms need also to be solved self-consistently.

From equation 29 in [BAR 01], we can show that if the perturbing potential of equation [2.51] acts only on the set of occupied states, no response of the charge density will occur. In order to have a non-vanishing variation of the electronic density in equation [2.49], the perturbing potential must couple occupied states to unoccupied ones [BAR 01, GON 95]. This can be shown by inspection of the left-hand side of equation [2.53]: it is singular because the determinant of $-\nabla^2 + v_{eff}(\mathbf{r}) - \varepsilon^m$ vanishes, as ε^m is one of the eigenvalues. Thus, the variation of the wavefunction is determined up to a multiple of $\varphi^m(\mathbf{r})$, which is written as [GIA 91]:

$$[-\nabla^2 + v_{eff}(\mathbf{r}) + \alpha\hat{P}_v - \varepsilon^m]\Delta\varphi^m(\mathbf{r}) = -\hat{P}_c\Delta v_{eff}(\mathbf{r})\varphi^m(\mathbf{r}), \qquad [2.54]$$

where α is a constant and $\hat{P}_v = \sum_v |\varphi_v\rangle\langle\varphi_v|$ is the projector onto the valence states, $\hat{P}_v + \hat{P}_c = \hat{1}$.

2.3.6.2. *DFPT on lattice-periodic functions*

The perturbing potential is a real functional and contains components of various wavelengths. It reads

$$\Delta v_{eff}(\mathbf{r}) = \sum_{\mathbf{q}} \Delta v^{\mathbf{q}}_{eff}(\mathbf{r}) = \sum_{\mathbf{q}} e^{i\mathbf{q}.\mathbf{r}}\delta v^{\mathbf{q}}_{eff}(\mathbf{r}), \qquad [2.55]$$

where $\Delta v_{eff}^{\mathbf{q}}(\mathbf{r})$ is the Bloch component of the perturbing potential, and $\delta v_{eff}^{\mathbf{q}}$ is the lattice periodic part:

$$\delta v_{eff}^{\mathbf{q}}(\mathbf{r}+\mathcal{R}) = \delta v_{eff}^{\mathbf{q}}(\mathbf{r}), \qquad [2.56]$$

where $\mathcal{R}$ is a Bravais lattice vector. Its consists of the variation of the external and the screening potential:

$$\delta v_{eff}^{\mathbf{q}}(\mathbf{r}) = \delta v_{ext}^{\mathbf{q}}(\mathbf{r}) + \delta^{\mathbf{q}} v_H(\mathbf{r}) + \delta v_{xc}^{\mathbf{q}}[\delta\rho^{\mathbf{q}}(\mathbf{r})]. \qquad [2.57]$$

We note that $\delta v_{eff}^{\mathbf{q}}$ is a complex function, and so is the charge density variation $\delta\rho^{\mathbf{q}}$ in response to $\delta v_{eff}^{\mathbf{q}}$.

In a linear response, responses to perturbations of different wavelengths can be decoupled [PIC 70], and equation [2.49] reads

$$\Delta\rho(\mathbf{r}) = \sum_{\mathbf{q}} \Delta\rho^{\mathbf{q}}(\mathbf{r}) = \sum_{\mathbf{q}} e^{i\mathbf{q}.\mathbf{r}}\delta\rho^{\mathbf{q}}(\mathbf{r}), \qquad [2.58]$$

where each component of the charge density response reads:

$$\delta\rho^{\mathbf{q}}(\mathbf{r}) = 4\sum_{\mathbf{k}} u_v^{\mathbf{k}^*}(\mathbf{r})\,\delta u_v^{\mathbf{k}}(\mathbf{r}). \qquad [2.59]$$

where $u_v^{\mathbf{k}}$ is the lattice periodic part of the Kohn–Sham state, $\varphi_v^{\mathbf{k}} = e^{i\mathbf{k}.\mathbf{r}}u_v^{\mathbf{k}}$, with $u_v^{\mathbf{k}}(\mathbf{r}+\mathcal{R}) = u_v^{\mathbf{k}}(\mathbf{r})$. The same definition applies for the lattice-periodic part of the perturbed state, $\Delta\varphi_v^{\mathbf{k}+\mathbf{q}} = e^{i\mathbf{k}+\mathbf{q}.\mathbf{r}}\delta u_v^{\mathbf{k}+\mathbf{q}}$. Though mathematically well-defined, $\delta\rho^{\mathbf{q}}$ is physically not well-defined, in the sense that only the total real charge density response $\Delta\rho(\mathbf{r})$ of equation [2.58] can be observed/measured.

The periodic part of the perturbed wavefunction, $\delta u_v^{\mathbf{k}}$, obeys an equation derived from equation [2.54]:

$$[-\nabla^2 + v_{eff}(\mathbf{r}) + \alpha\hat{P}_v^{\mathbf{k}+\mathbf{q}} - \varepsilon_v^{\mathbf{k}}]\delta u_v^{\mathbf{k}+\mathbf{q}}(\mathbf{r}) = -\hat{P}_c^{\mathbf{k}+\mathbf{q}}\delta v_{eff}^{\mathbf{q}} u_v^{\mathbf{k}}(\mathbf{r}), \qquad [2.60]$$

where the action of $\delta v_{eff}^{\mathbf{q}}$ onto the unperturbed wavefunction yields a function of the conduction manifold at $\mathbf{k}+\mathbf{q}$, $(\mathbf{k}+\mathbf{q}, c)$.

The power of equation [2.60] is that it involves lattice-periodic wavefunctions only: although it aims at computing charge density response for a wavelength incommensurate with the length of the unit cell, we work inside the crystal unit cell, and the workload to compute the linear response is in principle not much more than the computation of the ground-state density.

2.3.6.3. *Remarks about phonon calculations*

The computational workload of a phonon calculation comes from the evaluation of the first-order variation of the charge density, equations [2.60], [2.57] and [2.59]. The linear response and DFT have been implemented in several packages [GIA 09, GON 09]. Calculations implemented in some packages do not use linear response: instead, they rely on (less-efficient) finite difference of the forces with respect to atomic displacements [PAR 97, SOL 02].

2.3.7. *Beyond the harmonic approximation: computing the phonon–phonon interaction*

Vibrational properties of solids are nowadays routinely calculated from DFPT or the small displacement method (frozen phonon, section 2.3.5). High-order terms beyond the harmonic approximations have been calculated from perturbation theory to compute the lifetime [DEB 95] and frequency shift [LAZ 03, BON 07] of zone center phonons. The lifetime of a phonon at Γ is limited by its decay into two phonon of lower energies at $\mathbf{q}$ and $-\mathbf{q}$, and this phonon–phonon interaction is given by the third-order derivative of the total energy, while the phonon frequency shift is due to both third- and fourth-order terms. These terms are evaluated through the $2n + 1$ theorem [BAR 01, GON 95] or through the finite difference method applied to second-order derivatives [DEB 95, LAZ 03].

Recently, the calculation of the third-order derivative has been extended to the decay of a phonon at any point in the Brillouin zone (BZ) [PAU 13] and is the fundamental ingredient for the calculations of the bulk lattice thermal conductivity [FUG 13, LIN 11, MAR 16b].

Methods have also been designed for cases where the phonon–phonon interaction is so large that perturbation theory is not appropriate, a short introduction of which can be found in [ERR 14].

2.4. Electron–phonon coupling

In this section, we are going to introduce the notions that are necessary for the calculation from first principles of electron–phonon coupling in the valence or conduction band of semiconducting materials, but also to compute phonon-mediated superconductivity in metals. A more detailed review of the subject of *ab initio* calculations of the electron–phonon coupling in semiconductors can be found in [SJA 14].

2.4.1. *Electron–phonon matrix element*

A single electron–phonon scattering event involves an initial electronic Bloch state $|\varphi_n^{\mathbf{k}}\rangle$, characterized by its band number n and wave vector $\mathbf{k}$, a vibrational state (phonon) $|\nu, \mathbf{q}\rangle$, and a final electronic state $|\varphi_m^{\mathbf{k}'}\rangle$. An electron with wavevector $\mathbf{k}$ can scatter into a state $\mathbf{k}'$ by emission or absorption of a phonon of wavevector $\mathbf{q}$. The wave vectors of initial and final electronic states are constrained by the momentum conservation rule: $\mathbf{k}' = \mathbf{k} \pm \mathbf{q}$. Transitions must also satisfy the symmetry selection rules [BIR 66]. The energy-conservation condition must be fulfilled:

$$\varepsilon_{\mathbf{k}'}^{m} = \varepsilon_{\mathbf{k}}^{n} \pm \hbar\omega_{\mathbf{q}}^{\nu}, \qquad [2.61]$$

where $\omega_{\mathbf{q}}^{\nu}$ is the phonon frequency.

In presence of a phonon field, an electron in the Bloch state $|\varphi_n^{\mathbf{k}}\rangle$ of the crystal experiences a perturbation, which, to the first order, is proportional to the phonon amplitude. This perturbation induces coupling between Bloch states $|\varphi_n^{\mathbf{k}}\rangle$ and $|\varphi_m^{\mathbf{k}'}\rangle$.

DFPT [BAR 01], which has been introduced in the previous section, is the main state-of-the-art method to calculate the electron–phonon coupling. As for the calculation of vibrational properties, the first main hypothesis that underlies most of the *ab initio* calculations of electron–phonon coupling is the adiabatic approximation, which allows disentangling the ionic and electronic motions (section 2.2.2). The second hypothesis is the harmonic approximation for the dynamics of the ions (section 2.3.1).

In the case of phonons, the perturbation of the self-consistent crystal is induced by displacements of the atoms: $\mathbf{R}_I = \mathbf{R}_I^{eq} + \mathbf{u}_I$. Here, $\mathbf{u}_I$ is the deviation from the equilibrium position of the s^{TH} atom in L^{TH} unit cell: $I = \{L, s\}$. In the framework of DFPT, the perturbation of the self-consistent crystal potential Δv_{eff} is yielded by solution of equation [2.54]. Once equation [2.54] is solved, the electron–phonon matrix element can be calculated in the framework of DFPT as [MAU 96, SJA 14, CAL 10]:

$$g_{nm}^{\nu}(\mathbf{k}, \mathbf{k} \pm \mathbf{q}) = \langle \varphi_{\mathbf{k},n} | \frac{\delta v_{eff}}{\delta \mathbf{u}_I} | \varphi_{\mathbf{k}\pm\mathbf{q},m} \rangle. \qquad [2.62]$$

Here, $\varphi_{\mathbf{k},n}$ and $\varphi_{\mathbf{k}\pm\mathbf{q},m}$ are Bloch wavefunctions of the initial and final electronic states, $\mathbf{k}$ is the electronic wavevector, $\mathbf{q}$ is phonon wavevector and n and m are the initial and final band numbers. $\frac{\delta v_{eff}}{\delta \mathbf{u}_I}$ is the perturbation of the self-consistent crystal potential introduced by phonon displacement $\mathbf{u}_I$.

One of the greatest advantages of DFPT is that responses to perturbations of different wavelengths can be decoupled, and the solution of DFPT equations can be found for any given $\mathbf{q}$ vector (see equation [2.60]) [BAR 01]. Equation [2.60] yields the perturbation of the periodic part of the Bloch wavefunction, $|u_{n,\mathbf{k}}\rangle$, with respect to $\delta v^{\mathbf{q}}_{eff}$, the periodic part of the perturbing potential (see equations [2.55]–[2.60] for details). One can express the electron–phonon coupling in terms of monochromatic perturbations in the following way:

$$\mathbf{d}^{s}_{mn}(\mathbf{k}, \mathbf{k} \pm \mathbf{q}) = \langle u_{n,\mathbf{k}}|\frac{\delta v^{\mathbf{q}}_{eff}}{\delta \mathbf{u}_{\mathbf{q}s}}|u_{m,\mathbf{k}\pm\mathbf{q}}\rangle. \qquad [2.63]$$

Here, $\mathbf{u}_{\mathbf{q}s}$ is the Fourier transform of the phonon displacement $\mathbf{u}_{\mathbf{I}}$, and the electron–phonon matrix element [2.62] reads:

$$g^{\nu}_{nm}(\mathbf{k}, \mathbf{k} \pm \mathbf{q}) = \sum_{s} \mathbf{e}^{s\nu}(\mathbf{q})\mathbf{d}^{s}_{mn}(\mathbf{k}, \mathbf{k} \pm \mathbf{q})/\sqrt{2M_s\omega_{\mathbf{q}\nu}}, \qquad [2.64]$$

where $\mathbf{e}^{s\nu}(\mathbf{q})$ is the phonon eigenvector, $\omega_{\mathbf{q}\nu}$ is phonon frequency and M_s is the atomic mass of the s^{TH} atom in the unit cell.

Calculations from first principles of the electron–phonon coupling matrix elements in metals exist since the late 1990s [SAV 96, MAU 96] and were applied with success for the calculation of the superconducting properties of many materials [CAL 04, GIU 07, CAL 08, BRU 09, CAL 11]. More recently, electron–phonon coupling matrix elements were calculated *ab initio* in semiconductors [SJA 07a, MUR 08]. For reviews, see [SJA 14, GIU 16].

2.4.2. *Wavefunctions of the initial and final electronic states*

As already mentioned, DFPT was first successfully used to calculate electron–phonon coupling in metals [SAV 96, MAU 96]. In the case of semiconductors, an important advance was to show that the electronic wavefunctions of the conduction band in semiconductors obtained in DFT, $\varphi_{n\mathbf{k}}$, are very close to the quasi-particle orbitals calculated with more elaborated methods, even when the latter are computed self-consistently [HYB 86, BRU 06].

Indeed, in Figure 2.2, we can see that the wavefunction of the first conduction band in silicon, calculated within the DFT-LDA approximation, is very close to the one calculated within self-consistent GW approximation, which takes into account part of the electron–electron correlation effects discussed at the end of section 2.2.4.

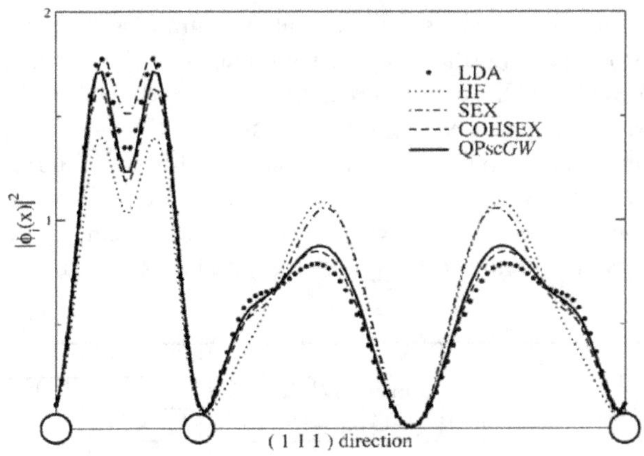

First conduction band of silicon: squared modulus of the wave function along the direction (1 1 1) at **k**=(−1/8,−3/8,1/4). The white circles represent the location of the silicon atoms.

Figure 2.2. *First conduction band of silicon: squared modulus of the wavefunction along the (111) direction calculated within DFT, Hartree–Fock theory and three implementations of the GW method. The white circles represent the location of the silicon atoms [BRU 06]*

The finding of [BRU 06] opened the way to the calculation of the electron–phonon matrix elements involved in the scattering of electrons excited in the conduction band. It should be noted, however, that in the case of surfaces, the difference between the DFT-LDA wavefunction and the quasi-particle one is larger [PUL 99].

2.4.3. *Selection rules and strong dispersion of electron–phonon matrix elements*

As already mentioned, the electron–phonon transitions must satisfy the symmetry selection rules: the phonon must belong to the space-group representation given by the Kronecker product of space-group representations of the initial and final electronic states [BIR 66]. Indeed, if we consider electron–phonon transitions between high-symmetry points in the BZ of a crystal, many electron–phonon scattering channels turn out to be forbidden by symmetry selection rules, and their electron–phonon matrix element is strictly equal to zero.

This is the case, for example, for the intervalley electron–phonon transitions between equivalent minima of the conduction band of Si (Table 2.3). Silicon is an indirect band gap semiconductor, with the lowest point of the conduction band situated close to the X point in BZ, in Δ direction. Thus, electron–phonon “intervalley” transitions between six equivalent Δ minima of the conduction band (Δ valleys) have to be taken into account in simulations of the electronic transport in silicon [JAC 83, WAN 11]. According to the phonon $\mathbf{q}$-vector direction, we can distinguish between g-type and f-type processes [YU 01] (see upper panels of Figure 2.3). The electron–phonon matrix elements for the intervalley transitions between the equivalent Δ minima of the conduction band of silicon are shown in the second column of Table 2.3. As we can see, most channels are forbidden by symmetry. The channels allowed by symmetry are also called zero-order transitions [FER 76].

			Ab initio data		
		adjusted D	Δ-Δ D	averaged D	
	phonon	[JAC 83]	[SJA 13]	[WAN 11]	[SJA 13]
g-processes					
	TA	0.5	0	0.6	0.8
	LA	0.8	0	1.2	1.7
	TO		0	1.3	1.2
	LO	11.0	4.7	4.2	4.0
f-processes					
	TA_1	0.3	0	0.2	
	TA_2		0	0.2	
	LA	2.0	0	1.1	
	LO		2.5	2.1	
	TO_1	2.0	0	2.4	
	TO_2		4.5	4.2	

Table 2.3. *Deformation potentials in eV/Å for the intervalley scattering in the conduction band: effective deformation potentials of [JAC 83] (column 1), and ab initio deformation potentials. Column 2: transitions between high-symmetry points. Columns 3 and 4: average deformation potentials. The value of D has been averaged over states in the final valley in [WAN 11] (column 3) and over both initial and final valleys for g-processes in [SJA 13] (column 4)*

We should also note that in Table 2.3, electron–phonon matrix elements are presented in the form of *deformation potentials*. The deformation potential for individual transition is defined as a quantity proportional to the absolute value of the electron–phonon matrix element of equation [2.62]:

$$D^{\nu}_{nm}(\mathbf{k}, \mathbf{k} \pm \mathbf{q}) = \frac{\sqrt{2\rho\Omega\omega_{\mathbf{q}\nu}}}{\hbar}|g^{\nu}_{nm}(\mathbf{k}, \mathbf{k} \pm \mathbf{q})|. \quad [2.65]$$

Here, ρ is the mass density of the crystal and Ω is the crystal volume.

As we can see from the comparison with the data fitted on the transport measurements in the framework of a Monte-Carlo simulation ([JAC 83], column 1 of

Table 2.3), the number of channels allowed by symmetry selection rules for transitions between high-symmetry points is insufficient to describe transport properties in Si. Moreover, the deformation potential values that have been averaged over many transitions are different from the high-symmetry ones. We can fix the energy close to the minima of the CB and average the deformation potentials over $\mathbf{k}$ and $\mathbf{k}+\mathbf{q}$ points of the initial and final states. The results of such calculations are shown in columns 3 and 4 of Table 2.3, which are very different from the ones in column 1. We can see that when *average* deformation potential values are considered, non-zero values are present in column 3, which is the signature that all scattering channels become allowed, and not only the zero-order transition ones.

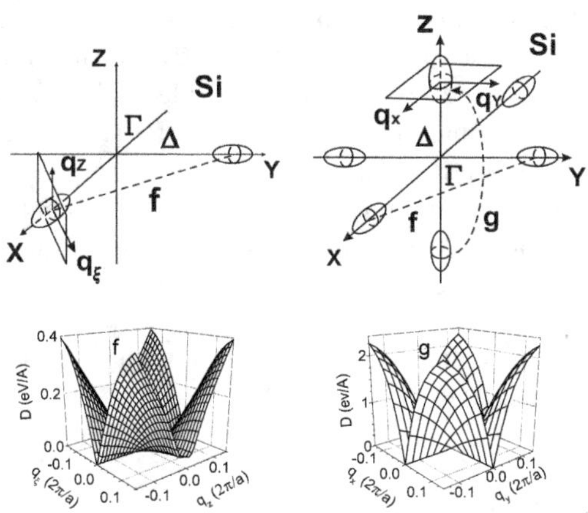

Figure 2.3. *Silicon. Upper panels: f and g type intervalley transitions (dashed lines) between the minima (Δ valley) of the CB. The planes perpendicular to the vectors linking the CB minima are defined for use in lower panels. Lower panels: Deformation potentials from equation [2.65], D (eV/Å) in silicon, as a function of q vector varying in planes defined in the upper panels. Left panels: TA f-processes. Right panels: LA g-processes [TYU 10]*

In fact, electron–phonon matrix elements depend strongly on the wavevectors of the initial and final electronic states: $\mathbf{k}$ and $\mathbf{k}\pm\mathbf{q}$, and change significantly over the BZ [ZOL 90b, SJA 06, SJA 07a, SJA 07b]. In the case of silicon (lower panels of Figure 2.3), the electron–phonon matrix elements that are forbidden by symmetry along high-symmetry directions grow very rapidly when the wavevectors of the initial and final electronic state deviate from high-symmetry directions. These transitions are called first-order transitions [FER 76]. As we can see in Table 2.3 (columns 3 and 4), first-order transitions contribute significantly to electron–phonon scattering due to the strong dependence of electron–phonon matrix elements on the wavevectors of the initial and final electronic states (strong dispersion of the electron–phonon coupling in the BZ).

2.4.4. *Electron–phonon transition probability: use of the Fermi golden rule*

From the electron–phonon matrix element of equation [2.62], one can calculate the probability for a single electron–phonon transition between one initial and one final electronic state:

$$W^{\nu}_{\mathbf{k}'m,\mathbf{k}n} = \frac{2\pi}{\hbar}|g^{\nu}_{\mathbf{k}'m,\mathbf{k}n}|^2 \left(N_{\mathbf{q}\nu} + \frac{1}{2} \pm \frac{1}{2}\right) \delta_{\mathbf{k}',\mathbf{k}\pm\mathbf{q}} \delta(\varepsilon_{\mathbf{k}n} - \varepsilon_{\mathbf{k}'m} \pm \hbar\omega_{\mathbf{q}\nu}), \quad [2.66]$$

where $g^{\nu}_{\mathbf{k}'m,\mathbf{k}n}$ is the electron–phonon scattering matrix element defined in equation [2.62] and ν is the phonon branch. $\hbar\omega_{\mathbf{q}\nu}$ is the phonon energy, with $\mathbf{q} = \pm(\mathbf{k}' - \mathbf{k})$ and $N_{\mathbf{q}\nu}$ the temperature-dependent phonon occupation number, which can be given, for example, by Bose–Einstein statistics. The upper (lower) sign in the delta function corresponds to phonon emission (absorption). The terms $\delta_{\mathbf{k}',\mathbf{k}\pm\mathbf{q}}$ and $\delta(\varepsilon_{\mathbf{k}n} - \varepsilon_{\mathbf{k}'m} \pm \hbar\omega_{\mathbf{q}\nu})$ represent the momentum and energy conservation laws, respectively, which were already discussed in section 2.4.1. Equation [2.66] is known as the *Fermi golden rule* for electron–phonon scattering.

The total probability of scattering via interaction with phonons gives a finite width $\Gamma_{n\mathbf{k}}$ to the initial electronic state $|n, \mathbf{k}\rangle$:

$$\Gamma_{n\mathbf{k}} = \sum_{\mathbf{k}',m,\nu} W^{\nu}_{\mathbf{k}'m,\mathbf{k}n}, \quad [2.67]$$

where the sum over $\mathbf{k}'$ is done over the BZ.

Quantity $\Gamma_{n\mathbf{k}}$ can be regarded either as a finite width of the electronic level due to the interaction with the phonon bath, or, alternatively, as a measure of the electron–phonon interaction strength. The relaxation of electronic states via electron–phonon coupling, which is described by Fermi's golden rule, assumes the weak coupling regime of the electron–phonon coupling, in which an electronic state irreversibly decays to a less excited state by emitting a phonon.

It turns out that the idea of electronic relaxation due to electron–phonon coupling is wrong in systems where electron–phonon interaction is very strong. That is the case, for example, of quantum dots where Fröhlich coupling is very strong, so that strong coupling regime occurs, and mixed electron-LO phonon states are formed [HAM 99]. These states are quasiparticles called polarons. To the best of our knowledge, few descriptions of polarons within DFT-based methods exists in the scientific literature today [KAV 16, FEN 13], as large electron–phonon coupling precludes the use of a perturbative approach like DFPT.

Further limitations of the use of Fermi's golden rule were discussed, for example, in [COM 01]. The Fermi golden rule is the long-time limit of the time-dependent transition probability from one initial state to a set of final states close in energy. On very short time scales, quantum effects such as uncertainty principle [COM 01] or a finite collision time for each collision event should be taken into account [LIP 91, REG 99, BOR 96].

2.4.5. *Measurable quantities and comparison with DFPT calculations*

The electron–phonon matrix element of equation [2.62] for one initial and one final electronic state is a purely theoretical quantity that cannot, in fact, be determined experimentally.

The quantities that can be determined experimentally are related to processes involving many electron–phonon transitions in the same time, such as electron–phonon scattering times and broadenings of electronic levels related to total probability of the electron–phonon scattering given by equation [2.67] (scattering rate). Experimental determination of such quantities allows for extracting the average, or effective, electron–phonon matrix elements.

Electron–phonon relaxation time (scattering time) is given by the inverse of the total electron–phonon scattering probability [2.67]:

$$\tau_{n\mathbf{k}} = \frac{\hbar}{\Gamma_{n\mathbf{k}}}. \qquad [2.68]$$

As only scattering times or scattering rates can be determined experimentally, the electron–phonon coupling constants (or deformation potentials) obtained in experiments are in fact effective values representing not a single one, but many transitions [ZOL 90a, SJA 07a]. This fact should not be neglected when one compares the *ab initio* calculated electron–phonon coupling characteristics with experimentally obtained ones. For example, in [WAN 11], *averaged* deformation potentials were compared to the ones fitted to experiment, as has been already discussed above with respect to the results presented in Table 2.3.

Experimentally, electron–phonon scattering rates and times can be estimated with the help of a wide range of experimental techniques, such as transport experiments, pump–probe experiments and spectroscopy experiments. Transport [PRU 05] and magnetotransport [YU 08] measurements provide a wealth of information on the electron–phonon scattering times and rates. The information on electron–phonon scattering can be extracted from mobility, power density, Hall mobility or other experimentally measured transport characteristics, as contributions from different scattering mechanisms to such quantities have different temperature dependencies.

The difficulty arises from the presence of different scattering mechanisms at the same time, and interpretation of experimental results requires to fit experimental data with models accounting for all of the phenomena, and even to perform Monte Carlo simulations [JAC 83]. Thus, scattering times and rates for a particular electron–phonon process depend on the particular model used to interpret experimental data, and their values are often distributed over a wide interval.

The uncertainty in the experimentally determined electron–phonon scattering constants, due to a large number of fitting parameters in the interpretation of experiments, reinforces the importance of *ab initio* calculations of the electron–phonon coupling constants. Indeed, such calculations allow for reducing the number of fitting parameters.

2.4.6. *Different electron–phonon scattering mechanisms and their role in the transport properties of semiconductors*

Different kinds of electron–phonon scattering mechanisms exist, and, depending on the properties of the materials, we have to take them into account in the calculation of the transport coefficients, such as electronic mobility or Seebeck coefficient. For instance, in a semiconductor with several local minima in the conduction band (valleys), we have to distinguish between intervalley scattering, i.e. scattering of electrons by short-wavelength phonons that connect different valleys, and intravalley scattering by long-wavelength zone-center phonons, where initial and final electronic states belong to the same valley. In polar semiconductors, other important scattering mechanisms appear, such as Fröhlich interaction with dipolar field created by zone-center longitudinal optical phonon [YU 01]. This dipolar-optical coupling has been computed in gallium arsenide [SJA 15, TAN 16] and titanium dioxide [VER 15]. A detailed description of different kinds of the electron–phonon scattering mechanisms can be found in [JAC 83, YU 01] as well as in [SJA 14].

The question about which one, among different electron–phonon scattering mechanisms, will yield the main effect on the transport properties of a semiconductor, depends on the temperature and the applied electric field, as well as on the semiconductor properties [JAC 83]. For instance, we have to know whether the semiconductor in question is polar or not, and whether its band gap is direct or indirect.

In non-polar semiconductors, at low applied electric fields and low temperatures, interaction with acoustic intravalley phonons (LA phonons) is the main momentum dissipation process [JAC 83]. Indeed, in Figure 2.4, we show the phonon-limited electronic mobility of Si, calculated in this work using BTE coupled with *ab initio* DFT data, as a function of temperature at low field. As we can see in Figure 2.4, at low temperatures, the mobility is mainly limited by interaction with acoustic intravalley phonons. At ambient temperatures, however, the relative importance of the acoustic intravalley phonons and intervalley scattering between equivalent minima of the conduction band are comparable. Indeed, in indirect semiconductors, such as Si and Ge, intervalley scattering between equivalent Δ minima in the conduction band is the main energy dissipation process [JAC 83]. At higher applied electric fields, however, intervalley scattering to the adjacent L, Δ and Γ minima of the conduction band becomes possible, and this scattering process dominates over all the other electron–phonon scattering mechanisms for both energy and momentum dissipation at very high fields, as it was shown, for example, in [JAC 83] for Ge.

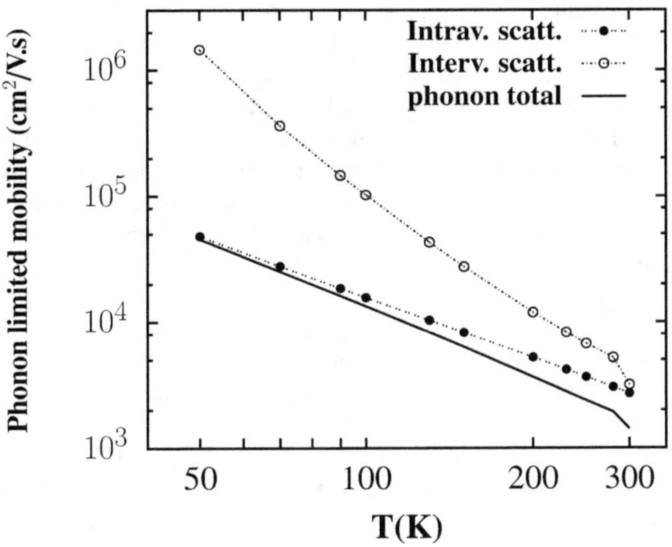

Figure 2.4. *Phonon limited mobility μ in Si as a function of temperature for two different electron–phonon scattering mechanisms (intravalley and intervalley electron–phonon scattering). The calculations have been done at low field by coupling Boltzmann transport equation and the first principles calculations of the electron–phonon scattering constants (see Table 2.3). At low temperatures, the mobility is essentially determined by the acoustic intravalley scattering. At room temperature, both scattering mechanisms (intravalley and intervalley) are important in the determination of the mobility*

In polar semiconductors at low applied electric fields, the scattering by acoustic intravalley phonons dominates at low temperatures. At high temperatures, however, Fröhlich interaction becomes the main scattering process that determines the transport characteristics such as mobility. This was shown, for example, in [STI 70] for GaAs, which is a polar semiconductor with a direct band gap. This holds for electrons near the bottom of the conduction band (low electric fields). However, at high enough electron excitation energies, it has been shown that intervalley scattering is the dominant process at room temperature [TAN 16].

2.4.7. *Electron–phonon coupling beyond standard approximations*

2.4.7.1. *Beyond BO approximation*

In state-of-the-art DFPT calculations, the adiabatic (BO) approximation is used (section 2.2.2), and the variation of Kohn–Sham self-consistent potential is calculated with respect to static ionic displacement. In some particular metallic systems, however, such as graphite, graphene, CaC_6, carbon nanotubes, a combined treatment of electrons and phonons is necessary in order to obtain correct phonon frequencies at particular points of the BZ and phonon linewidths due to electron–phonon and phonon–phonon scattering [PIS 04, LAZ 06b, CAL 10]. Such a dynamic approach was developed in [LAZ 06b, PIS 07], where phonon frequencies were computed with electronic population out of equilibrium, in graphene and carbon nanotubes. Recently, time-dependent linear response theory was formulated in [CAL 10], and its implementation was discussed in the framework of Wannier interpolation of dynamical matrix and electron–phonon matrix elements.

2.4.7.2. *Beyond DFT*

In some special cases, such as, for example, graphene, graphite, and fullerene, DFT and DFPT in LDA or GGA approximations turn out to be insufficient to describe some special phonon modes [LAZ 08, FAB 11]. The reason is that electron–electron correlations play an important role in these systems, and quasiparticle corrections obtained with the GW method have to be taken into account in calculations of phonon frequencies and electron–phonon coupling matrix elements [LAZ 08].

Another example of systems where we have to go beyond the standard DFT description is provided by chalcopyrites $CuInSe_2$ and $CuInS_2$ [VID 10]. These materials are attracting attention due to the possibility of photovoltaïc applications. It was shown in [VID 10] that the change in the photoemission band gap versus anion displacement was very different in DFT calculations with respect to GW methods. This indicates that for such systems, vibrational properties and electron–phonon coupling may also need a treatment beyond DFT.

2.5. Semiclassical transport properties

In what follows we are going to introduce some methods used to understand the transport properties of materials. In particular, we will show how the physical quantities defined above (bandstructure, electron–phonon coupling) are involved in the understanding of the carrier transport phenomena and investigate more precisely the role of DFT calculation to achieve this aim.

2.5.1. *Semiclassical transport: BTE*

We introduce here BTE, a semiclassical approach to carrier transport in materials. The first term (left-hand side of equation [2.69]) of BTE is based on the classical dynamics of carriers which, in absence of collision processes, describes the motion of carriers in the presence of external electric and magnetic fields [ASH 02]. The second term of BTE describes the collisional processes (right-hand side of equation [2.69]).

To determine the carrier dynamics, a distribution function $f_n(\mathbf{r}, \mathbf{k}, t)$ is defined. $f_n(\mathbf{r}, \mathbf{k}, t)$ is the probability to find a carrier with wavevector $\mathbf{k}$ at the position $\mathbf{r}$, and at the time t. n is the band index. By using the distribution function f_n, the Heisenberg uncertainty principle is not considered, and the position and momentum are always treated at the same time. BTE reads [ASH 02, ZIM 60, LUN 09]:

$$\frac{\partial f_n}{\partial t} + \mathbf{v}_n(\mathbf{k}) \cdot \nabla_{\mathbf{r}} f_n + \frac{\mathbf{F}}{\hbar} \cdot \nabla_{\mathbf{k}} f_n = \left. \frac{\partial f_n}{\partial t} \right|_{\text{coll}}, \qquad [2.69]$$

where $\mathbf{F} = -e\left[\mathbf{E} + \mathbf{v}_n(\mathbf{k}) \wedge \mathbf{B}\right]$ is the Lorentz force, e is the absolute value of the electron charge. $\mathbf{E}$ and $\mathbf{B}$ are, respectively, the external electric and magnetic field. $\mathbf{v}_n$ represents the group velocity and is given by:

$$\mathbf{v}_n = \hbar^{-1} \nabla \varepsilon_{\mathbf{k}n}, \qquad [2.70]$$

where $\varepsilon_{\mathbf{k}n}$ is the carrier energy. Note that in the absence of scattering events, an electron initially in the band n remains in it. The carrier motion between two scattering events obeys the semiclassical dynamics, but the collision processes, which are microscopic mechanisms, require a quantum treatment.

In absence of external applied fields, the distribution function at equilibrium is given by the Fermi–Dirac distribution f^o [ASH 02]:

$$f^o(\varepsilon_{\mathbf{k}n}) = \left[1 + \exp\left(\frac{\varepsilon_n(\mathbf{k}) - \mu}{k_B T}\right)\right]^{-1}, \qquad [2.71]$$

where μ is the chemical potential or Fermi energy and T is the temperature. Once equation [2.69] is solved, we can calculate the physical quantities such as the carrier density $n(\mathbf{r})$ and the current density $\mathbf{J}$, respectively, given by:

$$n(\mathbf{r}) = \frac{g_s}{L^d} \sum_{\mathbf{k},n} f_n(\mathbf{r}, \mathbf{k}, t), \qquad [2.72]$$

$$\mathbf{J}(\mathbf{r}) = -\frac{eg_s}{L^d}\sum_{\mathbf{k},n} \mathbf{v}_{n}, (\mathbf{k}) f_n(\mathbf{r},\mathbf{k},t) \qquad [2.73]$$

where g_s is the spin degeneracy. d is the dimension of the system ($d = 1, 2, 3$) and L^d is the normalization volume. In the following, we will omit the position and time dependence of the distribution function for clarity, i.e $f_n(\mathbf{r},\mathbf{k},t) \equiv f_{\mathbf{k}n}$.

2.5.2. *Scattering mechanisms*

Now let us focus on the right-hand side of equation [2.69]. As mentioned above, $\partial f_n/\partial t|_{\text{coll}}$ describes all collision processes. The scattering mechanisms that occur in the system can be classified into two types.

2.5.2.1. *Elastic scattering*

In an elastic process, the carrier is scattered without loss of energy. However, the direction of the incident wave vector changes from an initial state $|\mathbf{k}n\rangle$ to a final state $|\mathbf{k}'m\rangle$. At sufficiently low temperatures, scattering by impurities and defects is the main source of diffusion in a solid. The probability $W^{\text{el}}_{\mathbf{k}'m,\mathbf{k}n}dt$ for a transition from state $|\mathbf{k}n\rangle$ to $|\mathbf{k}'m\rangle$ in the time interval dt is obtained by using the Fermi golden rule. Assuming that the sample contains a small density of n_i randomly distributed immobile impurities, the quantum transition probability is given by [HAM 10]:

$$W^{\text{el}}_{\mathbf{k}'m,\mathbf{k}n} = \frac{2\pi}{\hbar} n_i |\langle \mathbf{k}'m|V_i|\mathbf{k}n\rangle|^2 \delta(\varepsilon_{\mathbf{k}n} - \varepsilon_{\mathbf{k}'m}), \qquad [2.74]$$

where $\langle \mathbf{k}'m|V_i|\mathbf{k}n\rangle$ is the impurity scattering matrix elements and V_i is the scattering potential. The collision integral describing elastic scattering processes reads:

$$\left.\frac{\partial f_{\mathbf{k}n}}{\partial t}\right|^{\text{el}}_{\text{coll}} = -\frac{1}{L^d}\sum_{\mathbf{k}'m} \left[W^{\text{el}}_{\mathbf{k}n,\mathbf{k}'m} f_{\mathbf{k}'m}(1 - f_{\mathbf{k}n}) - W^{\text{el}}_{\mathbf{k}'m,\mathbf{k}n} f_{\mathbf{k}n}(1 - f_{\mathbf{k}'m})\right]. \qquad [2.75]$$

Here, we explicitly write the momentum and band index to underline the fact that the electron can make a transition from one band (n) to another one (m) in the collision process. In equation [2.75], the factor $f_{\mathbf{k}n}(1 - f_{\mathbf{k}'m})$ gives the probability that an electron initially at the state $|\mathbf{k}n\rangle$ repopulates an empty state $|\mathbf{k}'m\rangle$. $W^{\text{el}}_{\mathbf{k}'m,\mathbf{k}n} f_{\mathbf{k}n}(1 - f_{\mathbf{k}'m})$ represents the outscattering or the flow of electrons from state $|\mathbf{k}n\rangle$ to state $|\mathbf{k}'m\rangle$ while $W^{\text{el}}_{\mathbf{k}n,\mathbf{k}'m} f_{\mathbf{k}'m}(1 - f_{\mathbf{k}n})$ is the inscattering or the flow of electrons from state $|\mathbf{k}'m\rangle$ to state $|\mathbf{k}n\rangle$.

Because V_i is Hermitian, we have $W^{\text{el}}_{\mathbf{k}n,\mathbf{k}'m} = W^{\text{el}}_{\mathbf{k}'m,\mathbf{k}n}$, and the elastic collision integral reduces to:

$$\left.\frac{\partial f_{\mathbf{k}n}}{\partial t}\right|^{\text{el}}_{\text{coll}} = \frac{1}{L^d}\sum_{\mathbf{k}'m} W^{\text{el}}_{\mathbf{k}'m,\mathbf{k}n}(f_{\mathbf{k}'m} - f_{\mathbf{k}n}). \qquad [2.76]$$

Examples of calculations of impurity scattering within the DFT can be found in [MUR 06, RES 09].

2.5.2.2. *Electron–phonon scattering*

In the case of the diffusion of an electron by a phonon, the transition probability, within the Fermi golden rule, is given by equation [2.66]. The electron–phonon collision integral has the general form of equation [2.75], but in addition we have to sum over the phonon branch ν:

$$\begin{aligned}\left.\frac{\partial f_{\mathbf{k}n}}{\partial t}\right|^{\text{e-ph}}_{\text{coll}} = \frac{1}{L^d}\sum_{\mathbf{k}'m,\nu} \Big[& W^{\nu}_{\mathbf{k}n,\mathbf{k}'m} f_{\mathbf{k}'m}(1 - f_{\mathbf{k}n}) \\ & - W^{\nu}_{\mathbf{k}'m,\mathbf{k}n} f_{\mathbf{k}n}(1 - f_{\mathbf{k}'m})\Big]. \end{aligned} \qquad [2.77]$$

In an electron–phonon scattering process, we can distinguish four terms corresponding to various transitions (see Figure 2.5): an electron with wavevector $\mathbf{k}$ can scatter into a state with wavevector $\mathbf{k}' = \mathbf{k} + \mathbf{q}$ by absorption of a phonon of wavevector $\mathbf{q}$ or emission of a phonon of wavevector $-\mathbf{q}$. Similarly, an electron of wavevector $\mathbf{k}'$ can scatter into the state $\mathbf{k}$ by emission of a phonon of wavevector $\mathbf{q}$ or absorption of a phonon of wavevector $-\mathbf{q}$. By taking in consideration the different transition processes mentioned above, the collision integral becomes [HAM 10]

$$\begin{aligned}\left.\frac{\partial f_{\mathbf{k}n}}{\partial t}\right|^{\text{e-ph}}_{\text{coll}} = -\frac{2\pi}{\hbar}\frac{1}{L^d}\sum_{\mathbf{k}'m,\nu} & |g^{\nu}_{\mathbf{k}'m,\mathbf{k}n}|^2 \{ f_{\mathbf{k}n}(1 - f_{\mathbf{k}'m}) N^{\nu}_{\mathbf{q}} \\ \times\delta(\varepsilon_{\mathbf{k}n} + \hbar\omega_{\mathbf{q}\nu} - \varepsilon_{\mathbf{k}'m}) & + f_{\mathbf{k}n}(1 - f_{\mathbf{k}'m})(1 + N^{\nu}_{-\mathbf{q}}) \\ \times\delta(\varepsilon_{\mathbf{k}n} - \hbar\omega_{-\mathbf{q}\nu} - \varepsilon_{\mathbf{k}'m}) & - (1 - f_{\mathbf{k}n}) f_{\mathbf{k}'m}(1 + N^{\nu}_{\mathbf{q}}) \\ \times\delta(\varepsilon_{\mathbf{k}n} + \hbar\omega_{\mathbf{q}\nu} - \varepsilon_{\mathbf{k}'m}) & - (1 - f_{\mathbf{k}n}) f_{\mathbf{k}'n} N^{\nu}_{-\mathbf{q}} \\ & \times\delta(\varepsilon_{\mathbf{k}n} - \hbar\omega_{-\mathbf{q}\nu} - \varepsilon_{\mathbf{k}'m})\}. \end{aligned} \qquad [2.78]$$

Electron–phonon scattering is a relevant scattering mechanism to understand, for example, the current saturation in metallic carbon nanotubes [LAZ 06a] and therefore the coupling constant $|g^{\nu}_{\mathbf{k}'m,\mathbf{k}n}|$ has to be well described for a better understanding of the transport properties. To this end, the first principles calculation (DFT) has proved to be a powerful tool [MAU 96, SJA 07a].

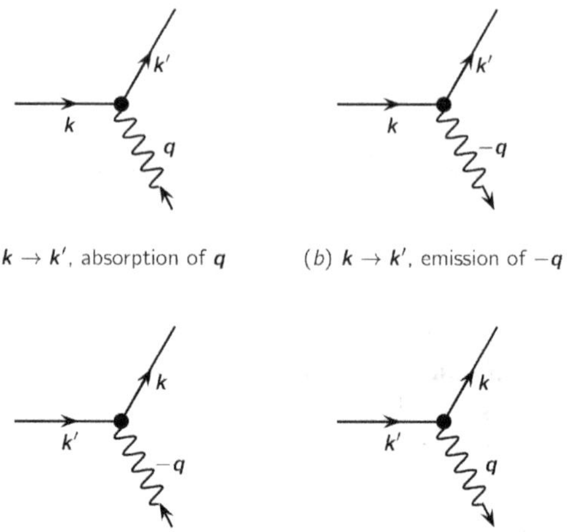

(a) $\mathbf{k} \to \mathbf{k}'$, absorption of $\mathbf{q}$ (b) $\mathbf{k} \to \mathbf{k}'$, emission of $-\mathbf{q}$

(c) $\mathbf{k}' \to \mathbf{k}$, emission of $\mathbf{q}$ (d) $\mathbf{k}' \to \mathbf{k}$, absorption of $-\mathbf{q}$

Figure 2.5. *Electron–phonon scattering processes*

2.5.2.3. *Electron–electron scattering*

In addition to the two scattering processes defined above, we may also have the electron–electron (e-e) scattering, which is an inelastic scattering event. In bulk materials, relaxation times for e-e interactions are typically larger than electron–phonon scattering times for a wide range of temperatures [ASH 02], which suggests that the e-e scattering can be neglected in the linear regime (low-field transport) for low values of the electron concentration. However, when increasing electron temperature or for large electron concentrations, the order of magnitude of the e-e interaction becomes relevant and must be treated. In some nanoscale systems such as semiconductor quantum wells, when the excited electrons do not have enough energy to emit a phonon (particularly an optical phonon), the e-e interaction has been suggested to be the most dominant relaxation process [TAV 01]. In silicon, for instance, e-e interaction has been treated empirically through an electron–plasmon term [FIS 91].

2.5.2.4. *Other scattering mechanisms*

Depending on the material under consideration, other mechanisms than electron–phonon and electron–electron scatterings come into play. The most used are the scattering by alloy disorder and by interface roughness. For these mechanisms, the potential of interaction is not well known and it is usually described within an empirical formalism. For examples of these scatterings, the interested reader can consult [UNU 03].

2.5.3. *Linear and nonlinear transport*

In this section, we define two regimes of transport depending on the value of the applied electric field: the linear and the nonlinear transport.

We assume that the system is under an applied static and uniform electric field ($\mathbf{B} = 0$). From equation [2.69], the steady-state distribution ($\partial f/\partial t = 0$) is given by:

$$\mathbf{v_k} \cdot \nabla_\mathbf{r} f_\mathbf{k} - \frac{e\mathbf{E}}{\hbar} \cdot \nabla_\mathbf{k} f_\mathbf{k} = \left.\frac{\partial f_\mathbf{k}}{\partial t}\right|_{\text{coll}}, \qquad [2.79]$$

where we have condensed both the wave vector and the band index. When the applied external field $\mathbf{E}$ is small, the steady-state distribution function is in general written as:

$$f_\mathbf{k} = f_\mathbf{k}^o + f_\mathbf{k}^1, \qquad [2.80]$$

where $f_\mathbf{k}^1$ is a small correction (first-order correction linear with the electric field) to the equilibrium distribution $f_\mathbf{k}^o$ (equation [2.71]). Therefore, finding $f_\mathbf{k}$ reduces to determine $f_\mathbf{k}^1$ by solving equation [2.79]. This situation is called the linear regime.

When the applied field is strong enough, the steady-state distribution $f_\mathbf{k}$ cannot be written as in equation [2.80] because the correction $f_\mathbf{k}^1$ is no longer small. In this case, we are in the nonlinear regime.

Solving equation [2.79] can be a difficult task since the microscopic expressions of the collision terms lead to an integro-differential equation. An approximation is to use the well-known relaxation time approximation (RTA) that consists in replacing the collision term by:

$$\left.\frac{\partial f_\mathbf{k}}{\partial t}\right|_{\text{coll}} = -\frac{f_\mathbf{k} - f_\mathbf{k}^o}{\tau_\mathbf{k}}. \qquad [2.81]$$

This equation describes the evolution of the non-equilibrium distribution function $f_\mathbf{k}$ toward the equilibrium distribution $f_\mathbf{k}^o$ of equation [2.71], with a characteristic transport time or relaxation time $\tau_\mathbf{k}$ given by [LUN 09]:

$$\tau_\mathbf{k} = \sum_{\mathbf{k}'} W_{\mathbf{kk}'}(1 - \frac{k'}{k}\cos\theta) \qquad [2.82]$$

with θ the angle between $\mathbf{k}$ and $\mathbf{k}'$. In Figure 2.6, we show the electronic mobility as a function of the carrier concentration N for n-doped Si at 300 K. The calculations (solid line) have been done by coupling BTE with first principles calculations of the electronic bandstructure and the deformation potentials for the electron–phonon scattering, and with the model of [WAN 11] for the impurity scattering. The BTE has been solved in the RTA approach. We compare the result with previous DFT calculations also based on RTA (crosses) [RES 09]. We can observe that the two calculations are close.

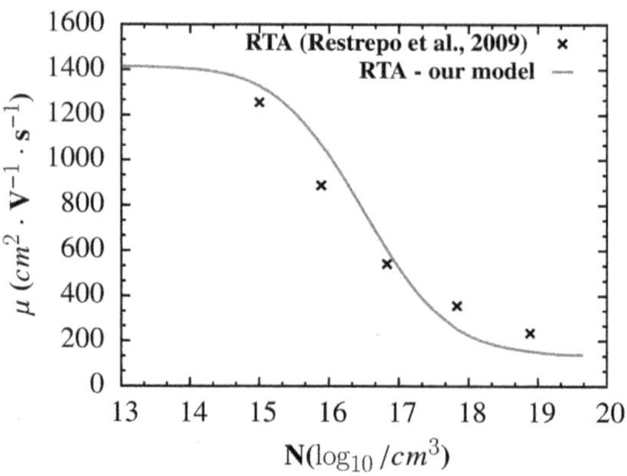

Figure 2.6. *Electronic mobility μ versus the carrier concentration N for n-doped Si at T=300 K. The solid line is our result, obtained by using the electronic bandstructure and the deformation potentials for the electron–phonon scattering from ab initio calculations (see Table 2.3). The BTE has been solved in the RTA approach. We compare the result with previous DFT calculations also based on RTA (crosses) [RES 09]*

As a remark, we should note that the RTA approach is only valid for elastic or isotropic scattering and does not work well under high fields (nonlinear regime) [ASH 02, LUN 09]. Therefore, if the scattering is not elastic nor isotropic, equation [2.79] must be solved iteratively (numerical treatment).

2.5.4. *Coupling BTE with DFT*

The *ab initio* calculation of the physical quantities (such as the electron–phonon coupling constants, the energy bandstructure) necessary for the understanding of the transport properties may reveal some underlying phenomenon that cannot be caught by using empirical parameters. In this context BTE has been, for instance, successfully combined with DFT calculations in [MUR 06, MUR 08]. Indeed in these studies, *ab initio* calculations of the electron–phonon coupling constants in

n-type SiGe alloy have revealed the importance of the f-type Δ intervalley scattering and L intervalley scattering for the understanding of the transport measurements. By using the approach described in [WAN 11], which consists of coupling first principles calculations to BTE, we have calculated the variation of the electronic mobility as a function of the carrier density in n-doped Si (see Figure 2.7). The transport calculation in this study is based on the BTE in the linear regime. The equations were solved iteratively beyond the RTA, and the electronic structure and electron–phonon scattering constants were obtained by first principles calculations (see Table 2.3).

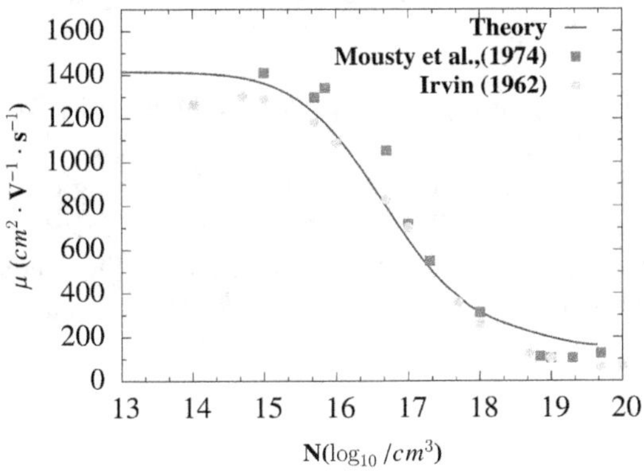

Figure 2.7. *Electronic mobility μ versus the carrier concentration N for n-doped Si at T = 300 K. The solid line is our result computed using BTE solved iteratively for 10 iterations, with the electronic bandstructure and the deformation potentials for the electron–phonon scattering calculated with DFT. Symbols stand for experimental data (circles [IRV 62], squares [MOU 74]). The experimental measurements are well described by our theoretical approach*

We compare our results with experimental measurements (circles: from [IRV 62], squares: from [MOU 74]). We can observe the good agreement between experiments and theory (solid line) in Figure 2.7.

2.5.5. *Limits of the semiclassical approach*

In the description above, the transport is supposed to be classical, which means we neglect the wave nature of the electron. For a system whose size is of the same order of the De Broglie wavelength, the BTE is no longer suitable since the wave-like nature of the electron can become important [SCH 98] and leads to coherence phenomena such as interference [TAK 99, SHE 99], weak localization [AVE 98] or tunneling [OHY 07]. In this case, we need new approaches to the transport like those described in the following section.

2.6. Quantum transport

2.6.1. *Quantum ballistic transport: the Landauer-Büttiker approach*

When the region of interest for the transport is smaller than the mean free path of the electron, the electron can go through the junction without experiencing any collision. The transport is then fully coherent and can be calculated by evaluating the probability that an electron can be transmitted through the junction. If, furthermore, the typical size of the junction is of the order of the De Broglie wavelength of the electron, the description of the system can be achieved only on the level of quantum mechanics.

We follow [CHO 99] and consider a system sandwiched between two contacts at zero temperature. We will consider that the system is infinite in two directions of the space and that we apply a voltage difference in the remaining direction (called the z direction). Hence, the transport will occur only in the z direction and the difference between the left and right electrochemical potential of the leads (μ_L and μ_R) is imposed via the bias (see Figure 2.8). In the following, equations are one dimensional in space.

$$\mu_L - \mu_R = -eV. \qquad [2.83]$$

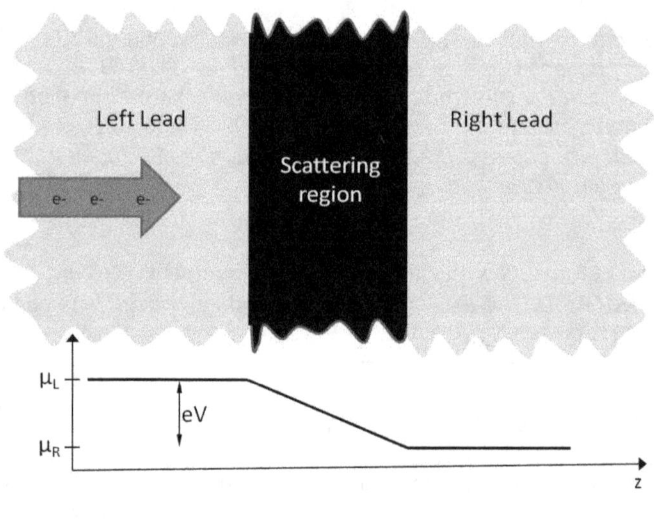

Figure 2.8. *Schematic representation of the transport in the system. Here, the electric field in the junction is considered as constant for the sake of simplicity of the representation*

The current flowing through the junction from left to right is:

$$I_{L\to R} = -e \sum_{kn} \mathrm{v}_{kn} f_L^o(\varepsilon_{kn}) = \frac{-e}{h} \sum_{kn} \int_0^\infty T(\varepsilon_{kn}) f_L^o(\varepsilon_{kn}) \frac{\partial \varepsilon_{kn}}{\partial k} dk, \quad [2.84]$$

$$I_{L\to R} = \frac{-e}{h} \int_0^\infty T(\varepsilon) f_L^o(\varepsilon) d\varepsilon, \quad [2.85]$$

where T is the probability of transmission of the electron, f_L^o is distribution for the left lead (see equation [2.71]), v_{kn} is the group velocity (see equation [2.70]) and k is the electron wavevector. Similarly, the current from right to left is:

$$I_{R\to L} = \frac{-e}{h} \int_0^\infty T(\varepsilon) f_R^o(\varepsilon) d\epsilon. \quad [2.86]$$

At zero temperature the Fermi–Dirac distribution becomes the Heaviside function, and only energy levels below the Fermi energy should be considered.

$$I = I_{L\to R} - I_{R\to L} = \frac{-e}{h} \int_{\mu_R}^{\mu_L} T(\varepsilon) d\varepsilon. \quad [2.87]$$

In the limit of an infinitely small voltage, we can easily perform the integration and we have for the total current:

$$I = \frac{-e}{h} T(\mu_L)(\mu_L - \mu_R) = \frac{e}{h} T(\mu_F) eV, \quad [2.88]$$

where μ_F is the common Fermi level at zero voltage. This yields the conductance in the Landauer–Büttiker approach:

$$G = \frac{I}{V} = \frac{e^2}{h} T(\mu_F). \quad [2.89]$$

The problem is now reduced to the calculation of the transmission $T(\mu_F)$. One divides the system into three parts, the two semi-infinite leads plus a central region, also called the scattering region because only in this part the electron can be reflected. In step 1, we have to calculate the propagating states in each lead. In step 2, we solve the Schrödinger equation in the central region with the boundary conditions that the wavefunctions and their derivatives must be continuous at the interfaces. Finally in step 3, we calculate the transmission by summing all of the channels available at the energy if the Fermi level. We then deduce the conductance.

2.6.1.1. *Transmission: example with the square potential barrier*

We take a very simple example in one dimension to illustrate how we can calculate the transmission for a square potential barrier $U(z)$ (see Figure 2.9). We detail the four steps of the calculation.

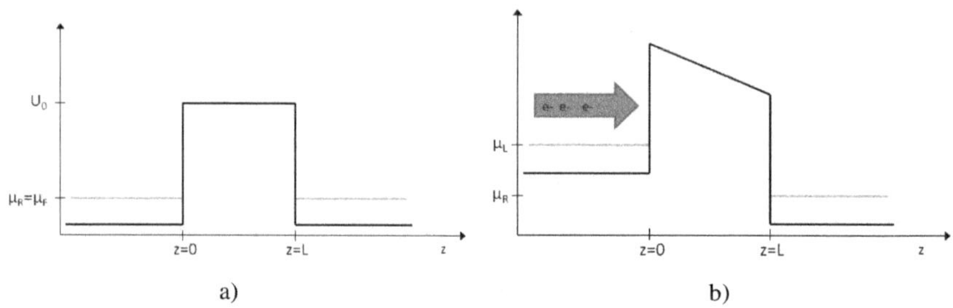

Figure 2.9. *The square potential: a toy model for calculating the conductance a) without applied voltage and b) with applied voltage*

The Schrödinger equation in one dimension is

$$-\frac{\hbar^2}{2m^*}\frac{d^2\psi(z)}{dz^2} + U(z)\psi(z) = E\psi(z), \qquad [2.90]$$

where m^* is the (effective) mass of the electron, ψ and E are, respectively, the wavefunction of the propagating state and its energy.

First, we solve equation [2.90] in each lead for a given E (step 1). The solution is $\psi(z) = Ae^{ikz}$ where the complex wavevector reads $k = \pm\sqrt{\frac{2m^*E}{\hbar^2}}$, and A is a constant. This expression is analogous to the solution of the Schrödinger equation for a free particle. The conceptual difference is that E is fixed and k is the solution, which we are looking for. For $E > 0$, k is real and the solution corresponds to a state propagating to the right or to the left depending on the sign of k. For $E < 0$, k is imaginary and the solution corresponds to a decaying state in the left (respectively, right) lead. We have thus obtained the complex band structure of the leads, which is similar to the usual band structure except that it is calculated for a given E and not a given k and that it can give complex values of k.

In the second step, we solve the generalized scattering problem. In the left lead, we search for a solution as the superposition of an incident and a reflected wave, so that only real values of k are relevant. In the right lead, the solution has the form of a transmitted wave:

$$\psi_L(z) = e^{ikz} + re^{-ikz},$$

$$\psi_R(z) = te^{ikz}, \qquad [2.91]$$

where r and t are the coefficients for the reflectance and the transmittance. In the scattering region, we solve equation [2.90] and find $\psi(z) = Be^{ik_1 z}$ where B is a constant and

$$k_1 = \pm\sqrt{\frac{2m(E-U_0)}{\hbar^2}}, \qquad [2.92]$$

where U_0 is the potential value in the scattering region. The general solution is:

$$\psi_S(z) = Ae^{ik_1 z} + Be^{-ik_1 z}, \qquad [2.93]$$

where A and B are the coefficients of the linear combination. The continuity of the solution and of its first derivative at $z = 0$ and $z = L$ determines the coefficients r, t, A and B (see [COH 97] for more details). This gives:

$$|r|^2 = \frac{(k^2 - k_1^2)^2 sinh^2(k_1 L)}{4k^2k_1^2 + (k^2 - k_1^2)^2 sinh^2(k_1 L)}, \qquad [2.94]$$

$$|t|^2 = \frac{4k^2k_1^2}{4k^2k_1^2 + (k^2 - k_1^2)^2 sinh^2(k_1 L)}. \qquad [2.95]$$

We are now in a position to determine the transmission T (step 3). The current in the right lead is $I_{\text{transmitted}} = |t|^2\hbar k/m$ and the incident current is $I_{\text{incident}} = \hbar k/m$. So that $T = I_{\text{transmitted}}/I_{\text{incident}} = |t|^2$. Finally, we deduce the conductance:

$$G = \frac{2e^2}{h}\frac{4k^2k_1^2}{4k^2k_1^2 + (k^2 - k_1^2)^2 sinh^2(k_1 L)}, \qquad [2.96]$$

where the factor 2 accounts for the spin degeneracy, $sinh$ is the sinus function when k_1 is real and it is the hyperbolic sinus when k_1 is imaginary. k and k_1 are calculated at an energy corresponding to the common Fermi level of the leads.

When $E > U_0$, k_1 is real, and the sinus in equation [2.96] will oscillate with the length of the barrier L. This is the resonant effect, very similar to what is seen in Fabry–Perot cavity for photons: when the length of the barrier is compatible with the wavelength of the electron, constructive interferences will be created in the barrier, resulting in a complete transmission. On the contrary, far from the resonance, the various electronic wave reflected onto the barrier interface will interfere destructively, resulting in a weak transmission. When $E < U_0$, the transmission will decrease exponentially with the length of the barrier, which is the famous tunneling effect.

Neither of these two phenomena can be addressed without the use of a quantum mechanical model. Another example of this kind of interference effect is the well-known resonant tunneling diode formed by two square potential barriers separated by a well. In this case, for some values of the energy, the double barrier can be totally transparent, because the incident electrons are in resonance with the energy of the states in the potential well.

2.6.1.2. *Transmission: real materials within DFT*

We now discuss the DFT evaluation of the conductance, which has been implemented in the PWCOND code that is part of the Quantum ESPRESSO package [GIA 09]. This code also provides the complex band structure calculation as shown in Figure 2.10. The method was first developed in [CHO 99], and was extended to ultra-soft-pseudopotential in [SMO 04] and to spin-orbit coupling in [DAL 06]. The interested reader can look into these papers for more details on the method.

For real materials, the potential taken as a square potential in the previous section is not known in the three regions. For this reason, an additional step is needed that consists in finding the self-consistent potential for each of the three regions separately. This is done from usual ground state DFT calculations (see section 2.2.4 and equations [2.17] and [2.34]), using the usual periodic boundary conditions. In practice, we define the central region to be the junction plus a part of the lead and we perform a super cell calculation. Deep into the lead the asymptotic form of the wavefunction becomes the wavefunction of the bulk. This allows for a description of the interface on the *ab initio* level and the matching procedure of the wavefunctions is made possible.

Once we have obtained the potentials in the three regions, we calculate the transmission in a way very similar to what was done for the square potential barrier. The main difference is that we are now in a three-dimensional case. We will keep the periodicity in x, y and search for the wavefunction as a Bloch state in these directions. Another difference is that we can have more than one propagating state at a given E as determined by the complex structure. For example, in Figure 2.10 at the fixed energy value of 0 eV (Fermi level), we find five states corresponding to five complex k vector.

For step 1, the complex band structure is calculated for fixed values of E and $k_\perp$. Instead of using the Schrödinger equation, we will use the Kohn–Sham equation [2.17] and periodic boundary conditions in the z direction for the wavefunction and its derivative with respect to z:

$$\varphi_k(r_\perp, z+d) = e^{ik_z d}\varphi_k(r_\perp, z), \qquad [2.97]$$

$$\varphi_k'(r\perp, z+d) = e^{ik_z d}\varphi_k'(r_\perp, z), \qquad [2.98]$$

where d the length of the unit cell in the z direction.

For step 2, we solve the scattering problem. We have to solve a system of equations that is very similar to the one for the complex band calculation. To this end, we solve the Kohn–Sham equation (equation [2.17]) for fixed values of E and $k_\perp$, but the boundary conditions are no longer periodic. Instead, we search for a solution that matches continuously with the following form in the leads:

$$\varphi = \varphi_j + \sum_{i \in L} r_{ij}\varphi_i, \quad z < 0 \qquad [2.99]$$

$$\varphi = \sum_{i \in R} t_{ij}\varphi_i, \quad z > L \qquad [2.100]$$

the transmittance t_{ij} and the reflectance r_{ij} are the quantities we are looking for. The matching is done at all atoms crossing the $z = 0$ or $z = L$ plane.

Finally for the step 3, we evaluate the transmission coefficients as:

$$T_{ij} = \sqrt{I_i/I_j t_{ij}} \qquad [2.101]$$

$$T = \sum_{ij} |T_{ij}|^2 \qquad [2.102]$$

where I_j is the current of the propagating state φ_j, evaluated from standard quantum mechanics. The sum runs for all propagating states in the leads at the Fermi energy.

An example of the *ab initio* complex band structure for the ferroelectric phase of $BaTiO_3$ is shown in Figure 2.10. The imaginary part of k_z controls the exponential penetration of the metal wavefunction into the barrier. The lowest value of $\mathrm{Im}(k_z)$ is the more pertinent one for the tunneling, and only the region of energy within the gap will contribute to the transport. We can also note that imaginary bands connect the two real band of same symmetry [BUT 01]. We can see clearly in Figure 2.10, that at the Γ point, the channel of conductance change at 1.7 eV. In fact, above 1.7 eV, the lowest value of k_z are coming from the first valence and conduction band, whereas below 1.7 eV, it comes from the first valence band and from the second conduction band, reflecting a change in the symmetry of the solution. At energy below 0.5 eV, we have a further change and the lowest $\mathrm{Im}(k_z)$ comes from a complex wavevector, which is neither purely real nor purely imaginary.

2.6.1.3. *Limits of the ab initio Landauer–Büttiker approach*

As we can see in equation [2.92], the imaginary part of k_z (called k_1 in this equation) are linked very tightly to the value of the gap of the barrier, because in the square potential model, U_0 is the energy of the bottom of the conduction band of the barrier. It is also notable in the complex band structure in Figure 2.10 that increasing the gap will result in larger value for $\mathrm{Im}(k_z)$ and lower transmission, as complex bands coming from the coupling of bands with greater energy differences have greater value of $\mathrm{Im}(k_z)$. Because DFT-LDA and GGA suffer from the well-known gap problem (see section 2.2.5), the same limitation will be encountered for the band gap of the complex bands structure: it will be strongly underestimated, and consequently the transmission will be overestimated.

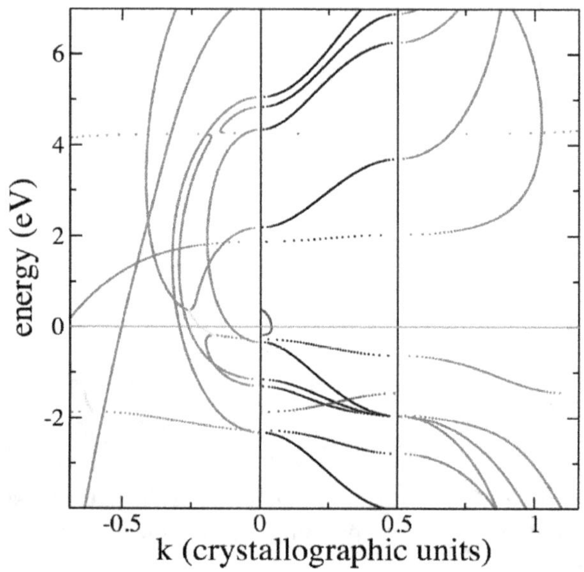

Figure 2.10. *Example of a complex band structure: BaTiO$_3$ at $k_\perp = 0$. Central panel: black points are for k_z that are purely real. Left panel: red points are purely imaginary k_z points. Right panel corresponds to k_z points imaginary part, where the real part equal to 0.5. Green points are for complex k_z, neither purely real nor purely imaginary, the green points are the imaginary part and the corresponding real part are in blue in the central panel. For a color version of this figure, see www.iste.co.uk/dollfus/nanodevices.zip*

Furthermore, when performing transport calculations with a realistic supercell structure for a ferroelectric barrier, the predicted gap can be so erroneously small that the conduction band of the barrier lies under the Fermi level energy of the leads. This can produce a severe problem in the simulation, because the electron of the leads will be transferred unphysically to the barrier and even the ground electronic density will be false [STE 11].

2.6.2. *The master equation*

In the previous section (section 2.6.1), we introduced quantum transport within the Landauer–Büttiker approach. However, in this approach the transport is purely ballistic. When scattering processes (electron–electron, electron–phonon) effects become important, we need to go beyond the Landauer–Büttiker method.

In this section, we introduce the quantum master equation for the electronic system that describes the evolution of the electronic density matrix, allowing for the inclusion of interband coupling and all of the relative phenomena. More precisely, we will focus on the quantum master equation formalism for a dissipative system (electrons interacting with phonons). We first introduce definitions, among which the density matrix operator, which is the key variable in the master equation (see Table 2.1).

2.6.2.1. *Pure state, mixed state and density matrix operator*

An electronic system is called a *pure state* [COH 97] when it can be described by a single wavefunction $|\phi_e^m(t)\rangle$ that can be decomposed into a complete basis set of stationary states $\{|\varphi_e^{\alpha,m}\rangle\}$ (see equation [2.5]) as:

$$|\phi_e^m(t)\rangle = \sum_\alpha c_{\alpha,m}(t)|\varphi_e^{\alpha,m}\rangle, \quad [2.103]$$

where the coefficients $c_{\alpha,m}(t)$ defined by

$$c_{\alpha,m}(t) = \langle\varphi_e^{\alpha,m}|\phi_e^m(t)\rangle \quad [2.104]$$

are determined by means of the time-dependent Schrödinger equation:

$$i\hbar\frac{\partial}{\partial t}|\phi_e^m(t)\rangle = \hat{H}_e|\phi_e^m(t)\rangle. \quad [2.105]$$

In this case, the expectation value of any operator $\hat{\mathbf{O}}$ is given by:

$$\langle\hat{\mathbf{O}}\rangle = \langle\phi_e^m(t)|\hat{\mathbf{O}}|\phi_e^m(t)\rangle = \sum_{\alpha,\alpha'} c_{\alpha,m}^*(t)c_{\alpha'm}(t)\hat{\mathbf{O}}_{\alpha,\alpha'}^m, \quad [2.106]$$

where $\hat{\mathbf{O}}_{\alpha,\alpha'}^m = \langle\varphi_e^{\alpha,m}|\hat{\mathbf{O}}|\varphi_e^{\alpha',m}\rangle$. In most of practical situations, for instance in the case of an electronic system coupled to a thermal bath, the system of interest cannot be described by a pure state. In this case, we must use a mixed state defined as a linear combination of states $|\phi_e^m(t)\rangle$ and perform statistical averages over the states $|\phi_e^m(t)\rangle$ involved in order to describe the quantum observables. A practical tool to reach this goal is the density matrix operator $\hat{\rho}_M$ defined by:

$$\hat{\rho}_M = \sum_m f_m |\phi_e^m(t)\rangle\langle\phi_e^m(t)|, \qquad [2.107]$$

where f_m denotes the probability of the total system to be in the state $|\Psi_i\rangle$ and $0 \leq f_m \leq 1$ with $\sum_m f_m = 1$. The density operator $\hat{\rho}_M$ obeys the following properties:

1) hermiticity: $\hat{\rho}_M^\dagger = \hat{\rho}_M$;

2) normalization: $\mathrm{Tr}\{\hat{\rho}_M\} = 1$;

3) semipositivity: $\langle\phi_e^m(t)|\hat{\rho}_M|\phi_e^m(t)\rangle \geq 0$ for all wavefunctions $|\phi_e^m(t)\rangle$.

$\mathrm{Tr}\{\hat{\rho}_M\}$ is the trace of the density matrix operator. For a pure state $\hat{\rho}_M^2 = \hat{\rho}_M$, $\mathrm{Tr}\{\hat{\rho}_M^2\} = 1$. The expectation value for any operator in a mixed state is given by:

$$\langle\hat{\mathbf{O}}\rangle \equiv \mathrm{Tr}\{\hat{\mathbf{O}}\hat{\rho}_M\} = \sum_m f_m \langle\phi_e^m(t)|\hat{\mathbf{O}}|\phi_e^m(t)\rangle. \qquad [2.108]$$

2.6.3. *Dynamics of the density matrix operator: closed and open system*

2.6.3.1. *Master equation for closed systems*

If one considers an electronic system (electrons) that does not interchange neither energy nor matter with its environment (a closed system), the time evolution of the density matrix operator $\hat{\rho}_M$ is given by the so-called Liouville–von Neumann master equation [COH 98]:

$$\frac{d\hat{\rho}_M}{dt} = \frac{1}{i\hbar}\left[\hat{H}_e, \hat{\rho}_M\right] \equiv \hat{\mathcal{L}}\hat{\rho}_M, \qquad [2.109]$$

where $\hat{H}_e$ is the many body Hamiltonian of the system (equation [2.6]) and where the "^" symbol underlines that the equation above deals with operators. $\hat{\mathcal{L}}$, the Liouville operator, acts on operators and, for this reason, is called a superoperator. The Liouville-von Neumann equation has the formal solution

$$\hat{\rho}_M(t) = \mathcal{U}(t, t_0)\hat{\rho}_M(t_0)\mathcal{U}^\dagger(t, t_0), \qquad [2.110]$$

with

$$\mathcal{U}(t, t_0) = \exp\left[-\frac{i}{\hbar}\hat{H}_e(t - t_0)\right]. \qquad [2.111]$$

$\hat{\rho}_M(t_0)$ defines the initial condition and $\mathcal{U}(t, t_0)$ is the time evolution operator. When the electronic system (electrons) is in permanent contact with its environment or a thermal bath (phonons), the system is open and the time evolution of the density matrix operator corresponding to the electronic system cannot be described by equation [2.109]. However, by assuming the whole system (electron + thermal bath) to be a closed system, it is possible, under some approximations, to derive an equation describing the time evolution of the density matrix operator of the electronic subsystem.

2.6.3.2. *Master equation for a dissipative system*

In this section, we give the main steps leading to the equation describing the time evolution of the density matrix operator corresponding to the electronic subsystem (electrons) interacting with the thermal bath (phonons bath) (Figure 2.11). A detailed derivation can be found in [COH 98, BRE 02, ROS 11]. The density matrix of the total system is denoted by $\hat{\rho}_M^{tot}$ and the total Hamiltonian by $\hat{H} = \hat{H}_e + \hat{H}_\nu + \hat{H}_{\nu e}$ (see sections 2.1–2.3). The total density matrix obeys the Liouville-von Neumann equation:

$$\frac{d\hat{\rho}_M^{tot}}{dt} = \frac{1}{i\hbar}\left[\hat{H}, \hat{\rho}_M^{tot}\right]. \qquad [2.112]$$

Since we are only interested in the electronic subsystem, we define the "reduced density matrix" $\hat{\rho}_M$ that reads:

$$\hat{\rho}_M = \mathrm{Tr}_\nu\{\hat{\rho}_M^{tot}\}, \qquad [2.113]$$

where we are taking the trace over all the phonon variables of the thermal bath. To achieve an equation of motion for the reduced density matrix $\hat{\rho}_M$, a number of assumptions are made. First, it is assumed that no correlations exist between the electronic system and the phonon bath at $t = t_0$. Hence, the initial density matrix factorizes as [COH 98, BRE 02, VEN 08, ROS 11]:

$$\hat{\rho}_M^{tot}(t_0) = \hat{\rho}_M(t_0)\hat{\rho}_\nu(t_0), \qquad [2.114]$$

where ρ_ν is the reduced density matrix of the bath. The interaction $\hat{H}_{\nu e}$ between the two subsystems (electron and phonon bath) is assumed to be weak, which allows to make the Born approximation [VEN 08] and write, when the interaction is turned on for $t > 0$, the total density as

$$\hat{\rho}_M^{tot}(t) \approx \hat{\rho}_M(t)\hat{\rho}_\nu(t_0)\cdot \quad [2.115]$$

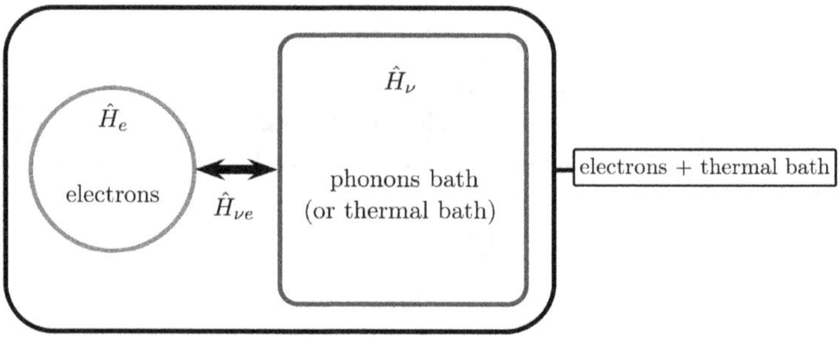

Figure 2.11. *Open system: electrons system coupled to a phonons bath*

The last approximation consists of considering that the future evolution of the electronic system is entirely determined by its present and not by its past [GEB 04a]. This approximation is known as the Markov approximation.

By means of the approximations mentioned above, we find the following equation, giving the time evolution of the reduced electronic density matrix also called Lindblad quantum master equation [COH 98, BRE 02, VEN 08, ROS 11]:

$$\frac{d\hat{\rho}_M}{dt} = -\frac{i}{\hbar}\left[\hat{H}_e, \hat{\rho}_M\right] + \hat{\mathcal{D}}\left[\hat{\rho}_M\right], \quad [2.116]$$

where the term involving the commutator with $\hat{H}_e$ describes the coherent evolution. $\hat{\mathcal{D}}$ describes the collisions with the phonons and produces the irreversible evolution of the system. Because of the presence of this operator, the electronic Hamiltonian appearing in equation [2.116] may not necessarily coincide with the Hamiltonian in equation [2.109] in the absence of bath. Indeed the bath, via its interaction with the system, may also contribute to the coherent part of the system evolution by shifting (or renormalizing) its energy states [VEN 08, p. 28]. $\hat{\mathcal{D}}\left[\hat{\rho}_M\right]$, also known as the dissipator, has the Lindblad form [HUB 96, GEB 04a]:

$$\hat{\mathcal{D}}\left[\hat{\rho}_M\right] = \sum_{mn} W_{mn}\left(2\hat{\mathcal{Q}}_{mn}\hat{\rho}_M\hat{\mathcal{Q}}_{mn}^\dagger - \hat{\mathcal{Q}}_{mn}^\dagger\hat{\mathcal{Q}}_{mn}\hat{\rho}_M - \hat{\rho}_M\hat{\mathcal{Q}}_{mn}^\dagger\hat{\mathcal{Q}}_{mn}\right) \quad [2.117]$$

where $\hat{\mathcal{Q}}_{mn} = |m\rangle\langle n|$ is an operator that describes the decay channel of the system (transition from state $|n\rangle$ to state $|m\rangle$: the operator $\hat{\mathcal{Q}}_{mn}$ destroys an electron in the state $|n\rangle$ and creates one electron into the state $|m\rangle$), and W_{mn} is the transition probability defined by equation [2.66], which is proportional to the electron–phonon coupling constant g_{mn}, the key quantity which is calculated *ab initio*.

2.6.3.3. *Single-particle description*

The quantum master equation (equation [2.116]) describes a many-electron system. The computational cell typically contains from hundreds to thousands of electrons, and it is of course impossible to work in terms of the exact many-particle wavefunctions or density matrices [GEB 06]. Moreover, in practical situations, many of the electronic properties of interest in the analysis of materials are single-particle quantities, like the carrier drift velocity [ROS 11]. It is therefore required to work with the reduced one-particle density matrix. To this end, we proceed like in [HUB 96] and neglect the electron–electron correlations induced by electron–electron interaction and electron–phonon coupling.

By defining the reduced one particle density matrix on the basis of single states $|m\rangle$, $|n\rangle$ as:

$$\rho_{mn} = \mathrm{Tr}\left\{c_n^\dagger c_m \hat{\rho}_M\right\}, \qquad [2.118]$$

with $c_m^\dagger$ (c_m) the creator (annihilator) of the eigenstate $|m\rangle$, we can show that (see Appendix A, section 2.8) the single-particle master equation reads:

$$\frac{d\rho_{mn}}{dt} = \left[\frac{d\rho_{mn}}{dt}\right]_{\hat{H}_e} + \left[\frac{d\rho_{mn}}{dt}\right]_{\hat{\mathcal{D}}} \qquad [2.119]$$

where

$$\left[\frac{d\rho_{mn}}{dt}\right]_{\hat{H}_e} = -\frac{i}{\hbar}\sum_p \left[(H_e)_{mp}\rho_{pn} - \rho_{mp}(H_e)_{pn}\right] \qquad [2.120]$$

is the time variation induced by the non-interacting Hamiltonian $\hat{H}_e$, with $(H_e)_{mp}$ is the Hamiltonian element matrix between the single-particle states.

$$\begin{aligned}\left[\frac{d\rho_{mn}}{dt}\right]_{\hat{\mathcal{D}}} &= \frac{1}{2}\sum_p (W_{mp} + W_{np})\rho_{pp}(\delta_{mn} - \rho_{mn}) \\ &\quad -\frac{1}{2}\sum_p (W_{pm} + W_{pn})\rho_{mn}(1 - \rho_{pp})\end{aligned} \qquad [2.121]$$

This part describes the collision of the electrons with the phonons.

2.6.3.4. *Physical meaning of the single-particle master equation*

In this section, we take a close look on equation [2.119]. An interesting fact is that by neglecting the off-diagonal elements in equation [2.119], the first part of the right-hand side terms, $\left[\frac{d\rho_{mn}}{dt}\right]_{\hat{H}_e}$, vanishes and we can recover the semiclassical BTE (see equation [2.69]), the diagonal elements of the reduced single-particle density ρ_{nn} corresponding to the single-particle carrier distribution f_n of the semiclassical Boltzmann theory. If, in addition, there is no collision process ($\left[\frac{d\rho_{mn}}{dt}\right]_{\hat{\mathcal{D}}} = 0$), equation [2.119] tell us that a particle initially on the band n remains on it ($\frac{d\rho_{mn}}{dt} = 0$), see Figure 2.12. This is true for large band gap materials but it is no longer valid in the presence of an external electric field in a narrow gap or zero gap material (graphene, for instance). Indeed in such materials, the electric field can induce a transition of the electrons from one band to another one through the forbidden energy gap. This phenomenon is known as the Zener tunneling and it can influence the transport properties. Now if we consider the off-diagonal elements of the single-particle density matrix, equation [2.119] does not vanish even in the absence of collision processes ($\frac{d\rho_{mn}}{dt} \neq 0$). The carrier populations (diagonal elements of the density matrix) of the bands (n and m) are related via the off-diagonal elements. This means that an electron initially on band n can transit to band m. The first part of the right-hand side terms of equation [2.119], $\left[\frac{d\rho_{mn}}{dt}\right]_{\hat{H}_e}$, is thus responsible for the transition (tunneling) of carriers from one band to another in the absence of collision processes.

2.6.3.5. *Master equation and DFT*

It is possible to describe the dissipative many-electron system, evolving under the master equation (equation [2.116]) using a generalization of DFT to dissipative systems. In the same way as in standard DFT, we can prove that the potential is uniquely determined (up to an arbitrary additional constant) by the charge density $\rho(\mathbf{r})$, it is possible to show that for a dissipative quantum system, no two different one-body potentials can give rise to the same time-dependent density $\rho(\mathbf{r}, t)$, given the superoperator $\hat{\mathcal{D}}$ in equation [2.116] and an initial density matrix $\hat{\rho}_M(t_0)$ [GEB 06, BUR 05]. In this case, the electronic Hamiltonian $\hat{H}_e$ in equation [2.120] is evaluated within DFT (see section 2.2) and so the terms ε^m and $|m\rangle$ will be, respectively, the eigenvalues and eigenstates of Kohn–Sham Hamiltonian (see equation [2.16]).

The approach described above (equation [2.119]) was used to study the transport in heterojunctions: a model of double-barrier resonant tunneling structure [GEB 04b] and a realistic simulation of benzene dithiolate monolayer sandwiched between two Au(111) surfaces [GEB 06]. In order to reach the steady state, the dissipation was treated in the leads with an electron–phonon coupling strength much stronger than what would be given by a realistic calculation of the electron–phonon coupling. Indeed in the experimental case, the leads will be much larger and the number of scattering events would be increased. The increase in the electron–phonon coupling strength thus compensates the small size of the model of leads.

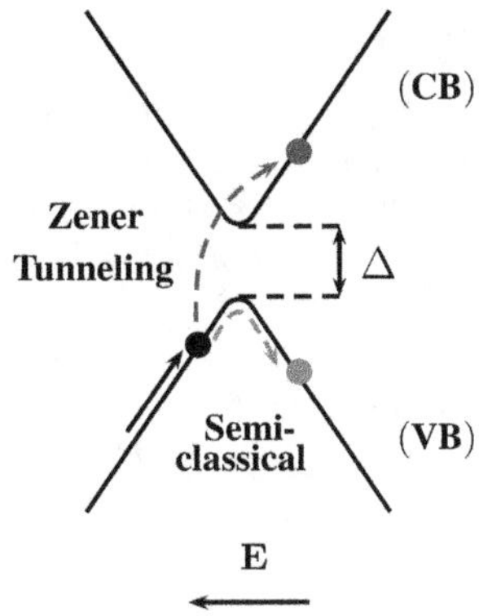

Figure 2.12. *Schematic of carrier dynamics in the presence of an external electric field* **E** *and in the absence of scattering processes.* Δ *represents the band gap. Under the applied electric field, an electron initially on the valence band (VB) can transit to the conduction band (CB) via Zener's tunneling or remain on the same band (semiclassical model)*

Another example based on equation [2.119] is the study of the Zener tunneling in quasi-metallic (Qm) carbon nanotubes (see Figure 2.13) [KAN 12]. In this case, the dissipation was not restricted to the leads. Instead, it was taken into account in the whole system by including the electron–phonon coupling of the nanotubes. The electron–phonon coupling was obtained by *ab initio* calculation [LAZ 06a]. In this study, it has been shown that the Zener tunneling can be made visible by defects in Qm nanotubes. Indeed, for small doping, the presence of Zener's tunneling leads to a finite value of the conductivity, which increases with decreasing elastic scattering length of the nanotube.

2.6.3.6. *Limitations of the method*

One limit of the master equation lies in an important approximation made to describe the collisional part, i.e. the dissipator term of equation [2.117], or alternatively, the right-hand side term of equation [2.119]. When developing this term, correlation terms are found out at all orders. To make the method tractable however, in its actual implementation, only first-order correlations of the density matrix are taken into account. Even second-order correlation are, for instance, neglected. This amounts to use a kind of mean-field (Hartree-like) approximation for the collisional part.

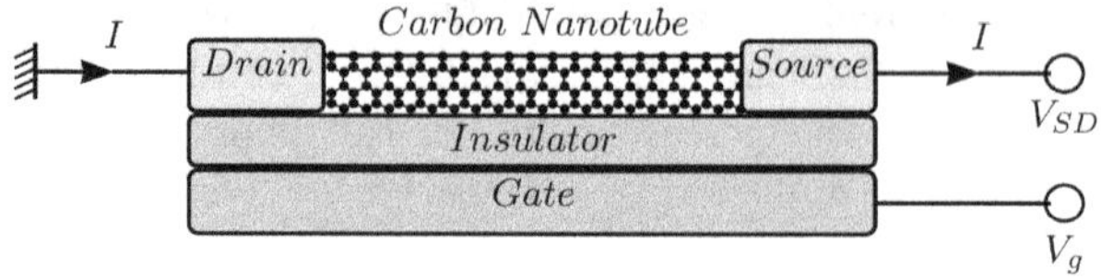

Figure 2.13. *Schematic of a carbon nanotube field-effect transistor (CNTFET). The electrical current I is obtained by applying a voltage V_{SD} (between the source and the drain leads) and a voltage V_g on the gate electrode. The gate voltage, V_g, allows for changing the density of the carriers in the nanotube*

We point out that contrastingly, in the commutator of equation [2.116], or, alternatively, equation [2.120], correlations in H_e are treated in the framework of density functional theory.

2.6.4. *Non-equilibrium Green's functions*

Another method that can be used to study transport properties in nanodevices is the NEGF approach. NEGF has been used to investigate nanoscale materials such as carbon-based materials [KIM 08] or single molecules [LIA 11]. In this formalism, the carrier dynamics and interaction are essentially described by means of Green's functions: a correlation function $G^<$ that plays a role similar to the distribution function in the Boltzmann transport or to the density matrix in the master equation, and the self-energy $\sigma^<$ that describes the effects of the leads or the scattering (electron defect, electron–phonon, etc.) on the carrier motion. For a detailed description of the NEGF formalism, reader should refer to [DAT 97, BUO 09] and Chapter 6 in this book.

The understanding of the transport properties of systems such as single molecules requires a good description of their electronic structure. To this end, first principles calculations do play a key role. The NEGF has been successfully combined to DFT [BRA 02, ROC 06, OZA 10] for the calculation of electronic transport properties of systems like a molecular transistor [SAH 10] and nanomolecular switch [GAN 08]. However, the excitation energies are not well described within DFT nor in time-dependent density functional theory (TDDFT) [ONI 02]. We will not develop this part further since NEGF formalism is largely discussed in detail in this book (see Chapter 6).

2.7. Conclusion

In this chapter, the basics of the description of the electronic structure within DFT, and of the interaction of electrons with the phonons, have been discussed. DFT and DFPT have proved to be well grounded and practical tools of high added value that allow for the study of electronic transport properties with a minimum (if any) of fitting parameters. Guidelines have been given about the coupling of the *ab initio* approaches with the BTE, the Landauer–Büttiker approach and the master equation. In each case, significant examples have been given that aim to edify the reader's mind in the handling of these tools. Finally, a special emphasis has been put in describing as clearly as possible the limitations of each method.

2.8. Appendix A

2.8.1. *Derivation of the single-particle master equation*

In this Appendix, we give the main steps leading to the single-particle master equation (equation [2.119]). We recall that the electronic subsystem Hamiltonian $\hat{H}_e$ can be written, in the second quantization scheme, as:

$$\hat{H}_e = \sum_m \varepsilon^m c_m^\dagger c_m. \qquad [2.122]$$

where $c_m^\dagger$, c_m are, respectively, creators and annihilators of the eigenstates $|m\rangle$ (single-particle state). The reduced one-particle density matrix is defined on the basis of single states $|m\rangle$, $|n\rangle$ by:

$$\rho_{mn} = \mathrm{Tr}\left\{c_n^\dagger c_m \hat{\rho}_M\right\}. \qquad [2.123]$$

The diagonal elements of the reduced single-particle density ρ_{nn} correspond to the single-particle carrier distribution f_n of the semiclassical Boltzmann theory (see equation [2.69]) and are called population terms, whereas the off-diagonal terms, also called interstate polarizations are responsible for the phase coherence effects between the single-particle states $|m\rangle$ and $|n\rangle$. The reduced one-particle density operator ρ is normalized to the number of electrons $\mathcal{N}$:

$$\sum_m \rho_{mm} = \mathcal{N}. \qquad [2.124]$$

To derive a corresponding equation of motion for the single-electron density matrix, we use:

$$\frac{d\rho_{mn}}{dt} = \mathrm{Tr}\left\{c_n^\dagger c_m \frac{d\hat{\rho}_M}{dt}\right\}, \qquad [2.125]$$

The electronic operator $\hat{Q}_{mn}$ in the collision operator [2.117] is written in the second quantization as $\hat{Q}_{mn} = c_n^\dagger c_m$. By developing equation [2.125], we find some expressions containing the two-particle density matrix $\mathrm{Tr}\left\{c_n^\dagger c_m^\dagger c_p c_q \hat{\rho}_M\right\}$. These terms are then approximated by products of elements of the single-particle density matrix $\mathrm{Tr}\left\{c_n^\dagger c_m^\dagger c_p c_q \hat{\rho}_M\right\} \approx \rho_{pm}\rho_{qn}$ (Wick theorem or Hartree approximation), in which we neglect the exchange term $(\rho_{qm}\rho_{pn})$ [GEB 04b]. After this simplification, the single-particle master equation reads

$$\frac{d\rho_{mn}}{dt} = \left[\frac{d\rho_{mn}}{dt}\right]_{\hat{H}_e} + \left[\frac{d\rho_{mn}}{dt}\right]_{\hat{\mathcal{D}}} \qquad [2.126]$$

where

$$\left[\frac{d\rho_{mn}}{dt}\right]_{\hat{H}_e} = -\frac{i}{\hbar}\sum_p \left[(H_e)_{mp}\rho_{pn} - \rho_{mp}(H_e)_{pn}\right] \qquad [2.127]$$

is the time variation induced by the non-interacting Hamiltonian $\hat{H}_e$, with $(H_e)_{mp}$ is the Hamiltonian element matrix between the single-particle states.

$$\begin{aligned}\left[\frac{d\rho_{mn}}{dt}\right]_{\hat{\mathcal{D}}} &= \frac{1}{2}\sum_p (W_{mp} + W_{np})\rho_{pp}(\delta_{mn} - \rho_{mn}) \\ &-\frac{1}{2}\sum_p (W_{pm} + W_{pn})\rho_{mn}(1 - \rho_{pp})\end{aligned} \qquad [2.128]$$

This part describes the collision of the electrons with the phonons.

2.9. Bibliography

[ABE 10] ABEDI A., MAITRA N.T., GROSS E.K.U., "Exact factorization of the time-dependent electron-nuclear wave function", *Physical Review Letters*, vol. 105, p. 123002, 2010.

[ALM 85] ALMBLADH C.-O., VON BARTH U., "Exact results for the charge and spin densities, exchange-correlation potentials, and density-functional eigenvalues", *Physical Review B*, vol. 31, p. 3231, 1985.

[AND 75] ANDERSEN O.K., "Linear methods in band theory", *Physical Review B*, vol. 12, p. 3060, 1975.

[ANT 88] ANTOGNETTI P., MASSOBRIO G. (eds), *Semiconductor Device Modeling with SPICE*, McGraw-Hill, New York, 1988.

[ASH 76] ASHCROFT N.W., MERMIN N.D., *Solid State Physics*, Holt-Saunders, Philadelphia,1976.

[ASH 02] ASHCROFT N., MERMIN N., *Physique des Solides*, EDP Sciences, 2002.

[AUL 99] AULBUR W.G., JÖNSSON L., WILKINS J.W., *Solid State Physics*, vol. 54, Academic Press, London, 1999.

[AVE 98] AVERKIEV N., GOLUB L., PIKUS G., "Weak localization in semiconductor structures with strong spin-orbit coupling", *Journal of Experimental and Theoretical Physics*, vol. 86, no. 4, pp. 780–789, 1998.

[BAC 82] BACHELET G.B., HAMANN D.R., SCHLÜTER M., "Pseudopotentials that work: from H to Pu", *Physical Review B*, vol. 26, p. 4199, 1982.

[BAL 83] BALDERESCHI A., RESTA R., *Ab initio Calculation of Phonon Spectra*, Plenum, New York, 1983.

[BAR 01] BARONI S., DE GIRONCOLI S., CORSO A.D. *et al.*, "Phonons and related crystal properties from density-functional perturbation theory", *Reviews of Modern Physics*, vol. 73, p. 515, 2001.

[BAR 07] BARTLETT R., MUSIA M., "Coupled-cluster theory in quantum chemistry", *Reviews of Modern Physics*, vol. 79, p. 291, 2007.

[BIR 66] BIRMAN J.L., LAX M., LOUDON R., "Intervalley-scattering selection rules in III-V semiconductors", *Physical Review*, vol. 145, p. 620, 1966.

[BLU 06] BLÜGEL S., BIHLMAYER G.,"Full potential linearized augmented planewave method", in GROTENDORST J. *et al.* (eds), *Computational Nanoscience: Do It Yourself!*, vol. 31, John von Neumann Institute for Computing, Jülich, 2006.

[BLU 09] BLUM V., GEHRKE R., HANKE F. *et al.*, "*Ab initio* molecular simulations with numeric atom-centered orbitals", *Computer Physics Communication*, vol. 180, p. 2175, 2009.

[BON 07] BONINI N., LAZZERI M., MARZARI N. *et al.*, "Phonon anharmonicities in graphite and graphene", *Physical Review Letter*, vol. 99, p. 176802, 2007.

[BOR 54] BORN M., HUANG K. (eds), *Dynamical Theory of Crystal Lattices*, Oxford University Press, New York, 1954.

[BOR 96] BORDONE P., VASILESKA D., FERRY D.K., "Collision-duration time for optical-phonon emission in semiconductors", *Physical Review B*, vol. 53, pp. 3846–3855, 1996.

[BRA 02] BRANDBYGE M., MOZOS J.-L., ORDEJÓN P. *et al.*, "Density-functional method for nonequilibrium electron transport", *Physical Review B*, vol. 65, p. 165401, 2002.

[BRE 02] BREUER H.P., PETRUCCIONE F., *The Theory of Open Quantum Systems*, Oxford University Press, Oxford, 2002.

[BRU 06] BRUNEVAL F., VAST N., REINING L., "Effect of self-consistency on quasiparticles in solids", *Physical Review B*, vol. 74, p. 045102, 2006.

[BRU 09] BRUN C., HONG I.-P., PATTHEY F. *et al.*, "Reduction of the Superconducting gap of ultrathin Pb islands grown on Si(111)", *Physical Review Letter*, vol. 102, p. 207002, 2009.

[BUO 09] BUOT F., *Nonequilibrium Quantum Transport Physics in Nanosystems: Foundation of Computational Nonequilibrium Physics in Nanoscience and Nanotechnology*, World Scientific, Singapore, 2009.

[BUR 05] BURKE K., CAR R., GEBAUER R., "Density functional theory of the electrical conductivity of molecular devices", *Phys. Rev. Lett.*, vol. 97, p. 146803, 2005.

[BUT 01] BUTLER W.H., ZHANG X.-G., SCHULTHESS T.C. *et al.*, "Spin-dependent tunneling conductance of f Fe|MgO|Fe sandwiches", *Physical Review B*, vol. 63, p. 054416, 2001.

[BYR 69] BYRON F., FULLER R., *Mathematics of Classical and Quantum Physics*, Dover Publications, New York, 1969.

[CAL 04] CALANDRA M., VAST N., MAURI F., "Superconductivity from doping Boron Icosahedra", *Physical Review B*, vol. 69, p. 224505, 2004.

[CAL 08] CALANDRA M., MAURI F., "High-T_c superconductivity in superhard diamondlike BC_5", *Physical Review Letters*, vol. 101, p. 016401, 2008.

[CAL 10] CALANDRA M., PROFETA G., MAURI F., "Adiabatic and nonadiabatic phonon dispersion in a Wannier approach", *Physical Review B*, vol. 82, p. 165111, 2010.

[CAL 11] CALANDRA M., MAURI F., "Charge-density wave and superconducting dome in $TiSe_2$ from electron-phonon interaction", *Phys. Rev. Lett.*, vol. 106, p. 196406, 2011.

[CEP 80] CEPERLEY D., ALDER B., "Ground state of the electron gas by a stochastic method", *Phys. Rev. Lett.*, vol. 45, p. 566, 1980.

[CHO 99] CHOI H.J., IHM J., "*Ab initio* pseudopotential method for the calculation of conductance in quantum wire", *Physical Review B*, vol. 59, p. 2267, 1999.

[CHR 84] CHRISTENSEN N.E., "Force Theorem and elastic constant of solids", *Solid State Communication*, vol. 49, p. 701, 1984.

[CIC 69] CICCO P.D., JOHNSON F., "The quantum theory of lattice dynamics", *Proceedings of the Royal Society of London Series A*, vol. 310, p. 111, 1969.

[COH 97] COHEN-TANNOUDJI C., DIU B., LALOË F., *Quantum Mechanics, Volume 1*, John Wiley and Sons, New York, 1997.

[COH 98] COHEN-TANNOUDJI C., DUPONT-ROC J., GRYNBERG G., *Atom-Photon Interactions*, John Wiley and Sons, New York, 1998.

[COM 01] COMBESCOT M., "On the generalized golden rule for transition probabilities", *J. Phys. A: Math Gen.*, vol. 34, p. 6087, 2001.

[DAL 06] DAL CORSO A., SMOGUNOV A., TOSATTI E., "*Ab initio* ballistic conductance with spin-orbit coupling: application to monoatomic wires", *Physical Review B*, vol. 74, p. 045429, 2006.

[DAT 97] DATTA S., *Electronic Transport in Mesoscopic Systems*, Cambridge University Press, Cambridge, 1997.

[DEB 95] DEBERNARDI A., BARONI S., MOLINARI E., "Anharmonic phonon lifetimes in semiconductors from density-functional perturbation theory", *Physical Review Letters*, vol. 75, p. 1819, 1995.

[DOB 98] DOBSOV J., VIGNALE G., DAS M. (eds), *Electronic Density Functional Theory: Recent Progress and New Directions*, Plenum, New York, 1998.

[DRE 90] DREIZLER R., GROSS E., *Density Functional Theory*, Spinger-Verlag, Berlin, Heidelberg, 1990.

[ERR 14] ERREA I., CALANDRA M., MAURI F., "Anharmonic free energies and phonon dispersions from the stochastic self-consistent harmonic approximation: application to platinum and palladium hydrides", *Physical Review B*, vol. 89, p. 064302, 2014.

[FAB 11] FABER C., JANSSEN J.L., CÔTÉ M. *et al.*, "Electron-phonon coupling in the C_{60} fullerene within the many-body GW approach", *Physical Review B*, vol. 84, p. 155104, 2011.

[FEN 13] FENG Z., TIMOSHEVSKII V., MAUGER A. *et al.*, "Dynamics of polar on formation in Li_2O_2 from density functional perturbation theory", *Physical Review B*, vol. 88, p. 184302, 2013.

[FER 76] FERRY D., "First-order optical and intervalley scattering in semiconductors", *Physical Review B*, vol. 14, p. 1605, 1976.

[FEY 39] FEYNMAN R., "Forces in molecules", *Physical Review*, vol. 56, p. 340, 1939.

[FIS 91] FISCHETTI M.V., "Effet of the electron-plasmons on the electronic mobility in silicon", *Physical Review B*, vol. 44, p. 5527, 1991.

[FUG 13] FUGALLO G., LAZZERI M., PAULATTO L. *et al.*, "*Ab initio* variational approach for evaluating lattice thermal conductivity", *Physical Review B*, vol. 88, p. 045430, 2013.

[GAN 08] GANJI M., NOUROZI F., "Density functional non-equilibrium Green's function (DFT-NEGF) study of the smallest nano-molecular switch", *Physica E: Low-dimensional Systems and Nanostructures*, vol. 40, pp. 2606–2613, 2008.

[GEB 04a] GEBAUER R., CAR R., "Current in open quantum systems", *Physical Review Letters*, vol. 96, p. 160404, 2004.

[GEB 04b] GEBAUER R., CAR R., "Kinetic theory of quantum transport at the nanoscale", *Physical Review B*, vol. 70, p. 125324, 2004.

[GEB 06] GEBAUER R., BURKE K., CAR R., "Kohn-Sham master equation approach to transport through single molecules", in MARQYES M. *et al.* (eds), *Time-Dependent Density Functional Theory*, Springer, Berlin/Heidelberg, 2006.

[GEN 08] GENOVESE L., NEELOV A., GOEDECKER S. *et al.*, "Daubechies wavelets as a basis set for density functional pseudopotential calculations", *Journal of Chemical Physics*, vol. 129, p. 014109, 2008.

[GEO 96] GEORGES A., KOTLIAR G., KRAUTH W. *et al.*, "Dynamical mean-field theory of strongly correlated fermion systems and the limit of infinite dimensions", *Reviews of Modern Physics*, vol. 68, p. 13, 1996.

[GIA 91] GIANNOZZI P., DE GIRONCOLI S., PAVONE P. *et al.*, "*Ab initio* calculation of phonon dispersions in semiconductors", *Physical Review B*, vol. 43, p. 7231, 1991.

[GIA 09] GIANNOZZI P., BARONI S., BONINI N. *et al.*, "QUANTUM ESPRESSO: a modular and open-source software project for quantum simulations of materials", *Journal of Physics: Condensed Matter*, vol. 21, p. 395502, 2009.

[GIU 07] GIUSTINO F., YATES J.R., SOUZA I. *et al.*, "Electron-phonon interaction via electronic and lattice Wannier functions: superconductivity in boron-doped diamond reexamined", *Physical Review Letters*, vol. 98, p. 047005, 2007.

[GIU 16] GIUSTINO F., "Electron-phonon interactions from first principles", available at: www.arxiv.org/pdf/1603.06965.pdf, 2016.

[GOL 13] GOLESORKHTABAR R., PAVONE P., SPITALER J. *et al.*, "ElaStic: a tool for calculating second-order elastic constants from first principles", *Computer Physics Communication*, vol. 184, pp. 1861–1873, 2013.

[GON 95] GONZE X., "Adiabatic density-functional perturbation theory", *Physical Review A*, vol. 52, p. 1096, 1995.

[GON 09] GONZE X., AMADON B., ANGLADE P.-M. *et al.*, "ABINIT: first-principles approach to material and nanosystem properties", *Computer Physics Communications*, vol. 180, p. 2582, 2009.

[GRO 95] GROSS E.K.U., DREIZLER R.M. (eds), *Density Functional Theory*, NATO ASI, Plenum, New York, 1995.

[GRU 06] GRÜNING M., MARINI A., RUBIO A., "Density functionals from many-body perturbation theory: The band gap for semiconductors and insulators", *Journal of Chemical Physics*, vol. 124, p. 154108, 2006.

[HAF 08] HAFNER J., "*Ab-initio* simulations of materials using VASP: density-functional theory and beyond", *Journal of Computational Chemistry*, vol. 29, p. 2044, 2008.

[HAM 99] HAMEAU S., GULDNER Y., VERZELEN O. *et al.*, "Strong electron-phonon coupling regime in quantum dots: evidence for everlasting resonant polarons", *Physical Review Letters*, vol. 83, p. 4152, 1999.

[HAM 10] HAMAGUCHI C., *Basic Semiconductor Physics*, Springer, Berlin/Heidelberg, 2010.

[HED 69] HEDIN L., LUNDQVIST B., *Solid State Physics*, vol. 23, Academic Press, New York, 1969.

[HEL 37] HELLMANN H., *Einführung in die Quantenchemie*, Deuticke, Leipzig, Germany, 1937.

[HOH 64] HOHENBERG P., KOHN W., "Inhomogeneous electron gas", *Physical Review*, vol. 136, p. B864, 1964.

[HUB 96] HÜBNER R., GRAHAM R., "Landau-Zener transitions and dissipation in a mesoscopic ring", *Physical Review B*, vol. 53, p. 4870, 1996.

[HYB 86] HYBERTSEN M., LOUIE S., "Electron correlation in semiconductors and insulators: Band gaps and quasiparticle energies", *Physical Review B*, vol. 34, p. 5390, 1986.

[IHM 79] IHM J., ZUNGER A., COHEN M., "Momentum-space formalism for the total energy of solids", *Journal of Physics C: Solid State Physics*, vol. 12, p. 4409, 1979.

[IMR 99] IMRY Y., LANDAUER R., "Conductance viewed as transmission", *Reviews of Modern Physics*, vol. 71, p. S306, 1999.

[IRV 62] IRVIN J.C., “Resistivity of bulk silicon and of diffused layers in silicon”, *Bell System Technical Journal*, vol. 41, pp. 387–410, 1962.

[JAC 83] JACOBONI C., REGGIANI L., “The Monte Carlo method for the solution of charge transport in semiconductors with applications to covalent materials”, *Reviews of Modern Physics*, vol. 55, p. 645, 1983.

[JAC 99] JACKSON J., *Classical Electrodynamics*, 2nd ed., John Wiley and Sons, New York, 1999.

[JON 89] JONES R.O., GUNNARSSON O., “The density functional formalism, its applications and prospects”, *Reviews of Modern Physics*, vol. 61, p. 689, 1989.

[KAN 12] KANÉ G., LAZZERI M., MAURI F., “Zener tunneling in the electrical transport of quasimetallic carbon nanotubes”, *Physical Review B*, vol. 86, p. 155433, 2012.

[KAT 51] KATO T., “Fundamental properties of Hamiltonian operators of Schrödinger type”, *Transactions of the American Mathematical Society*, vol. 70, pp. 195–211, 1951.

[KAV 16] KAVIANI M., STRAND J., AFANAS’EV V.V. *et al.*, “Deep electron and hole polarons and bipolarons in amorphous oxide”, *Physical Review B*, vol. 94, p. 020103(R), 2016.

[KIM 08] KIM W.Y., KIM K.S., “Prediction of very large values of magnetoresistance in a graphene nanoribbon device”, *Nature Nano*technology vol. 3, p. 408, 2008.

[KOH 65] KOHN W., SHAM L., “Self-consistent equations including exchange and correlation effects”, *Physical Review*, vol. 140, p. A1133, 1965.

[KOH 99] KOHN W., “Nobel lecture: electronic structure of matter – wave functions and density functionals”, *Reviews of Modern Physics*, vol. 71, p. 1253, 1999.

[KUB 57] KUBO R., “Statistical-mechanical theory of irreversible processes. I. General theory and simple applications to magnetic and conduction problems”, *Journal of the Physical Society Japan*, vol. 12, p. 570, 1957.

[KUB 66] KUBO R., “The fluctuation-dissipation theorem”, *Reports on Progress in Physics*, vol. 29, p. 255, 1966.

[KÜB 92] KÜBLER J., EYERT V., “Electronic structure calculations”, in BUSCHOW K.H.J. (ed.), *Electronic and Magnetic Properties of Metals and Ceramics*, VCH Verlagsgesellschaft, Weinheim, 1996.

[KUN 83] KUNC K., MARTIN R., *Ab initio Calculation of Phonon Spectra*, Plenum, New York, 1983.

[LAN 83] LANGRETH D., MEHL M., “Beyond the local-density approximation in calculations of ground-state electronic properties”, *Physical Review B*, vol. 28, p. 1809, 1983.

[LAN 84] LANGRETH D., MEHL M., “Erratum: Beyond the local-density approximation in calculations of ground-state electronic properties”, *Physical Review B*, vol. 29, p. 2310, 1984.

[LAZ 03] LAZZERI M., CALANDRA M., MAURI F., “Anharmonic phonon frequency shift in MgB2”, *Physical Review B*, vol. 68, p. 220509, 2003.

[LAZ 06a] LAZZERI M., MAURI F., "Coupled dynamics of electrons and phonons in metallic nanotubes: Current saturation from hot-phonon generation", *Physical Review B*, vol. 73, p. 165419, 2006.

[LAZ 06b] LAZZERI M., MAURI F., "Nonadiabatic Kohn Anomaly in a Doped Graphene Monolayer", *Physical Review Letters*, vol. 97, p. 266407, 2006.

[LAZ 08] LAZZERI M., ATTACCALITE C., WIRTZ L. *et al.*, "Impact of the electron-electron correlation on phonon dispersion: failure of LDA and GGA DFT functionals in graphene and graphite", *Physical Review B*, vol. 78, p. 081406(R), 2008.

[LEE 98] VAN LEEUWEN R., "Density gradient expansion of correlation functions", *Physical Review B*, vol. 87, p. 155142, 1998.

[LEV 79] LEVY M., "Universal variational functionals of electron densities, first-order density matrices, and natural spin-orbitals and solution of the v-repreisentability problem", *Proceedings of the National Academy of Science USA*, vol. 76, p. 6062, 1979.

[LIA 11] LIANG Y.Y., CHEN H., MIZUSEKI H. *et al.*, "Gate-controlled current and inelastic electron tunneling spectrum of benzene: a self-consistent study", *The Journal of Chemical Physics*, vol. 134, p. 144113, 2011.

[LIN 11] LINDSAY L., BROIDO D., REINECKE T., "First-principles determination of ultrahigh thermal conductivity of boron arsenide: a competitor for diamond?", *Physical Review Letters*, vol. 111, p. 025901, 2011.

[LIP 91] LIPAVSKÝ P., KHAN F.S., KALVOVÁ A.C.V. *et al.*, "High-field transport in semiconductors. II. Collision duration time", *Physical Review B*, vol. 43, pp. 6650–6664, 1991.

[LUN 09] LUNDSTROM M., *Fundamentals of Carrier Transport*, Cambridge University Press, 2009.

[MAR 16a] MARTIN R.M., REINING L., CEPERLEY D.M., *Interacting Electrons: Theory and Computational Approaches*, Cambridge University Press, 2016.

[MAR 16b] MARKOV M., SJAKSTE J., FUGALLO G. *et al.*, "Nanoscale mechanisms for the reduction of heat transport in bismuth", *Physical Review B*, vol. 93, p. 064301, 2016.

[MAU 96] MAURI F., ZAKHAROV O., DE GIRONCOLI S. *et al.*, "Phonon softening and superconductivity in tellurium under pressure", *Physical Review Letters*, vol. 77, p. 1151, 1996.

[MET 93] METHFESSEL M., VAN SCHILFGAARDE M., "Derivation of force theorems in density-functional theory: application to the full-potential LMTO method", *Physical Review B*, vol. 48, p. 4937, 1993.

[MOU 74] MOUSTY F., OSTOJA P., PASSARI L., "Relationship between resistivity and phosphorus concentration in silicon", *Journal of Applied Physics*, vol. 45, p. 4576, 1974.

[MUR 06] MURPHY-ARMANDO F., FAHY S., "*First-principles* Calculation of Alloy Scattering in $\mathrm{Ge}_x\mathrm{Si}_{1-x}$", *Physical Review Letters*, vol. 97, p. 096606, 2006.

[MUR 08] MURPHY-ARMANDO F., FAHY S., "*First-principles* calculation of carrier-phonon scattering in n-type Si_{1-x} Ge_x alloys", *Physical Review B*, vol. 78, p. 035202, 2008.

[NEE 10] NEEDS R.J., TOWLER M.D., DRUMMOND N. *et al.*, "Continuum variational and diffusion quantum Monte Carlo calculations", *Journal of Physics: Condensed Matter*, vol. 22, p. 023201, 2010.

[OHY 07] OHYA S., HAI P.N., MIZUNO Y. *et al.*, "Quantum size effect and tunneling magnetoresistance in ferromagnetic-semiconductor quantum heterostructures", *Physical Review B*, vol. 75, p. 155328, 2007.

[ONI 02] ONIDA G., REINING L., RUBIO A., "Electronic Excitations: Density Functional versus Many Body Green's Functions Approaches", *Reviews of Modern Physics*, vol. 74, p. 601, 2002.

[OZA 10] OZAKI T., NISHIO K., KINO H., "Efficient implementation of the nonequilibrium Green function method for electronic transport calculations", *Physical Review B*, vol. 81, p. 035116, 2010.

[PAR 89] PARR R., YANG W., *Density-Functional Theory of Atoms and Molecules*, Oxford University Press, 1989.

[PAR 97] PARLINSKI K., LI Z.-Q., KAWAZOE Y., "First-principles determination of the soft mode in cubic ZrO_2", *Physical Review Letters*, vol. 78, p. 4063, 1997.

[PAU 13] PAULATTO L., MAURI F., LAZZERI M., "Anharmonic properties from a generalized third-order ab initio approach: Theory and applications to graphite and graphene.", *Physical Review B*, vol. 87, p. 214303, 2013.

[PER 81] PERDEW J., ZUNGER A., "Self-interaction correction to density-functional approximations for many-electron systems", *Physical Review B*, vol. 23, p. 5048, 1981.

[PER 03] PERDEW J., KURTH S., "Density functionals for non-relativistic Coulomb systems in the new century", FIOLHAIS C., NOGUEIRA F., MARQUES M. (eds), *A Primer in Density Functional Theory*, Springer-Verlag, Berlin, vol. 620, 2003.

[PIC 70] PICK R., COHEN M., MARTIN R., "Microscopic theory of force constants in the adiabatic approximation", *Physical Review B*, vol. 1, p. 910, 1970.

[PIC 89] PICKETT W., "Pseudopotential methods in condensed matter applications", *Computer Physics Report*, vol. 9, p. 115, 1989.

[PIS 04] PISCANEC S., LAZZERI M., MAURI F. *et al.*, "Kohn anomalies and electron-phonon interactions in graphite", *Physical Review Letters*, vol. 93, p. 185503, 2004.

[PIS 07] PISCANEC S., LAZZERI M., ROBERTSON J. *et al.*, "Optical phonons in carbon nanotubes: Kohn anomalies, Peierls distortions, and dynamic effects", *Physical Review B*, vol. 75, p. 035427, 2007.

[PRU 05] PRUNNILA M., KIVINEN P., SAVIN A. *et al.*, "Intervalley-scattering-induced electron-phonon energy relaxation in many-valley semiconductors at low temperatures", *Physical Review Letters*, vol. 95, p. 206602, 2005.

[PUL 99] PULCI O., BECHSTEDT F., ONIDA G. *et al.*, "State mixing for quasiparticles at surfaces: nonperturbative GW approximation", *Physical Review B*, vol. 60, p. 16758, 1999.

[REG 99] REGGIANI L., STARIKOV E., SHIKTOROV P. *et al.*, "A fundamental time scale of collision duration", *Solid State Communications*, vol. 112, pp. 119–122, 1999.

[RES 09] RESTREPO O., VARGA K., PANTELIDES S., "*First-principles* calculations of electron mobilities in silicon: phonon and Coulomb scattering", *Applied Physics Letters*, vol. 94, p. 212103, 2009.

[ROC 06] ROCHA A.R., GARCÍA-SUÁREZ V.M., BAILEY S. *et al.*, "Spin and molecular electronics in atomically generated orbital landscapes", *Physical Review B*, vol. 73, p. 085414, 2006.

[ROS 11] ROSSI F., *Theory of Semiconductor Quantum Devices: Microscopic Modeling and Simulation Strategies*, Springer, Berlin, Germany, 2011.

[SAH 10] SAHA K.K., NIKOLI B.K., MEUNIER V. *et al.*, "Quantum-interference-controlled three-terminal molecular transistors based on a single ring-shaped molecule connected to graphene nanoribbon electrodes", *Physical Review Letters*, vol. 105, p. 236803, 2010.

[SAV 96] SAVRASOV S.Y., SAVRASOV D.Y., "Electron-phonon interactions and related physical properties of metals from linear-response theory", *Physical Review B*, vol. 54, pp. 16487–16501, 1996.

[SCH 95] SCHÖNHAMMER K., GUNNARSSON O. *et al.*, "Density-functional theory on a lattice: Comparison with exact numerical results for a model with strongly correlated electrons", *Physical Review B*, vol. 52, p. 2504, 1995.

[SCH 98] SCHÖLL E., *Theory of Transport Properties of Semiconductor Nanostructures*, Chapman and Hall, London, 1998.

[SCH 02] SCHWARZ K., BLAHA P., MADSEN G., "Electronic structure calculations of solids using the WIEN2k package for material sciences", *Computer Physics Communications*, vol. 147, p. 71, 2002.

[SEM 96] SEMINARIO J. (ed.), *Recent developments and Applications of Modern Density Functional Theory*, Elsevier, Amsterdam, 1996.

[SHE 99] SHEIK-BAHAE M., "Quantum interference control of current in semiconductors: universal scaling and polarization effects", *Physical Review B*, vol. 60, pp. R11257–R11260, 1999.

[SHI 00] SHIRAI K., "First-Principles PseudoPotential Electronic Structure Calculations OSAKA2000", 2000.

[SJA 06] SJAKSTE J., TYUTEREV V., VAST N., "*Ab initio* study of $\Gamma - X$ intervalley scattering in GaAs under pressure", *Physical Review B*, vol. 74, p. 235216, 2006.

[SJA 07a] SJAKSTE J., TYUTEREV V., VAST N., "Intervalley scattering in GaAs: *ab initio* calculation of the effective parameters for Monte Carlo simulations", *Applied Physics A*, vol. 86, p. 301, 2007.

[SJA 07b] SJAKSTE J., VAST N., TYUTEREV V., "*Ab initio* method for the electron-phonon scattering times in semiconductors: application to GaAs and GaP", *Physical Review Letters*, vol. 99, p. 236405, 2007.

[SJA 13] SJAKSTE J., VAST N., JANI H. *et al.*, "*Ab initio* study of the effects of pressure and strain on electron-phonon coupling in IV and III-V semiconductors", *Physica Status Solidi B*, vol. 250, p. 716, 2013.

[SJA 14] SJAKSTE J., TIMROV I., GAVA P. *et al.*, "First-principles calculations of electron-phonon scattering. *Annual Review of Heat Transfer*, Begell House Inc., Danbury, 2014.

[SJA 15] SJAKSTE J., VAST N., CALANDRA M. *et al.*, "Wannier interpolation of the electron-phonon matrix elements in polar semiconductors: Polar-optical coupling in GaAs", *Physical Review B*, vol. 92, p. 054307, 2015.

[SMO 04] SMOGUNOV A., DAL CORSO A., TOSATTI E., "Ballistic conductance of magnetic Co and Ni nanowires with ultrasoft pseudopotentials", *Physical Review B*, vol. 70, p. 045417, 2004.

[SOL 02] SOLER J., ARTACHO E., GALE J. *et al.*, "The SIESTA method for ab initio order-N materials simulation", *Journal of Physics: Condensed Matter*, vol. 14, p. 2745, 2002.

[STE 11] STENGEL M., AGUADO-PUENTE P., SPALDIN N.A. *et al.*, "Band alignment at metal/ferroelectric interfaces: insights and artifacts from first principles", *Physical Review B*, vol. 83, p. 235112, 2011.

[STI 70] STILLMAN G., WOLFE C., DIMMOCK J., "Hall coefficient factor for polar mode scattering in n-type GaAs", *Journal of Physics and Chemistry of Solids*, vol. 31, pp. 1199–1204, 1970.

[SUN 03] SUN G., KÜRTI. J., RAJCZY P. *et al.*, "Performance of the Vienna ab-initio simulation package (VASP) on chemical applications", *THEOCHEM – Journal of Molecular Structure*, vol. 624, p. 37, 2003.

[TAK 99] TAKAGAKI Y., PLOOG K.H., "Quantum interference effects in semiconductor-superconductor microjunctions", *Superlattices and Microstructures*, vol. 25, pp. 659–667, 1999.

[TAN 16] TANIMURA H., KANASAKI J., TANIMURA K. *et al.*, "Formation of hot-electron ensembles quasiequilibrated in momentum space by ultrafast momentum scattering of highly excited hot electrons photoinjected into the valley of GaAs", *Physical Review B*, vol. 93, p. 161203(R), 2016.

[TAV 01] TAVARES M.R.S., HAI G.-Q., SARMA D.S., "Carrier relaxation due to electron-electron interaction in coupled double quantum well structures", *Physical Review B*, vol. 64, p. 045325, 2001.

[TYU 10] TYUTEREV V., SJAKSTE J., VAST N., "Theoretical intrinsic lifetime limit of shallow donor states in silicon.", *Physical Review B*, vol. 81, p. 245212, 2010.

[UNU 03] UNUMA T., YOSHITA M., NODA T. *et al.*, "Intersubband absorption linewidth in GaAs quantum wells due to scattering by interface roughness, phonons, alloy disorder, and impurities", *Journal of Applied Physics*, vol. 93, no. 3, pp. 1586-1597, 2003.

[VAS 11] VASILESKA D., RALEVA K., GOODNICK S., *Applications of Monte Carlo Method in Science and Engineering*, InTech, 2011.

[VEN 08] VENTRA M.D., *Electrical Transport in Nanoscale Systems*, Cambridge University Press, Cambridge, 2008.

[VER 15] VERDI C., GIUSTINO F., "Fröhlich Electron-Phonon Vertex from First Principles", *Physical Review Letters*, vol. 115, p. 176401, 2015.

[VID 10] VIDAL J., BOTTI S., OLSSON P. *et al.*, "Strong interplay between structure and electronic properties in $CuIn(S,Se)_2$: a first-principles study", *Physical Review Letters*, vol. 104, p. 056401, 2010.

[VIT 96] VITERBO D., "Crystal lattices and crystal symmetry", in PISANI C. (ed.), *Quantum-Mechanical Ab-initio Calculation of the Properties of Crystalline Materials*, Springer-Verlag, Berlin, 1996.

[WAN 11] WANG Z., WANG S., OBUKHOV S. *et al.*, "Thermoelectric transport properties of silicon: Towards an *ab initio* approach", *Physical Review B*, vol. 83, p. 205208, 2011.

[WEI 35] VON WEIZSÄ C., "Zur Theorie der Kernmassen (On the theory of nuclear masses)", *Zeitschrift fü Physik (Journal of Physics)*, vol. 96, pp. 431–458, 1935.

[YAN 13] YAN J., NØRSKOV J.J., "Calculated formation and reaction energies of $3d$ transition metal oxides using a hierarchy of exchange-correlation functionals", *Physical Review B*, vol. 88, p. 245204, 2013.

[YU 01] YU P., CARDONA M., *Fundamentals of Semiconductors*, Springer-Verlag, Berlin/New York, 2001.

[YU 08] YU X.Z., YANG Y., PAN W. *et al.*, "Electron-phonon interaction in disordered semiconductors", *Applied Physics Letters*, vol. 92, p. 092106, 2008.

[ZIM 60] ZIMAN J.M., *Electrons and Phonons: The Theory of Transport Phenomena in Solids*, Oxford University Press, Oxford, 1960.

[ZOL 90a] ZOLLNER S., GOPALAN S., CARDONA M., "Effective intervalley deformation potentials in the description of time-resolved and hot-electron luminescence", *Solid State Communications*, vol. 76, p. 877, 1990.

[ZOL 90b] ZOLLNER S., GOPALAN S., CARDONA M., "Microscopic theory of intervalley scattering in GaAs: k-dependence of deformation potentials and scattering rates", *Journal of Applied Physics*, vol. 68, p. 1682, 1990.

3

Electronic Band Structure: Empirical Pseudopotentials, $\mathbf{k} \cdot \mathbf{p}$ and Tight-Binding Methods

While first principles calculations of material properties are computationally expensive and limited practically to systems of a few hundreds atoms, empirical methods can be used to calculate accurately and efficiently the full-band energy dispersion of semiconductors, including strain and confinement effects. In this chapter, we detail three frequently used empirical approaches, based on the pseudo-potential, $\mathbf{k} \cdot \mathbf{p}$, and tight-binding (TB) methods. The connection between these approaches is highlighted, and emphasis is placed on discussing how well these methods can reproduce experimental data and first principle simulations. The data provided by these techniques and the Hamiltonian they are based on can be used as inputs for accurate transport models such as the semiclassical Monte Carlo and the quantum nonequilibrium Green's function methods described in following chapters. Although they have important consequences in modern devices, confinement and strain effects will not be considered in detail in this chapter, which is intended to provide an introduction to the commonly used semiempirical methods for bulk semiconductors.

3.1. Band structure problem

3.1.1. *Introduction*

To study the electronic properties of any bulk material or nanostructure, we have to solve the Hamiltonian of the system as accurately as possible, while ensuring that

---

Chapter written by Denis RIDEAU, François TRIOZON and Philippe DOLLFUS.

the calculation remains computationally feasible. Material science computational methods for the calculation of electronic energy band structures can be classified in two general categories. The first category includes *ab initio* methods, such as those based on the Hartree–Fock approach or density functional theory (DFT), that deduce the electronic structure from first principles, i.e. without the use of empirical parameters. The empirical pseudopotential method, the TB method and the $\mathbf{k}\cdot\mathbf{p}$ method belong to the second category that encompasses semiempirical methods and involves fitting parameters.

Ab initio methods are self-consistent methods based on a variational approach to calculate the ground-state energy of an electronic system and therefore require a large computational effort. They can only be used in small structures with a limited number of atoms, depending on the technique used and in particular on whether many-body effects are included or not. They are not suitable for calculating the transport properties of large systems and devices with confined states. For this reason, the physics and electronic device community has developed much more computationally efficient semiempirical approaches.

Semiempirical approaches are relatively simple to implement and can be reduced to an eigenvalues problem, for which efficient algorithms have been developed. The most extensive use of semiempirical band structure models has been the computation of the optical properties of bulk crystals and of quantum wells, wires and dots. Because of their ability to describe high-energy effects and the impact of both mechanical strain and quantum confinement, "full-band" models have also been adopted in an increasing number of numerical simulation tools for electronic devices. Full-band Monte Carlo models [FIS 88, HES 91, JUN 03], whose validity in nanoscale transistors devices has received a general consensus are currently taken as a reference for simplified models, as for instance for the calibration of drift–diffusion tools. During the last decade, this method has been extensively used to determine both the performance of strained devices within the perspective of an industrial integration, and the intrinsic properties of new materials and alloys [MIC 09]. When modeling the electrical currents with Monte Carlo methods, a key ingredient of the solver is the electronic band structure itself. Monte Carlo transport simulations are usually based on empirical pseudopotential method (EPM), but full zone $\mathbf{k}\cdot\mathbf{p}$ [FER 06] and TB models [MIC 09] have been recently proposed. The combined effect of strain and confinement has been illustrated in Fully Depleted Silicon On Insulator (FD-SOI) structures [RID 09], but also in nanowires [ZHA 10, NIQ 12] and nanowire transistors [NGU 13], in which the channel mobility can be dramatically reduced by the subsequent effects of quantum confinement. Full quantum simulations have been reported to model the electronic and transport properties of entire nanodevices, such as fully depleted MOSFETs [RID 09] and silicon-based nanowires [LUI 06, NIQ 09a].

Most concepts of band structure calculation have been developed for bulk materials [KAN 66, CAR 66, CHE 76, JAN 98] and within the theory of semiconductor heterojunctions, in particular for the investigation of the electronic properties of two-dimensional structures and superlattices. A great number of reviews of properties and methods for such systems can be found in the literature [SMI 90, BUR 88, BUR 94, BUR 94, FOR 07a, FOR 07b, KLI 10]. Usually, the superlattice electronic band structure problem relies on a number of simplifying features. We typically care only about results in a small energy region near the superlattice zone center. Here, our objective is somewhat different: we mainly discuss the so-called "full zone" or "full band" methods that remain valid all over the entire first Brillouin zone and for a large energy domain. Thanks to various refinements of the pseudopotential, $\mathbf{k} \cdot \mathbf{p}$, and TB methods, it is currently possible to conduct full band calculations in confined and strained nanodevices. We would like to mention that such methods rely on a certain degree of "empiricism". Therefore, they can be limited by the accuracy of their parameterization. Hence, their ability to reproduce physical properties other than those which they were designed for can be limited. For that reason, even if we do not detail them, we put forth first-principle tools that are commonly adopted nowadays for the optimization of empirical models [NIQ 00, BOY 04, RID 06, NIQ 09, BOY 10], and that provide valuable benchmarks for empirical models.

In this chapter, for the sake of simplicity required for this introductory textbook, we mainly focus on the case of bulk semiconductors and just give some indications for the case of strained and confined systems. For more details about the treatment of such systems and for more details on the theories presented here, the reader can refer to more specialized textbooks or chapters of textbooks, such as for instance [YU 01, BAS 91, DEL 04, RID 12].

3.1.2. *Bulk semiconductors*

The band structure problem refers to the eigenvalue problem derived from the Schrödinger equation

$$H_0 \Psi_{b,\mathbf{k}} = E_{b,\mathbf{k}} \Psi_{b,\mathbf{k}}, \qquad [3.1]$$

where b is a band index. The single-body Hamiltonian of an electron of mass m_0 in a periodic crystal potential V_{cr} is given by:

$$H_0 = -\frac{\hbar^2}{2m_0}\nabla^2 + V_{cr}(\mathbf{r}). \qquad [3.2]$$

For the sake of simplicity, in what follows we will use units such that the Planck constant $\hbar$ and the mass m_0 are set to unity. In these units, [3.2] is rewritten as:

$$H_0 = -\frac{1}{2}\nabla^2 + V_{cr}(\mathbf{r}). \quad [3.3]$$

The total crystal potential $V_{cr}(\mathbf{r})$ can be written as a sum of the contributions of individual short-range potentials localized at the atomic positions $\mathbf{s}_\alpha$ in a primitive cell centered at the position $\mathbf{R}_l$, i.e.

$$V_{cr}(\mathbf{r}) = \sum_{l,\alpha} v_\alpha(\mathbf{r} - \mathbf{R}_l - \mathbf{s}_\alpha) \quad [3.4]$$

According to the Bloch theorem, the wave function is periodic as well, i.e.

$$\Psi_{b,\mathbf{k}}(\mathbf{r}) = \exp(i\mathbf{k}\cdot\mathbf{r})\, u_{b,\mathbf{k}}(\mathbf{r}) \quad [3.5]$$

with, for any $\mathbf{r}$, l,

$$u_{b,\mathbf{k}}(\mathbf{r} + \mathbf{R}_l) = u_{b,\mathbf{k}}(\mathbf{r}).$$

In such a periodic potential the Schrodinger equation becomes an eigenvalue problem and the solutions exist for some discrete values at $E = E_{b,\mathbf{k}}$. In the band structure calculation, the wave vector $\mathbf{k}$ of the reciprocal space is an input vector, and we aim to determine the energy values $E_{b,\mathbf{k}}$ that are solutions of this eigenvalue problem. In practice, a basis set $\{|\Phi_\mathbf{k}(n)\rangle\}$ is chosen to expand the wave function in the form:

$$|\Psi_{b,\mathbf{k}}\rangle = \sum_n A_{b,\mathbf{k}}(n)|\Phi_\mathbf{k}(n)\rangle, \quad [3.6]$$

where, depending on the chosen basis set, the index n will refer to a Fourier component, a band index, or an orbital quantum set. Equation [3.1] can then be expressed in the matrix form

$$H_{n'n}(\mathbf{k}) = \langle\Phi_\mathbf{k}(n')|H_0|\Phi_\mathbf{k}(n)\rangle. \quad [3.7]$$

The expansion coefficients $A_{b,\mathbf{k}}$ are the eigenvectors of this matrix, and $E_{b,\mathbf{k}}$ are the eigenvalues, i.e. the desired **k**-dependent energy dispersion.

In this section, we will detail three formalisms for electronic band structure calculation with examples for bulk silicon and germanium using different expansion functions as listed below:

1) For the empirical pseudopotential method, plane waves are considered, i.e.

$$\langle \mathbf{r} | \Phi_{\mathbf{k}}(\mathbf{G}) \rangle = \exp\left[i(\mathbf{k}+\mathbf{G}) \cdot \mathbf{r} \right] \qquad [3.8]$$

leading to matrix elements involving vectors **G** of the reciprocal space.

2) For the $\mathbf{k} \cdot \mathbf{p}$ method, **Γ**-centered Bloch functions are considered, i.e.

$$\langle \mathbf{r} | \Phi_{\mathbf{k}}(n) \rangle = \exp(i\mathbf{k} \cdot \mathbf{r})\, u_{n,\Gamma}(\mathbf{r}) \qquad [3.9]$$

3) For the TB method, we consider

$$\langle \mathbf{r} | \Phi_{\mathbf{k}}(\lambda) \rangle = \sum_{l,\alpha} \exp(i\mathbf{k} \cdot \mathbf{R}_{l,\alpha})\, \varphi_{\lambda}(\mathbf{r} - \mathbf{R}_{l,\alpha}), \qquad [3.10]$$

where the functions φ_{λ} are atomic orbitals of an isolated atom localized at position $\mathbf{R}_{l,\alpha}$, and $\lambda = \{s, p, d, ...\}$ refers to the orbital quantum states.

These approaches are equivalent in principle and would lead to the same results if complete basis sets were used. However, as shown in the following, they lead to different kinds of matrix equations and the optimization strategy for the "empirical" parameters is rather different.

3.1.3. *Confined systems*

Now, though it is outside the scope of this chapter, we briefly consider the calculation of energy bands under the presence of an additional external potential $U(\mathbf{r})$ that may induce confinement effects. The eigenfunctions ψ of a confined electron gas satisfy the Schrödinger equation:

$$\left[H_0 + U(\mathbf{r}) \right] \psi(\mathbf{r}) = E\, \psi(\mathbf{r}) \qquad [3.11]$$

where H_0 is the Hamiltonian [3.3] of the bulk semiconductor. To proceed further, we have to consider a complete set of functions $|n\mathbf{k}\rangle$ in which the wave function can be expanded, i.e.

$$|\psi\rangle = \sum_{n',\mathbf{k}'} A_{\mathbf{k}'}^{n'} |n'\mathbf{k}'\rangle \qquad [3.12]$$

Luttinger and Kohn demonstrated that any wave function may be expanded in one of the following sets of functions [LUT 54]:

$$\langle \mathbf{r}|n\mathbf{k}\rangle = \begin{cases} \exp(i\mathbf{k}\cdot\mathbf{r})\, u_{n,\mathbf{k}}(\mathbf{r}) \\ \exp(i\mathbf{k}\cdot\mathbf{r})\, u_{n,\Gamma}(\mathbf{r}) \end{cases} \qquad [3.13]$$

where $\mathbf{k}$ belongs to the first Brillouin zone. As in the bulk case, a matrix equation is obtained by applying the bra $\langle n\mathbf{k}|$ to the Schrödinger equation, with integration over the full crystal volume V, which gives

$$\sum_{n',\mathbf{k}'} \langle n\mathbf{k}|H_0|n'\mathbf{k}'\rangle_V A_{\mathbf{k}'}^{n'} + \sum_{n',\mathbf{k}'} \langle n\mathbf{k}|U|n'\mathbf{k}'\rangle_V A_{\mathbf{k}'}^{n'} = \varepsilon_{\mathbf{k}} A_{\mathbf{k}}^{n} \qquad [3.14]$$

The first term of the left-hand side of [3.14] contains the bulk Hamiltonian, while the second term contains the effect of the confinement potential. To evaluate this term, it is convenient to use the expansion of the wave function [3.12] with one of the expressions of the functions $|n\mathbf{k}\rangle$ given in [3.13]. The former one, referred to as the linear combination of bulk bands (LCBBs) approach is closely related to the empirical pseudopotential method, while the latter one, the basis of the envelope function approximation, is conveniently used in the $\mathbf{k}\cdot\mathbf{p}$ method.

3.2. Empirical pseudopotentials method

The development of the EPM is closely linked to the interpretation of the optical properties of semiconductors and insulators. As discussed in this section, the EPM relies on a small number of parameters, the so-called Fourier coefficients of the crystal potential or, more commonly, "form factors". The reduction of the band structure problem to the determination of a limited number of empirical parameters is one of the reasons that led to the success of this method.

Experimental quantities related to the band structure are of prime importance for the adjustment of the empirical parameters. In particular, in the interpretation of optical absorption/reflectivity spectra (precisely the structure of the imaginary part of the dielectric constant), it has been a common practice to associate critical points structure with high-symmetry points in the reciprocal space of the diamond lattice. Phillips was the first to note that a reasonable EPM band structure for silicon could be obtained using only two parameters [PHI 58]. A third parameter was added later, but only as a refinement of the method. Early EPM parameterizations proposed with three form factor parameters for silicon, germanium and diamond were performed on the basis of these assumptions using experimental reflectivity data [BRU 62, SAS 66, COH 66].

Some years later, additional characterization methods based on high-resolution photoemission spectroscopy (PS), e.g. X-ray photoemission spectroscopy (XPS) and ultraviolet photoemission spectroscopy (UPS), became available. These enabled access to important information such as the configuration of the lowest lying valence bands; a new series of EPM parameterizations have been proposed for a larger range of materials [CHE 73]. In contrast with reflectivity data that can be difficult to interpret [KAN 66], one of the striking features of XPS and UPS (in addition to an excellent agreement between the two) is that they both directly mimic the VB density of states. The latter quantity can be calculated using an EPM and directly compared to PS measurements.

Meanwhile, Brust re-examined the predictions of the EPM for the relevant band gaps in Si, Ge and Sn along with experiments and proposed substantial improvements to EPM using a nonlocal correction to the potential terms [BRU 71]. In a reference paper, Chelikowsky and Cohen extracted a new series of form factor parameters and nonlocal correction parameters, with a total of eight parameters, for 11 diamond-like and zinc-blende semiconductors [CHE 76]. These parameter sets are still commonly used nowadays in EPM-based calculations and unless stated otherwise, the EPM results shown in this chapter rely on these parameters.

With the recent interest in the semiconductor industry for strained Si as a major technological booster for the carrier mobility, adaptation of the EPM for strained semiconductors was needed. The calculation of deformation potentials with EPM was first demonstrated by Saravia and Brust [SAR 69] by applying the Bir and Pikus transformation [BIR 74]. Now, an interpolation function between the lowest order form factor coefficients is commonly used [FRI 76, RIE 93, FIS 96, RID 06].

3.2.1. *Principles of the model – basic equations*

The periodic part $u_{\mathbf{k}}(\mathbf{r})$ of the Bloch wave function [3.5] can be expanded in Fourier series in terms of wave vectors $\mathbf{G}'$ of the reciprocal lattice, which leads to the wave function

$$\psi_{\mathbf{k}}(r)=e^{i\mathbf{k}\cdot\mathbf{r}}\sum_{\mathbf{G}'}A_{\mathbf{k}}(\mathbf{G}')e^{i\mathbf{G}'\cdot\mathbf{r}} \quad [3.15]$$

By substituting this expression into the Schrödinger equation, premultiplying by $\exp(-i\mathbf{G}\cdot\mathbf{r})$ and integrating over the whole crystal volume V, we obtain the matrix representation in the form

$$\frac{1}{2}A_{\mathbf{k}}(\mathbf{G})(\mathbf{k}+\mathbf{G})^2+\sum_{\mathbf{G}'}V_{\mathbf{G}-\mathbf{G}'}A_{\mathbf{k}}(\mathbf{G}')=E_{\mathbf{k}}A_{\mathbf{k}}(\mathbf{G}) \quad [3.16]$$

where the potential term $V_{\mathbf{G}-\mathbf{G}'}$ is written

$$V_{\mathbf{G}-\mathbf{G}'}=\frac{1}{V}\int_V \exp(-i\mathbf{G}\cdot\mathbf{r})V_{cr}(\mathbf{r})\exp(i\mathbf{G}'\cdot\mathbf{r})d^3\mathbf{r} \quad [3.17]$$

We have to determine the eigenvectors $A_{\mathbf{k}}(\mathbf{G})$ and the eigenvalues $E_{\mathbf{k}}$ of system [3.16]. Inserting expression [3.4] of the crystal potential into the potential term [3.17] leads to

$$V_{\mathbf{G}-\mathbf{G}'}=\frac{1}{N\Omega}\sum_{l,\alpha}\int_V v_\alpha(\mathbf{r}-\mathbf{R}_l-\mathbf{s}_\alpha)\exp\left[-i(\mathbf{G}-\mathbf{G}')\cdot\mathbf{r}\right]d^3\mathbf{r} \quad [3.18]$$

where N is the number of primitive cells of volume Ω. Using the change in variables $\mathbf{G}''=\mathbf{G}-\mathbf{G}'$ and $\mathbf{r}'=\mathbf{r}-\mathbf{R}_l-\mathbf{s}_\alpha$ [3.18] may be rewritten as

$$V_{\mathbf{G}''}=\frac{1}{N\Omega}\sum_\alpha \exp[-i\mathbf{G}''\cdot\mathbf{s}_\alpha]\sum_l \exp[-i\mathbf{G}''\cdot\mathbf{R}_l]\int_V v_\alpha(\mathbf{r}')\exp[-i\mathbf{G}''\cdot\mathbf{r}']d^3\mathbf{r}'$$

Assuming the potential $v_\alpha(\mathbf{r}')$ to be short range, it decays rapidly from the atom position and the integral can be limited to a single primitive cell of volume Ω.

Additionally, recalling that $\exp[-i\mathbf{G}''\cdot\mathbf{R}_l]=1$ for any $\mathbf{G}''$ and $\mathbf{R}_l$, we finally obtain

$$V_{\mathbf{G}''}=\frac{1}{\Omega}\sum_{\alpha}\exp[-i\mathbf{G}''\cdot\mathbf{s}_\alpha]\int_{\Omega}v_\alpha(\mathbf{r}')\exp[-i\mathbf{G}''\cdot\mathbf{r}']d^3\mathbf{r}' \qquad [3.19]$$

This expression can be conveniently rewritten in the form

$$V_{\mathbf{G}''}=\sum_{\beta}S_{\mathbf{G}''\beta}\,v_\beta(\mathbf{G}'') \qquad [3.20]$$

where β indexes the atom type and $S_{\mathbf{G}''\beta}$ is summed over the positions of atoms of same type in a unit cell:

$$\begin{cases} S_{\mathbf{G}''\beta}=\sum_{\mathbf{s}_\beta}\exp\left[-i\mathbf{G}''\cdot\mathbf{s}_\beta\right] \\ v_\beta(\mathbf{G}'')=\frac{1}{\Omega}\int_{\Omega}v_\beta(\mathbf{r}')\exp[-i\mathbf{G}''\cdot\mathbf{r}']d^3\mathbf{r}' \end{cases} \qquad [3.21]$$

Hence, the potential term [3.20] can be written in terms of products of the structure factor $S_{\mathbf{G}\beta}$ that depends only on the atom positions in the cell, and of the atomic form factors $v_\beta(\mathbf{G})$ coming from the Fourier transform of the atomic potentials. In diamond and zinc-blende semiconductors with two atoms per unit cell at positions $\mathbf{s}_1=-\mathbf{s}=-(a/8,a/8,a/8)$ and $\mathbf{s}_2=\mathbf{s}=(a/8,a/8,a/8)$, respectively, where a is the lattice constant, $V_{\mathbf{G}}$ can be rewritten as a sum of symmetric and antisymmetric terms, i.e.

$$V_{\mathbf{G}}=V_S(\mathbf{G})\,\cos(\mathbf{G}\cdot\mathbf{s})+i\,V_A(\mathbf{G})\,\sin(\mathbf{G}\cdot\mathbf{s}) \qquad [3.22]$$

where

$$\begin{cases} V_S(\mathbf{G})=v_1(\mathbf{G})+v_2(\mathbf{G}) \\ V_A(\mathbf{G})=v_1(\mathbf{G})-v_2(\mathbf{G}) \end{cases} \qquad [3.22]$$

V_S and V_A are called symmetric and antisymmetric form factors, respectively. They can be considered as adjustable parameters that justify the name of EPM.

3.2.2. *The EPM equation in matrix form*

To solve [3.16] computationally, the sum over the infinite number of G vectors must be truncated to consider only the vectors with a magnitude smaller than a given cut-off value. We summarize in Table 3.1 the EPM matrix size of the reciprocal vectors of lowest magnitude in a cubic crystal.

In silicon and germanium, a good convergence of the energy band gap can be achieved when including all vectors up to $\mathbf{G} = 2\pi/a\,[2\,2\,4]$ of square modulus $G^2 = 24(2\pi/a)^2$, which also provides a satisfactory periodicity of the energy bands in the reciprocal space, i.e. $E_{\mathbf{k}+\mathbf{G}} \approx E_{\mathbf{k}}$.

Reciprocal vector **G** [in $2\pi/a$ units]	$\mathbf{G}^2$ [in $(2\pi/a)^2$ units]	Number of equivalent vectors	Size of EPM matrix
[0 0 0]	0	1	1
[1 1 1]	3	8	9
[2 0 0]	4	6	15
[2 2 0]	8	12	27
[3 1 1]	11	24	51
[2 2 2]	12	8	59
[4 0 0]	16	6	65
[3 3 1]	19	24	89
[2 2 4]	24	24	113

Table 3.1. *Reciprocal vectors in order of increasing magnitude in diamond and zinc blende crystals, number of equivalent vectors and size of the corresponding EPM matrix*

In principle, the form factors $V_S(\mathbf{G})$ and $V_A(\mathbf{G})$ should be determined for each vector listed in Table 3.1. However, the potential converges so rapidly in reciprocal space that it can be conveniently described by its lowest Fourier coefficients. The first-four series of G vectors are in the units of $2\pi/a$, the eight vectors $[\pm1\pm1\pm1]$ of square modulus $G^2 = 3$, the six vectors $[\pm2\,0\,0]$, $[0\pm2\,0]$ and $[0\,0\pm2]$ of square modulus $G^2 = 4$, the 12 vectors $[\pm2\pm2\,0]$,

$[\pm 2\,0 \pm 2]$ and $[0 \pm 2 \pm 2]$ of square modulus $G^2 = 8$, and the 12 vectors $[\pm 3 \pm 1 \pm 1]$, $[\pm 1 \pm 3 \pm 1]$ and $[\pm 1 \pm 1 \pm 3]$ of square modulus $G^2 = 11$. In an undistorted zinc-blende crystal, all directions *Ox*, *Oy* and *Oz* are equivalent and if we assume that the potential $v_\alpha(\mathbf{r})$ has a spherical symmetry, the form factors are real and depend only on the modulus of **G**. Additionally, in centrosymmetric crystals, only the symmetric form factors have to be considered and the contribution of $V_S(G^2 = 4)$ vanishes. Finally, in silicon and germanium only the three lowest form factors, i.e. $V_S(G^2 = 3)$, $V_S(G^2 = 8)$ and $V_S(G^2 = 11)$, have to be considered and adjusted according to experimental data [CHE 73]. In practice, the form factor of vectors with $G^2 > 11$ can be ignored. It has been shown that, including higher order, form factors may provide more accurate results in Si [MAS 83], but in most EPM computations reported so far, only the three lowest order form factors are considered for Si and Ge. In III–V compounds and SiGe alloys that are not centrosymmetric, the antisymmetric form factors must be included. Examples of pseudopotential form factors of some group – IV, III-V and II-VI – semiconductors are summarized in Table 3.2. It should be noted that it may be convenient to extract other series of form factor parameters with additional nonlocal correction parameters, leading to a total of eight parameters [CHE 76]. A full discussion of the method and a complete tabulation of the fitted potentials for the entire periodic table have been given by Bachelet *et al.* [BAC 82].

Material	$V_S(3)$	$V_S(8)$	$V_S(11)$	$V_A(3)$	$V_A(8)$	$V_A(11)$
Si	–0.211	0.04	0.08	0	0	0
Ge	–0.269	0.038	0.035	0	0	0
GaAs	–0.252	0	0.08	0.068	0.066	0.012
GaP	–0.249	0.017	0.083	0.081	0.055	0.003
InAs	–0.270	0.02	0.041	0.078	0.038	0.036
InSb	–0.250	0.01	0.044	0.049	0.038	0.01
ZnSe	–0.230	0.01	0.06	0.18	0.12	0.03
CdTe	–0.245	–0.015	0.073	0.089	0.084	0.006

Table 3.2. *Pseudopotential form factors of some group – IV, III–V and II-VI semiconductors, in units of Rydbergs = 13.6 eV; data from [YU 01, FAL 66, CHE 73, VAR 73]*

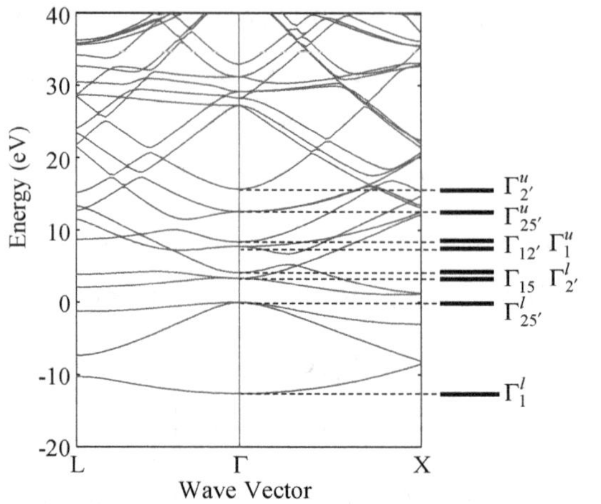

Figure 3.1. *Electronic band structure of Si calculated within EPM; the Γ_i states labeled in the irreducible representation of the O_h group are indicated on the right side*

In Figure 3.1, we plot the band structure of silicon. An indirect band gap appears around $E=0$. The free electron states corresponding to the reciprocal vectors $\mathbf{G}=[0\,0\,0]$, $\mathbf{G}=[1\,1\,1]$, $\mathbf{G}=[2\,0\,0]$, ... and equivalent vectors are degenerated at the center of the Brillouin zone Γ. It should be noted that these Γ-centered states, labeled in the figure in the irreducible representation of the O_h group, form the basis of the $\mathbf{k}\cdot\mathbf{p}$ method that will be presented in the following section.

3.2.3. *Effects of strain within EPM*

In the case of strained materials, we also need to determine the form factors for wave vectors $\mathbf{G}'$ different from the wave vectors $\mathbf{G}$ of the unstrained crystal, i.e.

$$\mathbf{G}'=\mathbf{G}/(1+\boldsymbol{\varepsilon}) \qquad [3.23]$$

where $\boldsymbol{\varepsilon}$ is the strain tensor.

A simple way consists of interpolating the form factors between the values $V_S(3)$, $V_S(8)$ and $V_S(11)$ of the unstrained crystal. The interpolation function is rather empirical and often chosen for its flexibility. We can mention the interpolation function, introduced by Friedel *et al.* for local EPM [FRI 89] and later used by Fischetti and Laux for nonlocal EPM [FIS 96], i.e.

$$V_S(\mathbf{G}) = \frac{a_1\left(\mathbf{G}^2 - a_2\right)}{1+\exp\left[a_3\left(\mathbf{G}^2 - a_2\right)\right]} \times \frac{1}{2}\tan\left[\left(\frac{a_5 - \mathbf{G}^2}{a_6}\right) + \frac{1}{2}\right] \qquad [3.24]$$

where the coefficients (a_i) are fitting parameters. Other interpolations have been proposed for Si and Ge. Rieger and Vogl used a cubic spline interpolation between the lowest form factors coefficients [RIE 93], and other authors have preferred a polynomial interpolation for the nonlocal EPM [RID 06]. In the latter case, the polynomial coefficients have been chosen by empirically fitting the slopes of $\partial V(\mathbf{G})/\partial \mathbf{G}$ in order to obtain correct band gap deformation potentials for strained semiconductor [WAL 86], and also to reproduce first principles nonlinear energy band shifts in the case of large crystal distortions ($\approx$ 5%) [RID 06].

3.3. The $\mathbf{k}\cdot\mathbf{p}$ method

The $\mathbf{k}\cdot\mathbf{p}$ method is based on the perturbation theory. In this method, the energy is calculated near a band extremum: the eigenvalues and eigenvectors of the single electron Hamiltonian are expanded as a function of $\mathbf{k}$ around the vector $\mathbf{k}_0$ corresponding to a band minimum or maximum. This method, first introduced by Bardeen [BAR 38], was applied by Dresselhaus *et al.* in silicon and germanium to discuss the connection between the cyclotron resonance frequencies and the $\mathbf{k}\cdot\mathbf{p}$ parameters [DRE 55] that define the warping of the energy surfaces, i.e. L, M and N parameters, closely related to the Luttinger parameters γ_1, γ_2 and γ_3. Meanwhile, Kane extracted another important $\mathbf{k}\cdot\mathbf{p}$ parameter from optical experiments [KAN 66], the so-called "Kane" parameter that characterizes the interaction of the valence and conduction band states.

Luttinger and Kohn have proposed a general formulation of the $\mathbf{k}\cdot\mathbf{p}$ method [LUT 54] that was later extended by Pikus and Bir to the case of mechanically strained crystals [PIK 60]. It has proven to be an efficient method to study the local properties of materials. Among other applications, we can mention the calculation of the confined states in Si/Ge heterostructures [RID 01] and the calculation of the energy levels of impurities in bulk Si and Ge, within the Luttinger and Kohn envelope function approximation that was originally developed for the calculation of shallow impurity levels [LUT 54].

The first $\mathbf{k}\cdot\mathbf{p}$ model for Si and Ge valid throughout the entire Brillouin zone, referred to as a "full zone" $\mathbf{k}\cdot\mathbf{p}$ model, was introduced by Cardona and Pollak [CAR 66]. It is worth mentioning that such a full zone $\mathbf{k}\cdot\mathbf{p}$ approach requires a large number of fitting parameters. While it is straightforward to determine the energy band structure at any point $\mathbf{k}$ of the Brillouin zone, once these parameters have been adjusted, it usually requires a non-negligible effort to come up with a satisfactory set of parameters valid in all the conditions. For this reason, the full-zone $\mathbf{k}\cdot\mathbf{p}$ method has been rarely used and only applied to some bulk semiconductors, including Si, Ge [MAS 83, CAR 66, HUM 81] and α-Sn [POL 70], while no extension of the method has been published for strained semiconductors until recently.

With the advent of first principle methods that can be very helpful to determine the unknown $\mathbf{k}\cdot\mathbf{p}$ parameters, the Cardona–Pollak Hamiltonian has been recently revisited to calculate the electronic states in strained Si and Ge [RIC 04, RID 06]. The $\mathbf{k}\cdot\mathbf{p}$ results presented here are based on the Cardona–Pollak Hamiltonian and the Bir–Pikus correction for strain, together with a recent first principle based model parameterization [RID 06].

3.3.1. *Principles of the model – basic equations*

Starting with the same single-body Hamiltonian [3.1], the full zone $\mathbf{k}\cdot\mathbf{p}$ method is built on a rather different approach than EPM. Indeed, the wave functions are not explicitly used in the determination of the matrix elements, which are actually directly fitted. As presented in this section, these matrix elements will be explicitly written as linear combinations of k_x, k_y and k_z using group theory considerations. However, even though there are only a limited number of independent matrix elements, the number of fitting parameters in a full zone $\mathbf{k}\cdot\mathbf{p}$ method remains fairly large in comparison with the EPM ones.

Let us recall the Schrödinger equation [3.1]

$$H_0 \Psi_{b,\mathbf{k}} = E_{b,\mathbf{k}} \Psi_{b,\mathbf{k}}$$

where $E_{b,\mathbf{k}}$ is the energy level for the band index b with a dependence on the wave vector $\mathbf{k}$ that spans the first Brillouin zone. The solutions of this equation have the form of Bloch functions [3.5], i.e.

$$\Psi_{b,\mathbf{k}}(\mathbf{r}) = \exp\left(i\mathbf{k}\cdot\mathbf{r}\right) u_{b,\mathbf{k}}\left(\mathbf{r}\right)$$

Introducing this form of wave function in [3.1] with the general Hamiltonian H_0 given by [3.2], the Schrödinger equation may be rewritten as

$$\left[-\frac{1}{2}\nabla^2 + V_{cr}(\mathbf{r}) - i\mathbf{k}\cdot\nabla + \frac{1}{2}\mathbf{k}^2\right] u_{b,\mathbf{k}}(\mathbf{r}) = \mathrm{E}_{b,\mathbf{k}}\, u_{b,\mathbf{k}}(\mathbf{r}) \qquad [3.25]$$

The first part of the left-hand side contains the Hamiltonian H_0 itself, while the other part vanishes when **k** tends to 0. Considering that the Γ-centered states form a complete set of functions [LUT 54], the function $u_{b,\mathbf{k}}(\mathbf{r})$ can be expanded as

$$u_{b,\mathbf{k}}(\mathbf{r}) = \sum_n C_n^b \times u_{n,\Gamma}(\mathbf{r}) \qquad [3.26]$$

It should be noted that a $\mathbf{k}\cdot\mathbf{p}$ Hamiltonian centered at a different location of the **k**-space can be constructed similarly. Substituting [3.26] into [3.25], applying the bra $\langle u_{n',\Gamma}|$ and integrating over the unit cell volume lead to the so-called $\mathbf{k}\cdot\mathbf{p}$ equation,

$$\sum_n \left[\left(E_n^\Gamma + \frac{1}{2}\mathbf{k}^2 - E_{b,\mathbf{k}}\right)\delta_{n'n} + \mathbf{k}\cdot\langle u_{n',\Gamma}|\mathbf{p}|u_{n,\Gamma}\rangle\right] C_n^b = 0 \qquad [3.27]$$

where E_n^Γ are the energy extrema at Γ and $\mathbf{p} = -i\nabla$ is the momentum operator. It is worth noting that the carrier effective mass at a band extremum is related not only to the energy E_n^Γ, but also to the matrix elements $\langle u_{n',\Gamma}|\mathbf{p}|u_{n,\Gamma}\rangle$. Considering the second-order perturbative treatment of the eigenvalues [DRE 55], [LAW 71], the quadratic approximation of the energy for non-degenerate bands is written

$$E_{n,\mathbf{k}}^\Gamma \approx E_n^\Gamma + \frac{1}{2}\mathbf{k}^2 + \sum_{i,j} k_i k_j \sum_{n'\neq n} \frac{\langle u_{n,\Gamma}|p_i|u_{n',\Gamma}\rangle\langle u_{n',\Gamma}|p_j|u_{n,\Gamma}\rangle}{E_n^\Gamma - E_{n'}^\Gamma} \qquad [3.28]$$

where i and j stand for *x*, *y* or *z*. This expression may be rewritten in the simple form

$$E_{n,\mathbf{k}}^\Gamma \approx E_n^\Gamma + \frac{1}{2}\sum_{i,j}\frac{k_i k_j}{m_{i,j}} \qquad [3.29]$$

where the components of the effective mass tensor are defined as

$$\frac{1}{m_{i,j}} = \delta_{i,j} + 2\sum_{n' \neq n} \frac{\langle u_{n,\Gamma} | p_i | u_{n',\Gamma} \rangle \langle u_{n',\Gamma} | p_j | u_{n,\Gamma} \rangle}{E_n^\Gamma - E_{n'}^\Gamma}. \quad [3.30]$$

3.3.2. *Solution of the eigenvalue problem*

The solution of [3.27] provides the energies $E_{b,\mathbf{k}}$ and the vectors C_n^b to be calculated. However, since the $u_{n,\Gamma}$ functions have not been explicitly determined, we cannot compute the wave function from the $\mathbf{k} \cdot \mathbf{p}$ equation. The sum in equation [3.27] runs over an infinite number of Γ-centered states. In practice, it must be reduced to the first lowest energy states. Usually, the number of states needed to obtain accurate results over the entire Brillouin zone depends on the considered material. In silicon the ionic potential interaction is not very strong and the 15 lowest Γ-centered states remain relatively close to their original degenerate positions (when the potential is ignored) and do not mix with the next [2 2 0] states. Hence, it makes sense to retain in the calculation these 15 lowest states corresponding to the reciprocal vectors G originated from the single [0 0 0], the eight [1 1 1] and the six [2 0 0] plane waves, as originally proposed by Cardona and Pollak [CAR 66]. In some other Zinc-blende materials, such as diamond, it is preferable to extend the $\mathbf{k} \cdot \mathbf{p}$ basis by including the next 12 vectors corresponding to the [2 2 0] plane waves.

Using the original Cardona and Pollak irreducible representation of the O_h group (FD3M), these 15 states correspond to the three threefold-degenerated representations Γ_{15}, $\Gamma_{25'}^l$ and $\Gamma_{25'}^u$, the twofold representation $\Gamma_{12'}$ and the four singlefold representations Γ_1^u, Γ_1^l, $\Gamma_{2'}^u$ and $\Gamma_{2'}^l$, i.e.

$$\Gamma \in \left\{ \Gamma_1^l, \Gamma_{25'}^l, \Gamma_{2'}^l, \Gamma_{15}, \Gamma_1^u, \Gamma_{25'}^u, \Gamma_{2'}^u, \Gamma_{12'} \right\} \quad [3.31]$$

The energy spectrum E_n^Γ consists of eight energy levels to span the Γ-centered states listed in [3.31]. The $\Gamma_{25'}^l$ energy corresponds to the topmost valence band. To reproduce the band offsets between Si and Ge, it is convenient to arbitrarily set the energy level of $\Gamma_{25'}^l$ to zero in Si and to 0.68 eV in Ge [NIQ 09b]. The other values of E_n^{Γ} can be directly determined from first-principle calculation or from measurements when available.

The momentum operator $\mathbf{p} = -i\nabla$ is a vector operator that belongs to the Γ_{15} irreducible representation of the O_h (FD3M) space group. The application of group theory shows the existence of only 10 independent momentum matrix elements between the 15 Γ-centered states [3.31]. For illustration, we can consider for instance the *s*+ states Γ_1^u and Γ_1^l. Since p_x transforms like *x*, the momentum matrix elements $\langle s+|p_x|\alpha\rangle$ involving these states vanish for any state $\alpha \neq \Gamma_{15}$. As a result, the matrix representation written simply,

$$\langle \Gamma_1^u|\mathbf{p}|\Gamma_{15}\rangle = T\left[k_x\ k_y\ k_z\right] \qquad [3.32]$$

where T is a fitting parameter and the three elements of the vector in the right-hand side are associated with the three basis vectors $|X\rangle$, $|Y\rangle$ and $|Z\rangle$ of representation Γ_{15}, which are summarized as $|\Gamma_{15}\rangle$ in the left-hand side. All the other blocks that can be calculated this way are displayed in Figure 3.2. The coupling of bands by spin-orbit interaction can be calculated in a similar way and will not be detailed here. The corresponding matrix terms may be found in [CAR 66, RID 06].

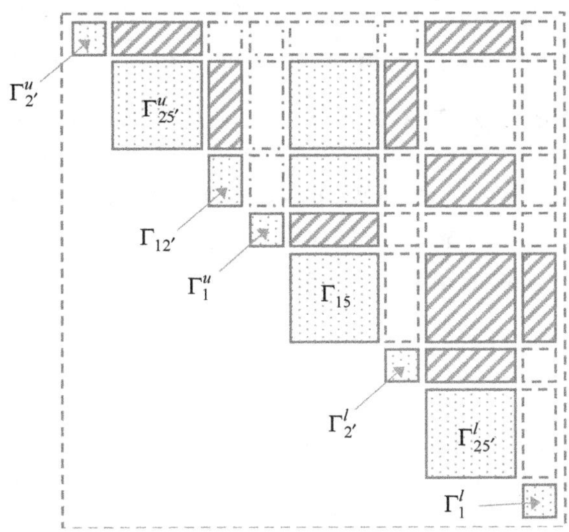

Figure 3.2. *Schematic representation of the* $k \cdot p$ *matrix; the non-vanishing blocks depend on eight Γ-centered energy levels (dotted blocks) and 10 matrix elements (dashed blocks); details can be found in [CAR 66, RID 06]*

3.3.3. *Effect of strain within the $k \cdot p$ model*

The analysis of the band structure of strained semiconductors using the $\mathbf{k} \cdot \mathbf{p}$ Hamiltonian was initially proposed by Pikus and Bir [PIK 60]. They derived the first-order $\mathbf{k} \cdot \mathbf{p}$ perturbation terms arising from strain effect. This general expression has been widely used in the six-level $\mathbf{k} \cdot \mathbf{p}$ Hamiltonian to analyze the effects of band deformation on the whole energy spectrum. An application of the Pikus–Bir formalism in the case of the previous 15-level $\mathbf{k} \cdot \mathbf{p}$ Hamiltonian can be found in [10]. According to the Pikus–Bir analysis, the perturbation term to be added to [3.27] can be written as:

$$W_{\mathbf{k}\cdot\mathbf{p}} = \sum_{i,j} k_i \, \varepsilon_{ij} \langle u_{n',\Gamma} | p_j | u_{n,\Gamma} \rangle + \varepsilon_{ij} \langle u_{n',\Gamma} | \Xi_{ij} | u_{n,\Gamma} \rangle \qquad [3.33]$$

where i and j stand for x, y and z and $\boldsymbol{\varepsilon}$ is the strain tensor. The first term of [3.33] accounts for the interaction between the strain tensor and the carrier momentum. Considering the same example as previously in [3.32], the matrix representation involving the s+ states Γ_1^u and Γ_{15} is written,

$$W_{\mathbf{k}\cdot\mathbf{p}}\left(\Gamma_1^u, \Gamma_{15}\right) = T \cdot \sum_i \left[\varepsilon_{ix} k_x \; \varepsilon_{iy} k_y \; \varepsilon_{iz} k_z \right] \qquad [3.34]$$

where T is the same fitting parameter as in the unstrained matrix.

The second term in [3.33] is referred to as the deformation potential operator that describes the change in the potential and the kinetic energy of carriers due to the strain itself. The unknown non-vanishing matrix elements $\langle u_{n',\Gamma} | \Xi_{ij} | u_{n,\Gamma} \rangle$ can be determined from group-theory selection rules. For instance, the strain matrix between states Γ_1^u and $\Gamma_{25'}^l$ takes the form

$$W_{\mathbf{k}\cdot\mathbf{p}}\left(\Gamma_1^u, \Gamma_{25'}^l\right) = f_\Gamma \cdot \left[\varepsilon_{yz} \; \varepsilon_{xz} \; \varepsilon_{xy} \right] \qquad [3.35]$$

where f_Γ is a deformation potential considered as a fitting parameter. All matrix elements can be constructed similarly. For details, see [CAR 66]. It should be pointed out that there are not less than 31 fitting parameters in the 15-level $\mathbf{k} \cdot \mathbf{p}$ strain model, which makes the optimization of this model quite difficult.

3.4. The TB method

The empirical TB method is based on an intuitive representation of the crystal originally proposed by Bloch [BLO 28], also known as the Linear Combination of Atomic Orbitals (LCAO). The method starts with atomic orbitals of isolated ions, and considers that the overlap between these orbitals should give rise to the band structure. In an anthological paper, Slater and Koster [SLA 54] suggested to treat the various overlap integrals between orbitals in a simplified way, arbitrarily discarding many of them. Instead of analytically computing the various integrals, it is possible to use them as fitting parameters to be chosen so that the method reproduces experimental data or the results of more accurate calculations like *ab initio* techniques. With this in mind, we should consider the empirical TB method as an interpolation method, rather than a primitive method. Nevertheless, many attractive qualitative features are present in this approach, since it gives solutions having all the correct symmetry properties of the system. The TB method is often called an "atomistic" method in the sense that its Hamiltonian can be constructed in a straightforward way once the positions of atoms that form the structure studied are known. It is thus suitable to describe the confined states in nanostructures such as quantum wells, nanowires and quantum dots.

For some materials, very simple TB Hamiltonians are sufficient to describe the band structure. A well-known example is graphene, which is a hexagonal lattice of carbon atoms. In principle, one s-basis and three p-basis orbitals for each atom should be necessary to describe the valence electrons. Choosing z as the normal direction to the graphene plane, the s, p_x and p_y orbitals account for the chemical bonds via sp^2 hybridization. The corresponding bonding states have energies well below the Fermi level. In the vicinity of the Fermi level, electronic states are very well described by a single orbital p_z on each atom [WAL 46]. This single orbital model is sufficient to describe transport properties with reasonable accuracy.

In the case of common semiconductors such as Si and Ge, the accuracy of the TB solution depends on how many orbitals are considered in the calculation. The simplest model is the original sp^3 Slater and Koster model [SLA 54], which features one s-basis and three p-basis orbitals for each atom. Although this approach yields an accurate description of the valence bands, it fails in accurately reproducing the conduction bands of indirect band gap semiconductors. Vogl *et al.* have introduced an excited s* orbital into the basis [VOG 83], which greatly improved the description of the conduction bands in indirect band gap semiconductors. However, this s* state (with s-symmetry) is mainly considered as a convenient numerical trick with the only purpose of temporarily "bending" the conduction band and obtaining a well-positioned indirect band gap. It is difficult to justify it physically since the conduction band states exhibit rather a d-symmetry [68]. Considering the importance of the d-excited states in silicon, Ren *et al.* [REN 97] and Jancu *et al.*

[JAN 98] developed a sp^3d^5 and a $sp^3d^5s^*$ TB model, respectively. As testified by the increasing number of models based on the latter $sp^3d^5s^*$ TB model, it has become a reference; and since the initial parameterization of this model [JAN 98], at least two other sets of parameter have been proposed for silicon [BOY 04, NIQ 09b].

3.4.1. *Principles of the model*

The total potential [3.4] of the crystal expressed as a combination of short-range potentials localized at the atomic positions is rewritten here for convenience in the form:

$$V_{cr}(\mathbf{r}) = \sum_n v_n(\mathbf{r} - \mathbf{R}_n) \qquad [3.36]$$

where the primitive cell index l and the atom position within this cell α have been merged into the single site index n, whose position is $\mathbf{R}_n = \mathbf{R}_l + \mathbf{s}_\alpha$. Let us start with the solution $\varphi_\lambda(\mathbf{r} - \mathbf{R}_n)$ of an isolated atom, i.e.

$$\left(-\frac{1}{2}\nabla^2 + v_n(\mathbf{r} - \mathbf{R}_n)\right)\varphi_\lambda(\mathbf{r} - \mathbf{R}_n) = \varepsilon_{n\lambda}\,\varphi_\lambda(\mathbf{r} - \mathbf{R}_n) \qquad [3.37]$$

and with the following expansion of the solution of the Schrodinger equation in terms of localized atomic orbitals:

$$\psi_b(\mathbf{r}) = \sum_{n,\lambda} C_{n\lambda}^b\,\varphi_\lambda(\mathbf{r} - \mathbf{R}_n). \qquad [3.38]$$

Here, the index λ refers to the orbitals of an isolated atom and b to the eigenvalues index.

3.4.2. *The $sp^3d^5s^*$ TB model for Si and Ge*

In the case of silicon, germanium and other conventional diamond-like or zinc-blende semiconductors, it is necessary to consider 4–10 orbitals per atom to reproduce accurately the band structure. In the $sp^3d^5s^*$ TB model, the index λ in [3.38] corresponds to one of the following orbitals:

$$\lambda \in \left\{s, s^*, p_x, p_y, p_z, d_{yz}, d_{xy}, d_{xz}, d_{x^2-y^2}, d_{3z^2-r^2}\right\}. \qquad [3.39]$$

In the most common form of TB models, the interaction between different orbitals is allowed only if the atomic sites are nearest neighbors and if the orbitals are orthogonal. This means that these so-called Löwdin orbitals differ from the usual atomic wave functions in that they have been constructed in such a way that the wave functions centered on different atomic positions are orthogonal to each other. From these assumptions, we can simplify the matrix elements of the TB Hamiltonian as:

$$\langle n,\lambda|H|n',\lambda'\rangle=\langle n,\lambda|-\frac{1}{2}\nabla^2+v_{n'}|n',\lambda'\rangle+\langle n,\lambda|\sum_{p\neq n'}v_p|n',\lambda'\rangle \quad [3.40]$$

Due to the orthogonality of wave functions, [3.40] can be rewritten

$$\langle n,\lambda|H|n',\lambda'\rangle=\varepsilon_{n\lambda}\delta_{n,n'}\delta_{\lambda,\lambda'}+\langle n,\lambda|\sum_{p\neq n'}v_p|n',\lambda'\rangle \quad [3.41]$$

The last term can be further simplified if the three site integrals corresponding to $n\neq n'\neq p$ are neglected. There are then two types of remaining terms:

– the so-called crystal field integrals (with $n=n'$), the main effect of which is to shift the energy $\varepsilon_{n\lambda}$ of the isolated atom:

$$\alpha_{n'}^{\lambda\lambda'}=\langle n',\lambda|\sum_{p\neq n'}v_p|n',\lambda'\rangle \quad [3.42]$$

– the interatomic hopping integrals (with $p=n$ and $n\neq n'$) that measure the ability of electrons to wander between two neighboring atoms:

$$\beta_{nn'}^{\lambda\lambda'}=\langle n,\lambda|v_n|n',\lambda'\rangle \quad [3.43]$$

The crystal field integrals are often neglected in the calculation, considering the onsite energies $\varepsilon_{n\lambda}$ as fitting parameters (i.e. four parameters, one for each orbital type). The interatomic hopping integrals have been evaluated for cubic crystals by Slater and Koster [SLA 54] and within the $sp^3d^5s^*$ basis, we have fourteen independent intersite parameters.

The TB matrix equation is finally written,

$$\left(\varepsilon_{n\lambda}-E_b\right)\cdot C_{n\lambda}^b+\sum_{n'\neq n,\lambda'}\left\{\beta_{nn'}^{\lambda\lambda'}\right\}\cdot C_{n'\lambda'}^b=0$$

the solutions of which are the energies E_b and the eigenvectors $C^b_{n\lambda}$. Within the $sp^3d^5s^*$ TB model, we have to deal with $10N \times 10N$ matrix, where N is the total number of atoms in the system. The size of the matrix doubles if the spin–orbit interaction is accounted for.

For the general case of a periodic system containing N_{cell} primitive cells (indexed by l) centered at the position $\mathbf{R}_l$, each containing N_{at} atoms localized at position $\mathbf{s}_\alpha$ in this primitive cell, the crystal potential [3.4] is rewritten here for convenience in the form

$$V_{cr}(\mathbf{r}) = \sum_{l,\alpha} v_\alpha(\mathbf{r} - \mathbf{R}_l - \mathbf{s}_\alpha)$$

According to the Bloch theorem, the wave function expansion [3.38] can be rewritten as:

$$\psi_{\mathbf{k},b}(\mathbf{r}) = \sum_{\alpha,\lambda} C^b_{\alpha\lambda}(\mathbf{k}) \frac{1}{\sqrt{N_{cell}}} \sum_l \exp(i\mathbf{k} \cdot \mathbf{R}_{l\alpha}) \varphi_\lambda(\mathbf{r} - \mathbf{R}_{l\alpha})$$

where $\mathbf{R}_{l\alpha} = \mathbf{R}_l + \mathbf{s}_\alpha$. This leads to the TB equation for the periodic case

$$\begin{aligned}(\varepsilon_{\alpha\lambda} - E_{\mathbf{k}b}) \cdot C^b_{\alpha\lambda}(\mathbf{k}) + \\ \sum_{l',\alpha',\lambda'} \left\{ \beta^{\lambda\lambda'}_{l'\alpha' l\alpha} \exp\left[i\mathbf{k} \cdot (\mathbf{R}_{l'\alpha'} - \mathbf{R}_{l\alpha}) \right] \right\} \cdot C^b_{\alpha'\lambda'}(\mathbf{k}) = 0\end{aligned} \qquad [3.44]$$

The solutions of this equation are the **k**-dependent $E_{\mathbf{k}b}$ energies and the $C^b_{\alpha'\lambda'}(\mathbf{k})$ eigenvectors. For the periodic zinc-blende crystals, where there are two atoms per unit cell, the size of the TB matrix is 20 × 20 or 40 × 40 when the spin orbit coupling is accounted for. This makes the TB eigenvalue problem almost as computationally efficient as the $\mathbf{k} \cdot \mathbf{p}$ one. The TB method has the utmost advantage of being perfectly periodic in reciprocal space, i.e. for each **k** and for any vector **G** of the reciprocal lattice, we have $E_{\mathbf{k+G}} = E_{\mathbf{k}}$.

3.4.3. *The simple case of graphene*

As mentioned above, a TB model with a single orbital p_z on each atom is sufficient to describe the electronic structure of graphene in the vicinity of the Fermi level [WAL 46]. These orbitals are indeed decoupled by symmetry from the s, p_x and p_y orbitals. The hexagonal lattice of a graphene layer is shown in Figure 3.3(a). In the simplest approximation, only nearest neighbor orbitals are coupled in the Hamiltonian by a hopping energy $\beta = -2.7$ eV. Since there is a single orbital per atom, the onsite energies are all equal and set to zero here, by convention. With the labeling of Figure 3.3(a), equation [3.44] becomes:

$$\begin{cases} \beta\left[e^{i\mathbf{k}.(\mathbf{R}_{12}-\mathbf{R}_{11})}+e^{i\mathbf{k}.(\mathbf{R}_{22}-\mathbf{R}_{11})}+e^{i\mathbf{k}.(\mathbf{R}_{32}-\mathbf{R}_{11})}\right]C_2^b(\mathbf{k})=E_{\mathbf{k}b}C_1^b(\mathbf{k}) \\ \beta\left[e^{i\mathbf{k}.(\mathbf{R}_{11}-\mathbf{R}_{12})}+e^{i\mathbf{k}.(\mathbf{R}_{41}-\mathbf{R}_{12})}+e^{i\mathbf{k}.(\mathbf{R}_{51}-\mathbf{R}_{12})}\right]C_1^b(\mathbf{k})=E_{\mathbf{k}b}C_2^b(\mathbf{k}) \end{cases} \quad [3.45]$$

We define:

$$f(\mathbf{k})=e^{i\mathbf{k}.(\mathbf{R}_{12}-\mathbf{R}_{11})}+e^{i\mathbf{k}.(\mathbf{R}_{22}-\mathbf{R}_{11})}+e^{i\mathbf{k}.(\mathbf{R}_{32}-\mathbf{R}_{11})}$$

By symmetry of the hexagonal lattice, the term in brackets in the second line of [3.45] is simply the complex conjugate of f:

$$f^*(\mathbf{k})=e^{i\mathbf{k}.(\mathbf{R}_{11}-\mathbf{R}_{12})}+e^{i\mathbf{k}.(\mathbf{R}_{41}-\mathbf{R}_{12})}+e^{i\mathbf{k}.(\mathbf{R}_{51}-\mathbf{R}_{12})}$$

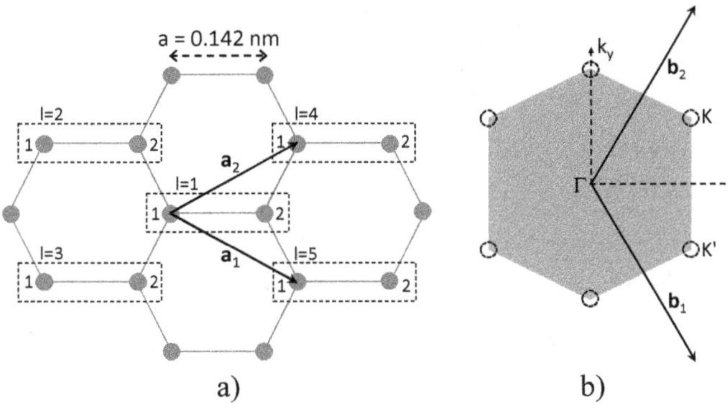

Figure 3.3. *a) Hexagonal lattice of graphene; $\mathbf{a}_1$ and $\mathbf{a}_2$ are the basis vectors of the Bravais lattice; each unit cell contains two carbon atoms labeled with $\alpha = 1$ and 2; the central unit cell and its neighboring unit cells have been arbitrarily labeled with index l = 1–5, in order to identify the Hamiltonian matrix elements and calculate the Bloch states; b) first Brillouin zone of graphene; $\mathbf{b}_1$ and $\mathbf{b}_2$ are the basis vectors of the reciprocal lattice; two corner points of the Brillouin zone are indexed by K and K'. The four other corners are equivalent to K or K' by reciprocal lattice translations*

Hence the eigenvalue problem [3.45] reads:

$$\begin{pmatrix} 0 & \beta f(\mathbf{k}) \\ \beta f^*(\mathbf{k}) & 0 \end{pmatrix} \begin{pmatrix} C_1^b(\mathbf{k}) \\ C_2^b(\mathbf{k}) \end{pmatrix} = E_{\mathbf{k}b} \begin{pmatrix} C_1^b(\mathbf{k}) \\ C_2^b(\mathbf{k}) \end{pmatrix}$$

The eigenvectors and eigenvalues are as follows:

$$\begin{pmatrix} C_1^1(\mathbf{k}) \\ C_2^1(\mathbf{k}) \end{pmatrix} = \begin{pmatrix} f(\mathbf{k}) \\ |f(\mathbf{k})| \end{pmatrix} \quad \text{with} \quad E_{\mathbf{k}1} = \beta |f(\mathbf{k})| \leq 0$$

and

$$\begin{pmatrix} C_1^2(\mathbf{k}) \\ C_2^2(\mathbf{k}) \end{pmatrix} = \begin{pmatrix} -f(\mathbf{k}) \\ |f(\mathbf{k})| \end{pmatrix} \quad \text{with} \quad E_{\mathbf{k}2} = -\beta |f(\mathbf{k})| \geq 0$$

The band structure is represented in Figure 3.4. The valence band (b = 1) and conduction band (b = 2) meet at the K points, defined as the six corners of the first Brillouin zone (see Figure 3.3(b)).

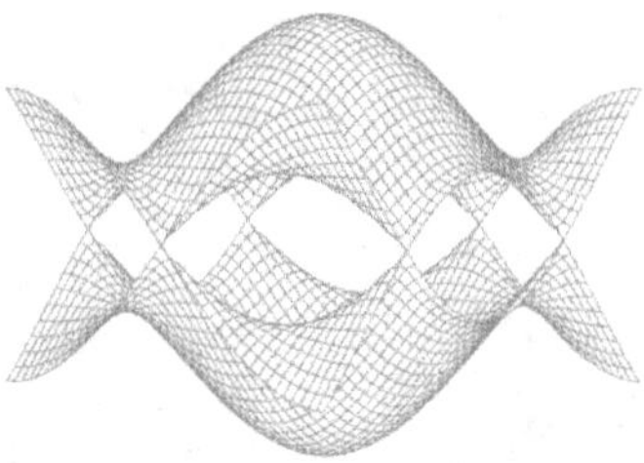

Figure 3.4. *Energy surfaces $E_{\mathbf{k}1}$ and $E_{\mathbf{k}2}$, representing, respectively, the valence band and the conduction band of graphene; they meet at the six K points of the first Brillouin zone*

The valence band is called the π band. It is formed by bonding linear combinations of p_z orbitals and contributes, with the lower energy orbitals s, p_x and p_y, to the cohesion of the graphene crystal. The conduction band is called the π* band, formed by antibonding linear combinations of p_z orbitals. The first order Taylor expansion of the dispersion relations around the K points yields:

$$E(\delta\mathbf{k}) \simeq \pm\frac{\sqrt{3}a\beta}{2}\|\delta\mathbf{k}\|$$

where a = 0.142 nm is the interatomic distance, the signs + and – stand for the valence and conduction bands, respectively, and $\|\delta\mathbf{k}\|$ is the distance to K point in reciprocal space. Hence, the energy surfaces are isotropic cones around K points, yielding an energy-independent carrier group velocity:

$$v = \frac{\sqrt{3}a|\beta|}{2\hbar} = 9.5\times10^5 \text{ m.s}^{-1}$$

The corresponding density of states cancels at the Fermi level, $E_F = 0$, and varies linearly with energy around E_F. Such a material with no band gap, but with a density of states that cancels at $E = E_F$, is called a semimetal.

3.4.4. *Strain effects in the TB model*

The effects of strain in TB models are accounted for through the bond-length dependence of the nearest-neighbor hopping integrals, which is usually fitted to a power law of the form

$$\beta(d) = \beta(d_0)\left(\frac{d}{d_0}\right)^{\eta} \qquad [3.46]$$

where d is the distance between two atoms in the strained crystal, d_0 is the equilibrium distance and η is a fitting parameter often called Harrison parameter. Although some hydrostatic and uniaxial deformation potentials can be reproduced by means of Harrison parameters only, Jancu *et al.* showed that much better accuracy can be achieved with the introduction of strain-dependent onsite parameters [JAN 98, JAN 07]. For a uniaxial [001] strain that lifts the degeneracy of the xy (z) and xz, yz (y, x) states, these authors suggested the on-site parameters:

$$\begin{cases} E_{xy} = E_d\left[1+2b_d\left(\varepsilon_{zz}-\varepsilon_{xx}\right)\right] \\ E_{yz} = E_{xz} = E_d\left[1-b_d\left(\varepsilon_{zz}-\varepsilon_{xx}\right)\right] \end{cases} \qquad [3.47]$$

where b_d is the shear parameter of the d states fitted to reproduce the uniaxial deformation of the valence-band edge [JAN 98]. However, the expressions of the strain tensor elements ε_{ij} as a function of atomic position are not univocal [NIQ 09b]. Boykin *et al.* therefore introduced position-dependent orbital energies in the $sp^3d^5s^*$ Hamiltonian, which improved the valence-band deformation potentials [BOY 02, 07], but not significantly the conduction-band deformation potentials, in particular for shear distortions. Recently, the same group generalized this approach to the case of [110]-oriented uniaxial strain [BOY 10], achieving a clear improvement of the model. Meanwhile, Niquet *et al.* proposed a model for the onsite matrix elements based on an explicit expression for the crystal field, assuming that the total potential is the sum of central, atomic contributions [NIQ 09b]. It features off-diagonal couplings between different orbitals and is able to reproduce the effects of arbitrary strains on the band energies and effective masses at all relevant **k** points for a large range of strain, up to 5%. In addition to the 14 usual Harrison parameters, this model introduced 13 new parameters that were fitted to reproduce accurately first-principle simulations of strained Si and Ge crystals.

3.5. Optimization of empirical models

As presented in the previous sections, empirical methods rely on adjusted parameters. The accuracy of the model strongly depends on the relevance of these parameters. In the development of an optimized empirical model, we have to fit experimental electronic properties as closely as possible. If in some cases a limited amount of experimental data can be enough, a rigorous optimization strategy is usually required. Currently, in silicon and germanium, some experimental information regarding the band energies are still missing to accurately determine all the model parameters, in particular at high energy (>4 eV). The situation can be even worse in certain materials in which important parameters, e.g. the effective masses and Luttinger parameters in SiC, have not been extensively characterized yet. For these reasons, a mixed approach, based on experimental data when available and *ab initio* results otherwise, is nowadays commonly used by the community for the optimization and the benchmark of empirical models.

3.5.1. *Reference data and methodology*

Over the past decades, the local-density approximation (LDA) variant of DFT has been established as a very powerful tool for studying the elastic properties and the deformation potentials of strained semiconductors [WAL 86, 89, NIE 85, JEV 03]. More recently, the GW many-body correction (where G stands for the Green's function and W for the screened Coulomb interaction) to the LDA DFT [HED 65] has led to semiconductor band structures with band gap values very close to the experimental data.

The empirical model parameters used below for Si and Ge have been optimized using the same first-principles references obtained from the LDA variant of the pseudopotential-based DFT. We will now briefly comment on three important aspects in order to generate reliable first principles DFT–LDA set of energy bands, masses and deformation potentials, which can be used later during the empirical model parameter optimization:

– the value of the equilibrium lattice parameter is calculated by minimizing the total energy. Any further LDA calculation have to use this theoretical value, instead of the experimental one, yielding thus a consistent set of zero-pressure reference data. In most semiconductor materials, the theoretical lattice parameter values usually agree within 1% with the experimental data [AUL 00, RID 06];

– it is known that the values of the band gap calculated with the LDA method are generally smaller than their experimental counterparts. However, the agreement can be greatly improved by the use of the Hedin's GW correction to include many-body effects [HED 65]. In practice this correction can be applied as a post-DFT scheme [AUL 00] in a non-self-consistent way. Nevertheless, in spite of the GW correction, the theoretical lowest conduction bands (CBs) typically lie within 0.05–0.2 eV of the range of experimental energies [AUL 00]. To obtain a reference set of energy bands in line with experimental data, a supplementary rigid "scissors" shift can be applied as a post-treatment [RID 06];

– a mechanical strain can be applied through a distortion of the Bravais vectors. However, before calculating the band structure, it is mandatory to perform a cell relaxation by minimizing the total energy of the strained crystal. For the determination of the equilibrium structure configuration under external applied strain, the total energy is minimized by varying the atomic position in the unit cell. This effect is characterized by the internal strain parameter ξ [KLE 62, KLE 63, GOR 63] that measures how the distance between the two atoms in the unit cell changes in response to the stress. It can be experimentally determined. For instance, in silicon we have $\xi = 0.54$ [COU 87]. The atomic positions in the cell are modified according to:

$$\mathbf{s}_\alpha{}' = (\mathbf{1}+\boldsymbol{\varepsilon})\cdot\mathbf{s}_\alpha \pm \frac{\xi}{4}\left[\varepsilon_{yz},\varepsilon_{xz},\varepsilon_{xy}\right] \qquad [3.48]$$

where the + and – signs refer to the atomic positions α = 1 and 2, respectively. These inner displacements of the atoms and the induced symmetry breaking have a strong effect on the electronic properties of strained semiconductors.

3.5.2. *Model parameters for Si and Ge*

Without extensive reference energy bands and deformation potentials, the full zone **k . p** model of [RID 06] and the $sp^3d^5s^*$ TB model of [NIQ 09b] that depend on more than 50 and 45 parameters, respectively (if the strain parameters are included), would be very difficult to optimize. The parameters of these models are briefly summarized in Table 3.3. Also shown are the parameters of the empirical non-local EPM of [RID 06].

During the last decade, several empirical models for silicon, germanium and their alloys have been efficiently fitted using conjugate-gradient procedures [RID 06], overall optimization procedures [BOY 04, BOY 10] or a mix of both [NIQ 09b] on first principle calculations. In general, satisfactory convergence is determined through a least-squares error function between empirical model predictions and experimental data as well as first principle results evaluated at relevant **k** points in the Brillouin zone. Considering the models presented in this chapter [RID 06], [NIQ 09b], it should be mentioned that they have been developed within a perspective of an application to silicon- or germanium-based devices, and particular attention has been paid to the near-Γ region and to the CB minima in order to obtain an accurate description of the curvature masses and Luttinger parameters. Some other aspects have also been considered. For instance, in the $\mathbf{k} \cdot \mathbf{p}$ model, particular care has been given at the edge of the Brillouin zone to ensure that the lowest conduction band and the highest valence band reach the edge of the Brillouin zone with zero slope (or average slope) as required by crystal symmetry.

Fitting the strain parameters can be more difficult than fitting the bulk ones. Indeed, strain can produce quite subtle effects on the band structure, ranging from energy band shifts to substantial modifications of the curvature masses. As highlighted in [RID 06, NIQ 09b], this effect has a deep impact on electron transport properties. Additionally, for all empirical models, a valence-band offset must be specified. This offset, combined with the bulk strain-dependent electronic structure model, determines the remaining part of the band line-ups.

Model	Number of bulk parameters	Matrix size (including spin–orbit)	Strain model parameters
Non local EPM	– 3 Form factors – 5 Non-local terms – Lattice parameter	>100 × 100	Interpolation function
Full-zone **k.p**	– 8 Γ-centered eigenvalues – 10 **k.p** matrix terms	30 × 30	– 31 Fitting parameters
$sp^3d^5s^*$ TB	– 4 Onsite terms – 14 Overlap terms – Lattice parameter	40 × 40	– 14 Harrison parameters – 13 Onsite parameters

Table 3.3. *Summary of fitting parameters to be used for empirical non-local EPM [RID 06], full zone k · p model [RID 10] and $sp^3d^5s^*$ model [NIQ 09b]*

3.5.3. *Comparison of the models for bulk Si and Ge*

Figure 3.5 compares GW results with those obtained with empirical full band models. The calculations have been performed along high-symmetry directions in reciprocal space. A good overall agreement is obtained, even at high energy. It also appears that the overall agreement between theoretical and experimental band gaps is good. Similar agreement can be found concerning the curvature masses and Luttinger parameters.

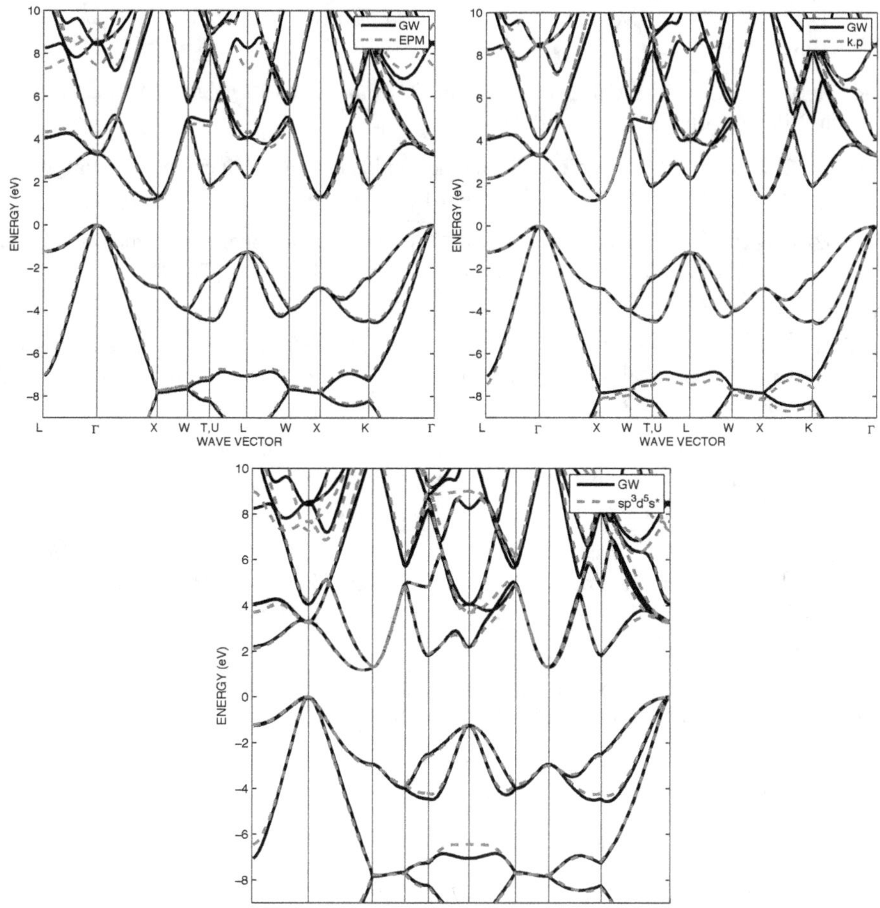

Figure 3.5. *Bulk silicon electronic band structure obtained using 30-level* **k·p** *model, EPM and sp3d5s* TB model, compared to ab initio DFT+GW calculation. For a color version of this figure, see www.iste.co.uk/dollfus/nanodevices.zip*

3.6. Bibliography

[AUL 00] AULBUR W.G., JÖNSSON L., WILKINS J.W., "Quasiparticle calculations in solids", *Solid State Physics*, vol. 54, pp. 1–218, 2000.

[BAC 82] BACHELET G.B., HAMANN D.R., SCHLÜTER M., "Pseudopotentials that work: from H to Pu", *Physical Review B*, vol. 26, no. 8, pp. 4199–4228, 1982.

[BAR 38] BARDEEN J., "An improved calculation of the energies of metallic Li and Na", *Journal of Chemical Physics*, vol. 6, pp. 367–372, 1938.

[BAS 91] BASTARD G., *Wave Mechanics Applied to Semiconductor Heterostructures*, Wiley Interscience, 1991.

[BIR 74] BIR G.L., PIKUS G.E., *Symmetry and Strain-Induced Effects in Semiconductors*, John Wiley & Sons, New York, 1974.

[BLO 28] BLOCH F., "Über die Quantenmechanik der Elektronen in Kristallgittern", *Zeitschrift für Physik*, vol. 52, pp. 555–600, 1928.

[BOY 02] BOYKIN T.B., KLIMECK G., BOWEN R.C. *et al.*, "Diagonal parameter shifts due to nearest-neighbor displacements in empirical tight-binding theory", *Physical Review B*, vol. 66, p. 125207, 2002.

[BOY 04] BOYKIN T.B., KLIMECK G., OYAFUSO F., "Valence band effective-mass expressions in the $sp^3d^5s^*$ empirical tight-binding model applied to a Si and Ge parametrization", *Physical Review B*, vol. 69, p. 115201, 2004.

[BOY 07] BOYKIN T.B., KHARCHE N., KLIMECK G., "Brillouin-zone unfolding of perfect supercells having nonequivalent primitive cells illustrated with a Si/Ge tight-binding parameterization", *Physical Review B*, vol. 76, p. 035310, 2007.

[BOY 10] BOYKIN T.B., LUISIER M., SALMANI-JELODAR M. *et al.*, "Strain-induced, off-diagonal, same-atom parameters in empirical tight-binding theory suitable for [110] uniaxial strain applied to a silicon parametrization", *Physical Review B*, vol. 81, p. 125202, 2010.

[BRU 62] BRUST D., COHEN M.L., PHILLIPS J.C., "Reflectance and photoemission from Si", *Physical Review Letters*, vol. 9, pp. 389–392, 1962.

[BRU 71] BRUST D., "Nonlocal corrections to the band structure of Si, Ge, and α-Sn", *Physical Review B*, vol. 4, pp. 3497–3501, 1971.

[BUR 88] BURT M.G., "An exact formulation of the envelope function method for the determination of electronic states in semiconductor microstructures", *Semiconductor Science and Technology*, vol. 3, pp. 739–753, 1988.

[BUR 94] BURT M.G., "Direct derivation of effective-mass equations for microstructures with atomically abrupt boundaries", *Physical Review B*, vol. 50, pp. 7518–7525, 1994.

[CAR 66] CARDONA M., POLLAK F.H., "Energy-band structure of germanium and silicon: the $k \cdot p$ method", *Physical Review*, vol. 142, pp. 530–543, 1966.

[CAR 88] CARDONA M., CHRISTENSEN N.E., FASOL G., "Relativistic band structure and spin-orbit splitting of zinc-blende-type semiconductors", *Physical Review B*, vol. 38, pp. 1806–1827, 1988.

[CHE 73] CHELIKOWSKY J.R., CHADI D.J., COHEN M.L., "Calculated valence band densities of states and photoemission spectra of diamond and zinc-blende semiconductors", *Physical Review B*, vol. 8, pp. 2786–2794, 1973.

[CHE 76] CHELIKOWSKY J.R., COHEN M.L., "Nonlocal pseudopotential calculations for the electronic structure of eleven diamond and zinc-blende semiconductors", *Physical Review B*, vol. 14, no. 2, pp. 556–582, 1976.

[COH 66] COHEN M.L., BERGSTRESSER T.K., "Band structures and pseudopotential form factors for fourteen semiconductors of the diamond and zinc-blende structures", *Physical Review*, vol. 141, pp. 789–796, 1966.

[COU 87] COUSINS C.S.G., GERWARD L., OLSEN J.S. *et al.*, "Surface effects in uniaxially stressed crystals: the internal-strain parameters of silicon and germanium revised", *Journal of Physics C: Solid State Physics.*, vol. 20, pp. 29–37, 1987.

[DEL 04] DELERUE C., LANNOO M., *Nanostructures: Theory and Modelling*, Springer-Verlag, Berlin Heidelberg New York, 2004.

[DRE 55] DRESSELHAUS G., KIP A.F., KITTEL C., "Cyclotron resonance of electrons and holes in silicon and germanium crystals", *Physical Review*, vol. 98, pp. 368–384, 1955.

[FAL 66] FALICOV L.M., *Group Theory and its Physical Applications*, University of Chicago Press, Chicago 1966.

[FER 06] FERAILLE M., RIDEAU D., GHETTI A. *et al.*, "Low-field mobility in strained silicon with full-band Monte Carlo simulation using k·p and EPM band structure", *Proceedings of the International Conference on Simulation of Semiconductor Processes and Devices (SISPAD 2006),* Monterey, pp. 264–266, 2006.

[FIS 88] FISCHETTI M.V., LAUX S.E., "Monte Carlo analysis of electron transport in small semiconductor devices including band-structure and space-charge effects", *Physical Review B*, vol. 38, no. 14, pp. 9721–9745, 1988.

[FIS 96] FISCHETTI M.V., LAUX S.E., "Band structure, deformation potentials, and carrier mobility in strained Si, Ge, and SiGe alloys", *Journal of Applied Physics*, vol. 80, pp. 2234–2252, 1996.

[FOR 07a] FOREMAN B.A., "Accurate quadratic-response approximation for the self-consistent pseudopotential of semiconductor nanostructures", *Physical Review B*, vol. 76, p. 045326, 2007.

[FOR 07b] FOREMAN B.A., "Valence-band mixing in first-principles envelope-function theory", *Physical Review B*, vol. 76, p. 045327, 2007.

[FRI 76] FRIEDEL P., HYBERTSEN M.S., SCHLÜTER M., "Local empirical pseudopotential approach to the optical properties of Si/Ge superlattices", *Physical Review B*, vol. 39, pp. 7974–7977, 1976.

[GOR 63] GOROFF I., KLEINMAN L., "Deformation potentials in silicon. III. Effects of a general strain on conduction and valence levels", *Physical Review*, vol. 132, p. 1080, 1963.

[HED 65] HEDIN L., "New method for calculating the one-particle green's function with application to the electron-gas problem", *Physical Review*, vol. 139, pp. A796–A823, 1965.

[HES 91] HESS K., *Monte Carlo Device Simulation: Full Band and Beyond*, Kluwer Academic Publishers, Boston,1991.

[HUE 08] HUET K., FERAILLE M., RIDEAU D. *et al.*, "Experimental and theoretical analysis of hole transport in uniaxially strained pMOSFETs", *Proceedings 38th European Solid-State Device Research Conference (ESSDERC 2008)*, Edinburgh, pp. 234–237, 2008.

[HUM 81] HUMPHREYS R.G., "Valence band averages in silicon: anisotropy and non-parabolicity", *Journal of Physics C: Solid State Physics*, vol. 14, pp. 2935–2942, 1981.

[JAN 07] JANCU J.M., VOISIN P., "Tetragonal and trigonal deformations in zinc-blende semiconductors: a tight-binding point of view", *Physical Review B*, vol. 76, p. 115202, 2007.

[JAN 98] JANCU J.M., SCHOLZ R., BELTRAM F. *et al.*, "Empirical spds* tight-binding calculation for cubic semiconductors: general method and material parameters", *Physical Review B*, vol. 57, pp. 6493–6507, 1998.

[JUN 03] JUNGEMANN C., MEINERZHAGEN B., *Hierarchical Device Simulation: The Monte Carlo Perspective*, Springer-Verlag, New York, 2003.

[KAN 66] KANE E.O., "Band structure of silicon from an adjusted Heine-Abarenkov calculation", *Physical Review*, vol. 146, pp. 558–567, 1966.

[KLE 62] KLEINMAN L., "Deformation potentials in silicon. I. Uniaxial strain", *Physical Review*, vol. 128, pp. 2614–2621, 1962.

[KLE 63] KLEINMAN L., "Deformation potentials in silicon. II. Hydrostatic strain and the electron-phonon interaction", *Physical Review*, vol. 130, pp. 2283–2289, 1963.

[KLI 10] KLIPSTEIN P.C., "Operator ordering and interface-band mixing in the Kane-like Hamiltonian of lattice-matched semiconductor superlattices with abrupt interfaces", *Physical Review B*, vol. 81, p. 235314, 2010.

[LAW 71] LAWAETZ P., "Valence-band parameters in cubic semiconductors", *Physical Review B*, vol. 4, pp. 3460–3467, 1971.

[LEV 03] LEVINE Z.H., BURNETT J.H., SHIRLEY E.L., "Photoelastic and elastic properties of the fluorite structure materials, LiF, and Si", *Physical Review B*, vol. 68, p. 155120, 2003.

[LUI 06] LUISIER M., SCHENK A., FICHTNER W. *et al.*, "Atomistic simulation of nanowires in the $sp^3d^5s^*$ tight-binding formalism: from boundary conditions to strain calculations", *Physical Review B*, vol. 74, p. 205323, 2006.

[LUT 54] LUTTINGER J.M., KOHN W., "Motion of electrons and holes in perturbed periodic fields", *Physical Review*, vol. 97, pp. 869–883, 1954.

[MAS 83] MASOVIC D.R., VUKAJLOVIC F.R., ZEKOVIC S. "Local-pseudopotential calculation for optical properties and photoemission valence-band spectrum of silicon", *Journal of Physics C: Solid State Physics*, vol. 16, p. 6731, 1983.

[MIC 09] MICHAILLAT M., RIDEAU D., ANIEL F. *et al.*, "Monte Carlo-based analytical models for electron and hole electrical parameters in strained SiGeC alloys", *Proceedings of the International Conference on Simulation of Semiconductor Processes and Devices (SISPAD 2009)*, pp. 55–58, 2009.

[NGU 13] NGUYEN V.H., TRIOZON F., BONNET F.D.R. *et al.*, "Performances of strained nanowire devices: ballistic versus scattering-limited currents", *IEEE Transactions on Electron Devices*, vol. 60, no. 5, pp. 1506–1313, 2013.

[NIE 85] NIELSEN O.H., MARTIN R.M., "Quantum-mechanical theory of stress and force", *Physical Review B*, vol. 32, pp. 3780–3791, 1985.

[NIQ 00] NIQUET Y.M., DELERUE C., ALLAN G. *et al.*, "Method for tight-binding parametrization: application to silicon nanostructures", *Physical Review B*, vol. 62, pp. 5109–5116, 2000.

[NIQ 09a] NIQUET Y.M., RIDEAU D., TAVERNIER C. *et al.*, "Onsite matrix elements of the tight-binding Hamiltonian of a strained crystal: application to silicon, germanium, and their alloys", *Physical Review B*, vol. 79, p. 245201, 2009.

[NIQ 09b] NIQUET Y.M., LHERBIER A., QUANG N.H. *et al.*, "Electronic structure of semiconductor nanowires", *Physical Review B*, vol. 73, p. 165319, 2009.

[NIQ 12] NIQUET Y.M., DELERUE C., "Carrier mobility in strained Ge nanowires", *Journal of Applied Physics*, vol. 112, p. 084301, 2012.

[PHI 58] PHILLIPS J.C., "Energy-band interpolation scheme based on a pseudopotential", *Physical Review*, vol. 112, pp. 685–695, 1958.

[PIK 60] PIKUS G.E., BIR G.L., "Effects of deformation on the hole energy spectrum of germanium and silicon", *Soviet Physics – Solid State*, vol. 1, pp. 1502–1517, 1960.

[POL 70] POLLAK F.H., CARDONA M., HIGGINBOTHAM C.W. *et al.*, "Energy-band structure and optical spectrum of grey tin", *Physical Review B*, vol. 2, pp. 352–363, 1970.

[REN 97] REN S.S., CHEN X., DOW J.D., "Tight-binding sp^3d^5 Hamiltonian for Si", *Journal of Physics and Chemistry of Solids*, vol. 59, pp. 403–410, 1997.

[RIC 04] RICHARD S., ANIEL F., FISHMAN G., "Energy-band structure of Ge, Si, and GaAs: A thirty-band k·p method", *Physical Review B*, vol. 70, p. 235204, 2004.

[RID 01] RIDENE S., BOUJDARIA K., BOUCHRIHA H. *et al.*, "Infrared absorption in $Si/Si_{1-x}Ge_x/Si$ quantum wells", *Physical Review B*, vol. 64, p. 085329, 2001.

[RID 06] RIDEAU D., FERAILLE M., CIAMPOLINI L. *et al.*, "Strained Si, Ge, and $Si_{1-x}Ge_x$ alloys modeled with a first-principles-optimized full-zone **k·p** method", *Physical Review B*, vol. 74, 195208, 2006.

[RID 09] RIDEAU D., FERAILLE M., MICHAILLAT M. *et al.*, "On the validity of the effective mass approximation and the Luttinger **k·p** model in fully depleted SOI MOSFETs", *Solid-State Electronics*, vol. 53, pp. 452–461, 2009.

[RID 12] RIDEAU D., "Full band models for strained silicon and germanium devices", in JIT S. (ed.), *Advances in Microelectronics and Photonics*, Nova Science Publishers, New-York, pp. 191–258, 2012.

[RIE 93] RIEGER M.M., VOGL P., "Electronic-band parameters in strained $Si_{1-x}Ge_x$ alloys on $Si_{1-y}Ge_y$ substrates", *Physical Review B*, vol. 48, pp. 14276–14287, 1993.

[SAR 69] SARAVIA L.R., BRUST D., "Strain-split energy bands in semiconductors: Ge", *Physical Review,* vol. 178, pp. 1240–1243, 1969.

[SAS 62] SASLOW W., BERGSTRESSER T.K., COHEN M.L., "Band structure and optical properties of diamond", *Physical Review Letter*, vol. 16, pp. 354–357, 1962.

[SLA 54] SLATER J.C., KOSTER G.F., "Simplified LCAO method for the periodic potential problem", *Physical Review*, vol. 94, pp. 1498–1524, 1954.

[SMI 90] SMITH D.L., MAILHIOT C., "Theory of semiconductor superlattice electronic structure", *Reviews of Modern Physics*, vol. 62, pp. 173–234, 1990.

[VAR 73] VAREA de ALVAREZ C., WALTER J.P., BOYD R.W. *et al.*, "Calculated band structures, optical constants and electronic charge densities for InAs and InSb", *Journal of Chemical Physics Solids*, vol. 34, pp. 337–345, 1973.

[VOG 83] VOGL P., HJALMARSON H.P., DOW J.D., "A semi-empirical tight-binding theory of the electronic structure of semiconductors", *Journal of Physics and Chemistry of Solids*, vol. 44, pp. 365–378, 1983.

[WAL 46] WALLACE P.R., "The band theory of graphite", *Physical Review*, vol. 71, pp. 622–634, 1946.

[WAL 86] VAN DE WALLE C.G., MARTIN R.M., "Theoretical calculations of heterojunction discontinuities in the Si/Ge system", *Physical Review B*, vol. 34, pp. 5621–5634, 1986.

[WAL 89] VAN DE WALLE C.G., "Band lineups and deformation potentials in the model-solid theory", *Physical Review B*, vol. 39, pp. 1871–1883, 1989.

[YU 01] YU P., CARDONA M., *Fundamentals of Semiconductors*, Springer-Verlag, Berlin, Heidelberg, 2001.

[ZHA 10] ZHANG W., DELERUE C., NIQUET Y.M. *et al.*, "Atomistic modeling of electron-phonon coupling and transport properties in n-type [110] silicon nanowires", *Physical Review B*, vol. 82, p. 115319, 2010.

4

Relevant Semiempirical Potentials for Phonon Properties

4.1. Introduction

This chapter provides a review of several key interaction potentials that are commonly used to estimate phonon properties – dispersion curves, density of states (DOS), group velocities and relaxation times – as well as their related macroscopic quantities – dilatation coefficient, heat flux, thermal conductivity and thermal conductance.

Computing those latter quantities using *ab initio* codes remains a challenge. Quantum molecular dynamics (MD), i.e. the integration of Newton's second law with interaction forces estimated by density functional theory at each time step, requires extensive resources. Lattice dynamics (LD) with harmonic and first-order anharmonic terms derived from *ab initio* calculations have recently been implemented and have provided reasonable predictions. However, they miss higher order terms and only apply to simple periodic crystals.

The use of pseudopotentials – potential energy functionals of the atomic positions – might appear as somehow rough to those aware of the details of the electronic cortege and the methods of estimating its potential energy. However, a pseudo-potential constitutes an adequate tool enabling the treatment of time and space scales specific to phonon transport and can reliably reproduce the complexity of electronic wave functions in terms of an average potential, be it in the long range, as for Coulombic interactions, or in the short-range ones when wave functions are merging. This reliability lies on the fact that a large set of pseudopotentials is in fact derived from exact *ab initio* calculations.

Chapter written by Sebastian VOLZ.

Pseudopotentials can be used in MD simulations, where atomic trajectories are computed according to time. All phonon properties are expressed in those atomic motions and can therefore be extracted. Pseudopotentials can also be used to set the force constant matrix and issue the characteristics of the phonon eigenmodes from LD.

This chapter is a guideline for those interested in the computation of phonon or vibrational properties with a selection of broadly used potentials in a vast variety of cases including semiconductors, metals and dielectrics but also polymers and water.

4.1.1. *Criteria for choosing the adequate potential in phonon computations*

Finding a relevant pseudopotential to compute phononic properties in a given system essentially relies on the dispersion curves this potential provides and how those curves are compared with experimental curves. Exact dispersion curves guarantee the accuracy of the quantities based on the harmonic interaction forces – or the second derivative of the potential energy – including eigenfrequencies, group velocity and elastic constants.

The correspondence between experimental and computed dispersion curves might be limited to a given frequency interval, which might still be relevant depending on the temperature under test. High temperatures correspond to the excitations of high frequencies while at lower temperatures low-frequency modes are populated. A rather accurate quantitative correspondence between temperature and frequency is deduced from the maximum of the Planck function at the given temperature.

The pertinence of the third derivative of the potential can be checked by the computation of the dilation coefficient or the Grüneisen coefficient, which are both deduced from the first-order anharmonic term. Relaxation times should also constitute a check but fewer reference data are available and those times usually include the contribution of higher order terms.

Quite unexpectedly, recent attempts to reconstruct the harmonic and the first anharmonic terms of a potential via density functional theory have revealed that fitting the dispersion curves on the whole Brillouin zone is a more difficult task than fitting the relaxation times. If confirmed, this fact might reduce the validity criteria for a potential to predict phononic properties to the sole dispersion curves constraint.

4.1.2. *Derivation of the phonon properties*

Dispersion curves can be derived by the conventional LD technique based on the force constant matrix. This matrix can be built from analytical derivations using the functional form of the pseudopotential, but also from the so-called "brute force" approach. This latter consists of establishing the static atomic system with interatomic forces set by the studied potential, and move one atom by a small fraction of the lattice constant. The change in force divided by the atomic displacement can be considered as a reasonable estimation of the force constants involving the modified atom. However, note that LD misses the anharmonic shift of the eigenfrequencies.

The second technique is based on MD where atomic motions include the dynamical expression of phonon modes. MD consists of solving atomic motion based on Newton's second Law and is a purely classical technique, which does not account for the low temperature quantum populations. In other words, MD will systematically provide the energy $k_B T/2$ to all degrees of freedom involved in the system. This will not affect the estimations of classical quantities such as eigenfrequencies, group velocities or Grüneisen coefficient, but will alter the predictions of the terms depending on the phonon populations such as the thermal conductivity, the heat capacity or the spectral free path and the spectral scattering rate. The validity of the classical description is of course defined by the relationship between temperature and Debye temperature, as high frequency modes are not fully populated when the temperature is lower than the Debye temperature. In this situation, for instance, the estimation of high-frequency relaxation times might generally be considered as dubious.

Phonon dispersion relation can be calculated from the projection of the kinetic energy on a given mode characterized by a wave vector k of the reciprocal space and a frequency of the Fourier space. For a given polarization, p = x, y, z, the reciprocal space representation of the velocity for atom i at time t can be written as:

$$v_{i,k}^{p}(t) = v_i^{p}(t) e^{-ik \cdot r_i^0} \qquad [4.1]$$

where r_i^0 refers to the equilibrium position of atom i. The per-atom kinetic energy in the reciprocal space then reads:

$$E_{i,c}^{p}(k,t) = \frac{m_i v_{i,k}^{p}(t) \cdot v_{i,k}^{p\,*}(t)}{2} \qquad [4.2]$$

By time Fourier transforming the total kinetic energy, the eigenfrequencies for a given wave vector and the corresponding amplitudes of the kinetic energy can be found:

$$E_c^p(k,\omega) = \frac{\left| \sum_i m_i^{1/2} \int_{-\infty}^{\infty} v_{i,k}^p(t) e^{i\omega t} dt \right|^2}{2} \quad [4.3]$$

When plotted in the frequency wave vector plane, this projected energy defines the dispersion curves from the resonance peak observed in the spectrum at a given wave vector. The mode relaxation time is derived from the reverse of the peak line width at mid-height multiplied by two. This data processing remains difficult to program due to the sharpness of the signal and should be often carried out in a per mode process. The DOS can also be deduced from the time Fourier transform of the atomic velocities as follows [DOM 04]:

$$g(\omega) \propto \sum_i^N \left| \int_{-\infty}^{+\infty} v_i(t) e^{i\omega t} dt \right|^2 \quad [4.4]$$

4.2. Generic pair potentials: the Lennard-Jones potential

The Lennard-Jones (L-J) potential (also referred to as the L-J potential, 6-12 potential, or 12-6 potential) is a model that approximates the interaction between a pair of neutral atoms or molecules [LEN 24].

The most common expression of the L-J potential is:

$$V_{LJ} = 4\varepsilon \left[\left(\frac{\sigma}{r}\right)^{12} - \left(\frac{\sigma}{r}\right)^{6} \right] \quad [4.5]$$

where ε is the depth of the potential well, σ is the finite distance at which the interparticle potential is zero and r is the distance between the particles. The r^{-12} term, which is the repulsive term, describes Pauli repulsion at short ranges due to overlapping electron orbitals and the r^{-6} term, which is the attractive term, describes van der Waals force, also called dispersion force. The L-J potential is commonly used to model weak interactions, as in rare gases such as Ar, Kr and Xe, but also stronger ones such as Cl–Cl and C–C (see section 4.6.2). The potential curve is shown in Figure 4.1.

The dispersion curve and phonon DOS of Ar, as shown in Figure 4.2, are calculated by choosing the parameter values: ε = 0.0103 eV and σ = 3.405 Å. The

MD phonon dispersion relations underestimate the experimental data measured by inelastic neutron scattering [FUJ 74]. The phonon DOS at very low temperature (10 K) reproduces accurately the calculations from LD. At higher temperature (50 K), the DOS deviates from LD calculations possibly due to the strong anharmonic effects appearing before melting.

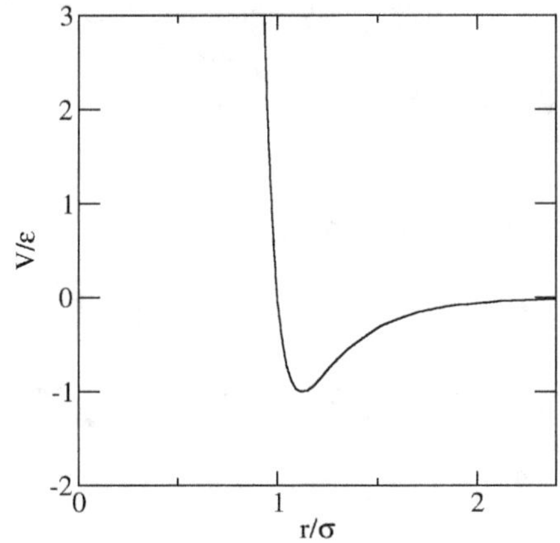

Figure 4.1. *Potential energy versus distance for the 12-6 Lennard-Jones potential*

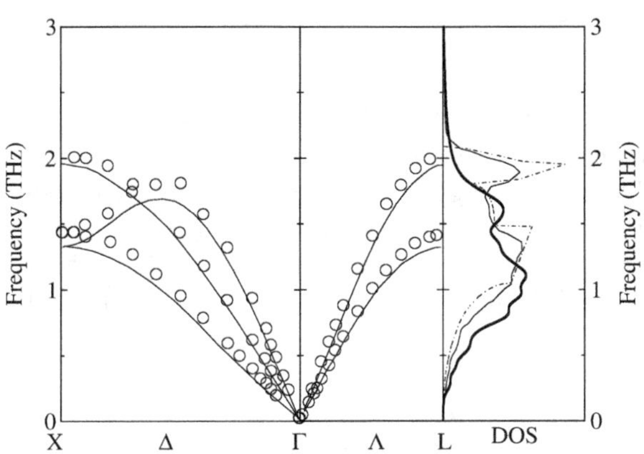

Figure 4.2. *(Left panel) Dispersion curves of Ar calculated by lattice dynamics (solid line) [HEI 07]. Experimental results are represented as open circles [FUJ 74]. (Right panel) Phonon density of states of Ar computed by molecular dynamics at 10 K (gray line) and 50 K (black line) and from lattice dynamics (dashed curve) [KAB 07]*

4.3. Semiconductors: Stillinger–Weber and Tersoff potentials

4.3.1. *Stillinger–Weber potential*

The Stillinger–Weber potential was developed as an interatomic potential for tetrahedrally bonded semiconductor materials [STI 85]. The potential energy E_p is composed of two parts: the two-body interaction $\phi_2(r_{ij})$ and the three-body interaction $\phi_3(r_{ij}, r_{ik}, \theta_{ijk})$, where r_{ij} represents the distance between two atoms i and j and θ_{ijk} refers to the angle between atoms i, j and k.

$$E_p = \sum_{\substack{i,j \\ j<i}} \phi_2(r_{ij}) + \sum_{\substack{i,j,k \\ k<j<i}} \phi_3(r_{ij}, r_{ik}, \theta_{ijk}) \quad [4.6]$$

$$\phi_2(r_{ij}) = \begin{cases} A\varepsilon\left[B\left(\frac{\sigma}{r_{ij}}\right)^p - \left(\frac{\sigma}{r_{ij}}\right)^q\right]\exp\left(\frac{\sigma}{r_{ij}-a\sigma}\right) \text{ if } r_{ij} < a\sigma \\ 0 \text{ if } r_{ij} \geq a\sigma \end{cases} \quad [4.7]$$

$$\phi_3(r_{ij}, r_{ik}, \theta_{ijk}) = \\ \begin{cases} \lambda\varepsilon\left(\cos\theta_{ijk} - \cos\theta_{ijk}^0\right)^2 \exp\left(\frac{\gamma\sigma}{r_{ij}-a\sigma}\right)\exp\left(\frac{\gamma\sigma}{r_{ik}-a\sigma}\right) \text{ if } r_{ij} < a\sigma \text{ and } r_{ik} < a\sigma \\ 0 \text{ if } r_{ij} \geq a\sigma \text{ or } r_{ik} \geq a\sigma \end{cases} \quad [4.8]$$

The two-body interaction $\phi_2(r_{ij})$ is similar to that of the L-J potential (Figure 4.1). It ensures that the covalent bond length at equilibrium, defined by the parameter σ, and the bond energy, referred as ε, are consistent with literature values. $a\sigma$ corresponds to the cutoff radius of the potential. The exponential term guarantees the continuity of the gradient of the interatomic potential, i.e. the interatomic forces. The three-body term $\phi_3(r_{ij}, r_{ik}, \theta_{ijk})$ tends to impose the angle between atoms at the value θ_{ijk}^0 in order to conserve a diamond-like structure.

As this potential was designed for tetrahedrally bonded crystals, it cannot handle other types of structure with a different coordination number such as surface diamond-like structures, graphite or semiconductor liquids. Silicon and germanium crystals should however be well described in terms of vibrational properties by implementing the Stillinger–Weber potential.

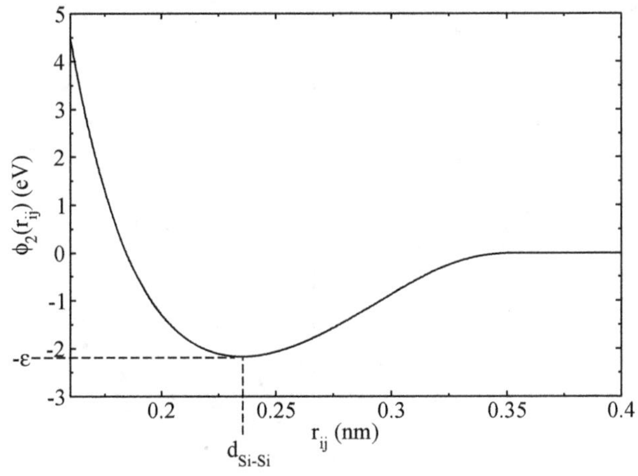

Figure 4.3. *Two-body interaction for the silicon crystal; the minimum value of $\phi_2(r_{ij})$ corresponds to the covalent bond length at equilibrium*

4.3.2. *Silicon crystal modeled with Stillinger–Weber potential*

The silicon crystal is arranged in a face centered cubic lattice with a pattern of two atoms at reduced coordinates (0 0 0) and (¼ ¼ ¼) and a lattice parameter of 0.5431 nm. The cohesive energy and the covalent bond length were among the main criteria to estimate the potential parameters given in Table 4.1. The minimum value of the two-body potential is reached at 0.235 nm, which is consistent with the experimental value. The cutoff radius is equal to 0.377 nm, which indicates that only the first-neighbor interactions are taken into account during MD simulations.

ε (eV)	σ (nm)	A	λ	γ
2.17	0.20955	1.80	21.0	1.20
θ_{ijk}^{0} (°)	A	B	P	Q
109.471	7.0495562	0.6022245	4.0	0.0

Table 4.1. *Stillinger–Weber parameters for silicon crystal*

Figure 4.4 provides the comparison between the vibrational properties computed by MD and the literature results.

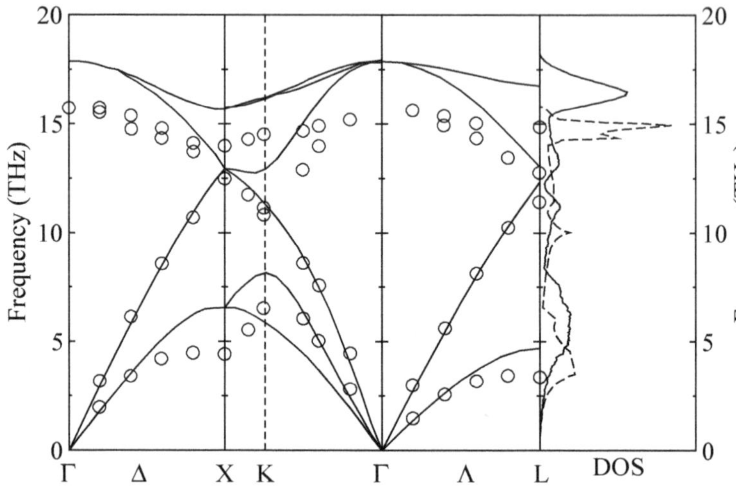

Figure 4.4. *Silicon phonon dispersion relation computed with molecular dynamics (solid lines) compared to inelastic neutron scattering experimental data (open circles) [DOL 62]; (left) phonon density of states (DOS) computed by molecular dynamics (solid line) compared to adiabatic bond charged model (dashed line) [WEB 77]*

The MD phonon dispersion relations overestimate the experimental data measured by inelastic neutron scattering [DOL 62] for certain branches. However, the longitudinal acoustic and the first half of the transverse acoustic branches are well modeled by the Stillinger–Weber potential. Consequently, the DOS are also overestimated compared to the one predicted by the adiabatic bond charge model [WEB 77]. As the modes of the acoustic branches are the main energy carrier in silicon, the frequency-dependent transport properties should be reasonably estimated by the Stillinger–Weber potential.

4.3.3. *Germanium crystal modeled with Stillinger–Weber*

As with carbon and silicon, germanium atoms are tetrahedrally bonded to form a diamond-like structure. The lattice parameter is equal to 0.5658 nm. The Stillinger–Weber potential was adapted for amorphous and crystalline germanium phases [DIN 86]. The cohesive energy at 0 K, the experimental bond length and the elastic constants are used as constraints to determine the potential parameters presented in Table 4.2.

ε (eV)	σ (nm)	A	λ	γ
1.93	0.2181	1.80	31.0	1.20
θ_{ijk}^{0} (°)	A	B	P	Q
109.471	7.0495562	0.6022245	4.0	0.0

Table 4.2. *Stillinger–Weber parameters for germanium amorphous and crystalline phases*

The phonon dispersion relations and the DOS were calculated using Stillinger–Weber potential in MD simulations and compared to reference results in Figure 4.5. The phonon dispersion relations have branches with higher frequency than the ones obtained from inelastic neutron scattering measurements [DOL 66], except for the longitudinal and the low k transverse acoustic modes. Therefore, the computed DOS is shifted to a higher frequency compared to DOS yielded by adiabatic charge model results [WEB 77]. Thus, the Stillinger–Weber potential overestimates in the same way the vibrational properties of silicon and germanium crystals, except for the majority of the acoustic modes, which remain preponderant in transport phenomena in the vicinity of room temperature.

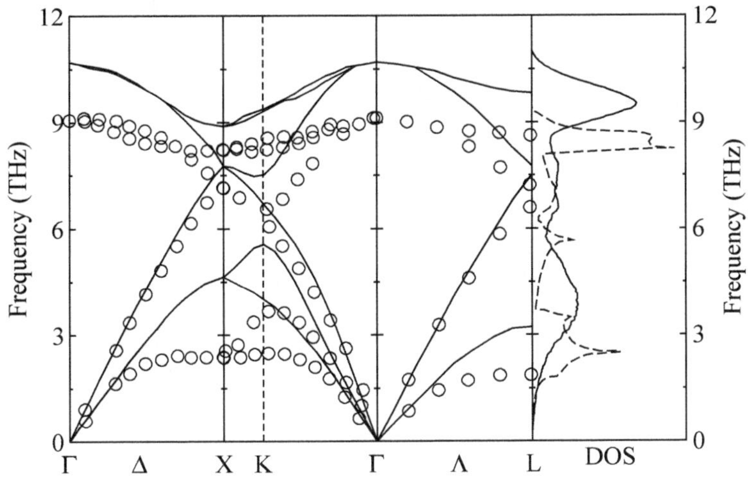

Figure 4.5. *Germanium phonon dispersion relation computed with molecular dynamics (solid lines) compared to inelastic neutron scattering experiment (open circles) [DOL 66]; phonon density of states (DOS) computed by molecular dynamics (solid line) compared to adiabatic bond charged model (dashed line) [WEB 77]*

4.3.4. *Tersoff potential*

First proposed for silicon by its namesake in 1986, the Tersoff potential takes the form of a Morse pair potential with a bond strength parameter that depends on the local bonding environment. This property results in the summation over neighbors to compute the bonding environment effects and is sometimes assimilated to a three-body potential. Tersoff *et al.* extended the potential to multicomponent systems in 1989 when they considered C-Si and Si-Ge systems. A simple method for mixing potential coefficients was also presented, which calculates pair interaction coefficients from elemental coefficients and a mixing parameter.

The Tersoff potential's primary form is unremarkable, being composed of an attraction term f_A, a repulsive term f_R and a cutoff function, all terms varying according to the distance between interaction atoms. The common expression is defined as follows:

$$E = \frac{1}{2}\sum_i \sum_{i \neq j}\left[f_C\left(r_{ij}\right)\left(f_R\left(r_{ij}\right) - b_{ij} f_A\left(r_{ij}\right)\right)\right] \quad [4.9]$$

$$f_C = \begin{cases} 1 & r < R - D \\ \frac{1}{2} - \frac{1}{2}\sin\left(\frac{\pi}{2}\frac{r - R}{D}\right) & R - D < r < R + D \\ 0 & r > R - D \end{cases} \quad [4.10]$$

$$f_R\left(r\right) = A\exp\left(-\lambda_1 r\right) \quad [4.11]$$

$$f_A\left(r\right) = B\exp\left(-\lambda_2 r\right) \quad [4.12]$$

The bond environment effects come into play in the multiplier of the attraction term, which is a function of the positions of atoms lying within the cutoff range of the reference atom. These expressions account for both the distance from the reference to the environmental atom as well as their relative angular positions.

$$b_{ij} = \left(1 + \beta^n \zeta_{ij}^n\right)^{-\frac{1}{2n}} \quad [4.13]$$

$$\zeta_{ij} = \sum_{k \neq i,j} \left[f_C(r_{ik}) g(\theta_{ijk}) \exp\left(\lambda_3^m (r_{ij} - r_{ik})^m \right) \right] \quad [4.14]$$

$$g(\theta) = 1 + \frac{c^2}{d^2} - \frac{c^2}{d^2 + (\cos\theta - h)^2} \quad [4.15]$$

The parameter h is sometimes noted as $\cos(\theta_0)$, but it should not be considered as a cosine function, as its value often falls outside the range of the cosine function.

The potential definition includes several coefficients to be fitted, and several studies have been dedicated to defining values for key materials. The Tersoff potential has found use in several areas of MD simulation, including the analysis of III–IV semiconductors, especially for III-As and III-N compounds. In 2007, Powell *et al.* fitted new coefficients that demonstrate the potential applicability to III-P and III-Sb zinc blende semiconductors as well, specifically to describe nonlinear strain effects.

Figures 4.6–4.8 report the dispersion curves computed using the Tersoff potential for Si, InAs and SiC in comparison to experimental data. An example of the parameters for silicon [TER 89] is shown in Table 4.3. The parameters for SiC can also be found from [TER 89]. The transverse acoustic branches are consistently found below the experimental values, and in the case of InAs, the problem extends to the longitudinal acoustic branches as well. In both Si and InAs cases, the optical branches are reproduced with a satisfying accuracy, while only the longitudinal optical branch remains in fair agreement with those of the experimental investigations. Similarly to transverse acoustic branches, the frequencies of the two transverse optical branches are overestimated.

M	λ_1	λ_2	λ_3	c	d	N
3	2.480	1.732	0	1.0039×10^5	16.217	0.78734
$\cos\theta_0$	β	B	R	D	A	
–0.5983	1.1×10^{-6}	471.18	2.85	0.15	1,830.8	

Table 4.3. *Tersoff parameters defined for silicon [TER 89]*

Figure 4.6 also compares the DOS as computed by the Tersoff potential with that of the experimental results. While the overall trends are reasonably similar, errors in the estimation of the acoustic branches are clearly revealed and lead to disagreement with experimental data. These results were computed using the

LAMMPS program from Sandia National Laboratories with potential coefficients provided in the work of Nordlund *et al.* [NOR 00].

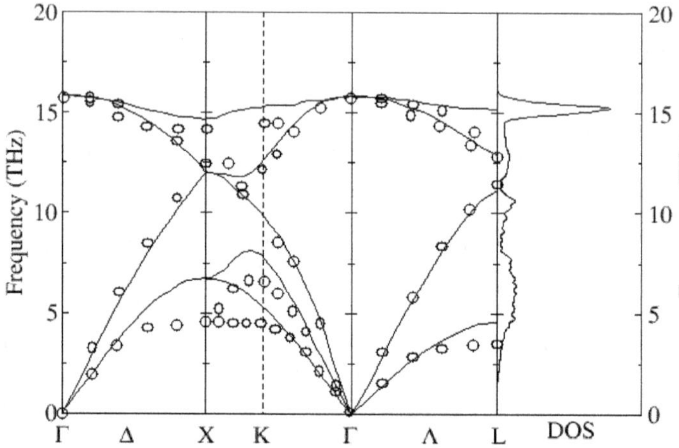

Figure 4.6. *Silicon dispersion relation from MD simulations based on Tersoff potential (solid lines) and experimental values (circles)*

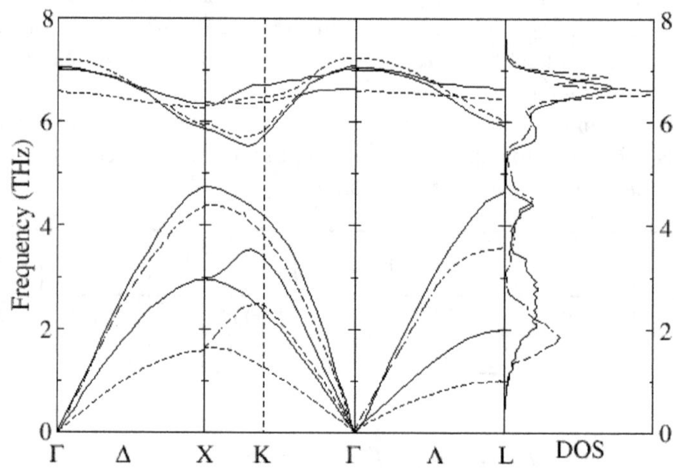

Figure 4.7. *InAs dispersion relation from MD simulations based on Tersoff potential (solid lines) and experimental values (dashed lines)*

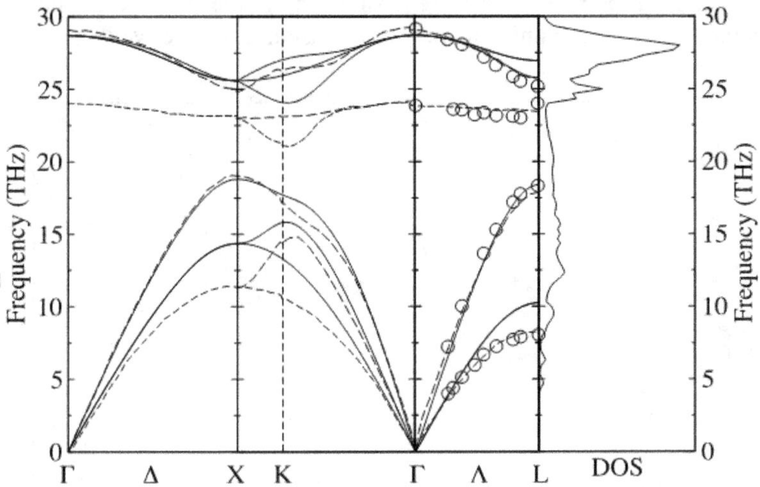

Figure 4.8. *Silicon carbide dispersion relation from MD simulations based on Tersoff potential (solid lines) and experimental values (circles)*

4.4. Oxydes: Van Beest, Kramer and van Santen potential

The van Beest, Kramer and van Santen potential consists of a two-body potential, which was originally developed in 1990 for silica and aluminophosphate polymorphs [BEE 90]. The potential is sometimes also known as the Buckingham potential as the name BKS would refer specifically to the parameterization of the functional form in the case of silica.

The functional form is relatively simple and is given by

$$\Phi_{ij} = \frac{q_i q_j}{r_{ij}} + A_{ij} e_{ij}^{-r/\rho_{ij}} - \frac{C_{ij}}{r_{ij}^6} \quad [4.16]$$

The first term describes point-charge interactions. This is usually the most computationally demanding portion of the potential since it is slow to converge, and will only do so under certain conditions. Ewald summation is usually employed to handle the long-range nature of this Coulombic interaction. The value q_i will be close to that of the formal charge of the atom i in units of e. Empirical fittings of the potential might cause deviations from the intuitive value. The remaining terms form what is known as the Buckingham potential. The second term describes repulsion at short range due to the overlap of closed electron shells (Born–Mayer type repulsion). It replaces the r^{-12} repulsion term in the standard L-J potential. The A and ρ parameters are in units of energy and distance, respectively.

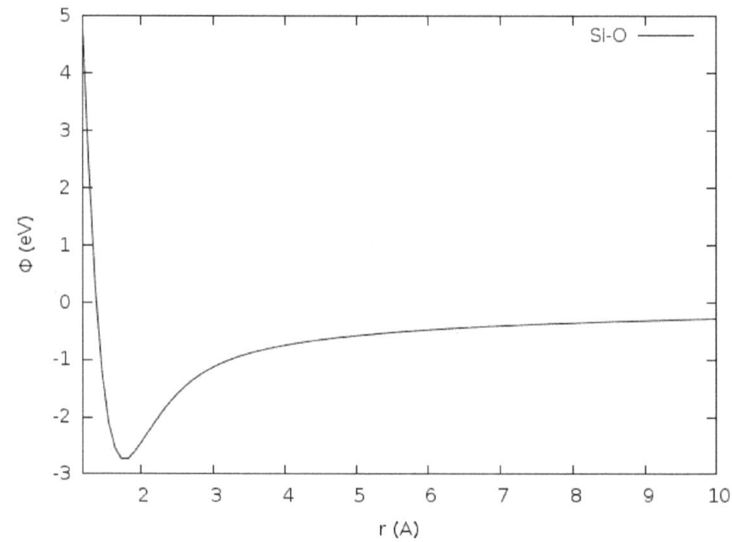

Figure 4.9. *BKS potential for Si-O; parameters taken from [BEE 90]*

The latter of these should be used to determine the required cutoff value for the potential. In van Beest *et al*'s original paper, these two parameters for the ij species were obtained by fitting the potential to *ab initio* data. Finally, the last term is the well-known London dispersive term and was originally fitted using bulk rather than *ab initio* data. However, the term also has a well-established physical meaning [BIN 93] and may be calculated using the Slater–Kirkwood formula [SLA 31] as was done for the case of ZnO.

Although this potential has a relatively simple form, for example, it does not contain any three-body terms, it has supplied, rather successfully, several important features of silica such as the $\alpha - \beta$ transition temperature of quartz, the unit cell parameters, elastic constants [BEE 90] and piezoelectric constants [HER 06] close to experimental values. Hence, the validity of the potential over a range of pressure values and over a number of different silica polymorphs made this functional form attractive for use in MD simulations. Van Beest's paper specifies that the potential could be applied to "... any other chemical elements that form tetrahedral network oxides ..." [BEE 90].

However, there still exist certain discrepancies when using this potential. For example, a recent paper has shown the advantage of other potentials incorporating extra degrees of freedom, and it has displayed the shortcomings of the BKS potential [HER 05] when it comes to fundamental quantities such as the vibrational DOS. Therefore, extensions to the potential are proposed and even modifications that might provide better accuracy. For example, we may extend the potential by a shell model that aims at modeling the polarizability of an atom using an additional degree of freedom represented by a massless particle harmonically coupled to the ionic core [DIC 58]. Another closely related potential to the BKS potential was proposed by Vashishta *et al.* It adds steric repulsion, Debye screening, charge–dipole interaction, and three-body terms to the functional form and has been applied to several materials including SiO, SiC and AlN [VAS 90, VAS 07, VAS 11].

4.4.1. *Application of BKS potential to ZnO crystal*

The wurtzite zinc oxide crystal belongs to the $P6_3mc$ space group – hexagonal lattice. Binks provided an empirical fit of the BKS potential for the case of ZnO using the lattice energy, cell parameters and elastic constants for several structures [BIN 94]. This was done under a shell model, but the results should be equally valid for the point ion model. The coefficients for ZnO are given in Table 4.4.

Species	A (eV)	ρ (Å)	C (eV Å^{-6})
O^{2-}–O^{2-}	9,547.96	0.21916	32.0
Zn^{2+}–O^{2-}	529.70	0.3581	0.0
Zn^{2+}–Zn^{2+}	0.0	0.0	0.0

Table 4.4. *BKS parameters for ZnO crystal*

We first calculate the DOS for a perfect ZnO crystal using the velocity autocorrelation function as shown in equation [4.16]. This autocorrelation is based on the atom velocities computed during a MD simulation. The MD system is defined by lattice cell parameters $a = 3.265$ Å, $c = 5.155$ Å and $u = 0.3882$ Å. We start with 15 unit cells in each direction, apply periodic boundary conditions and relax the system in a constant atom number, pressure and temperature ensemble for 10,000 timesteps (timestep of 1fs). After this period we obtain the vibrational DOS by taking the Fourier transform of the velocity autocorrelation. We compare our results with *ab initio* data from [SER 03].

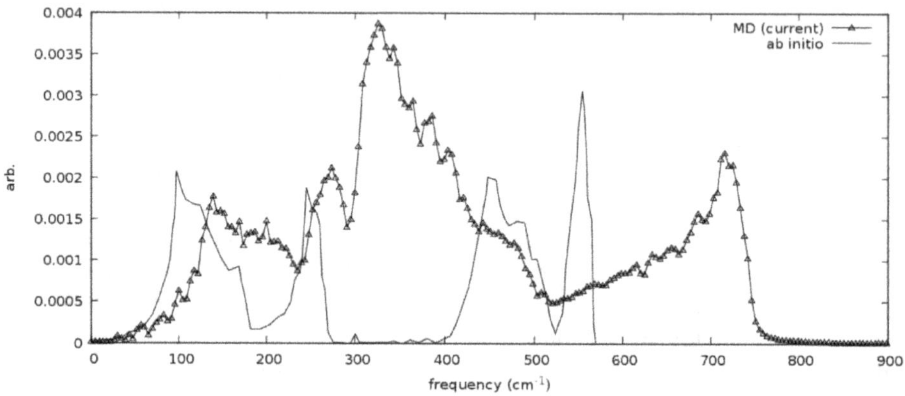

Figure 4.10. *Vibrational density of states for ZnO using BKS potential*

We also present the dispersion curves along two symmetry directions and compare them to *ab initio* data and neutron scattering data obtained from [SER 04]. The data were calculated using the code GULP for the same potential and cell parameters used in MD.

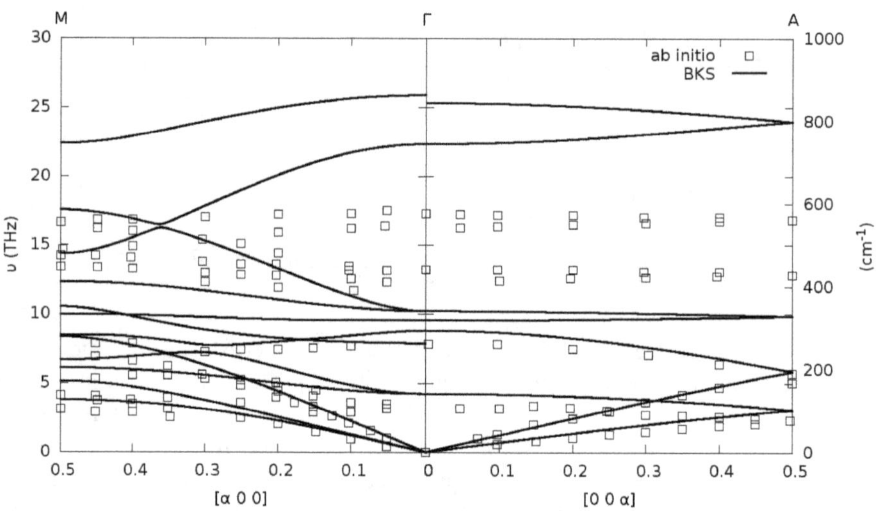

Figure 4.11. *Dispersion relation for ZnO along two symmetry directions*

While acoustic modes are fairly well predicted by MD, we note in both instances the large differences between *ab initio* calculated optical phonon states and those calculated using the BKS potential. The Debye temperature for ZnO (zincite) is given in the literature [LAN 99] to be approximately 440 at 300 K, which is equivalent to 9.16 THz, and is closer to the *ab initio* data than the BKS data. This discrepancy was to be expected as the potential was originally empirically fitted.

4.4.2. *Application of BKS potential to MgO crystal*

MgO crystal is an ionic crystal with a NaCl structure and a lattice constant of 0.421 nm. Matsui *et al.* provided several articles [MAT 87, MAT 88, MAT 89] on adapting BKS potential to MgO crystals. Charge neutrality, structure at 0 K and the measured bulk modulus at 1,000 K are considered as fitting constraints to determine the potential parameters that are presented in Table 4.4.

	A (eV)	ρ (nm)	C (eV·nm^6)
Mg^{2+}–Mg^{2}	1,310,148.44	0.0104	0.0
O^{2-}-O^{2}	2,147.023	0.0300	30.223627
Mg^{2+}–O^{2-}	9,898.297	0.0202	0.0

Table 4.4. *Parameters for MgO crystal using BKS potential, depending on the chemical nature of the ions*

The electronic charge of oxygen ions is equal to –1.4e and +1.4e for magnesium ions. The cutoff radius for Buckingham interactions is defined as 0.93 nm and for electrostatic forces as 1.0 nm. As electrostatic interactions are long range, setting a cutoff radius is not enough to guarantee that all non-zero interactions are considered. Long-range forces are computed using Ewald summation [EWA 21] with a relative error in forces equal to 1.0×10^{-4}.

A comparison between the DOS obtained by MD and *ab initio* calculations [SCH 94] is presented in Figure 4.12. The general structure of the phonon DOS agrees with *ab initio* calculations. Acoustic phonon frequencies are, however, a little underestimated by MD, contrary to optical phonons, which have higher frequency than *ab initio* calculations.

The phonon dispersion relations computed by MD are also compared to experimental data obtained by inelastic neutron scattering [PEC 67, SAN 70] in Figure 4.12. With the exception of longitudinal optical phonons, all the branches are modeled with a good agreement compared to *ab initio* results. This demonstrates the validity of BKS potential using Matsui parameters to model MgO crystals by MD.

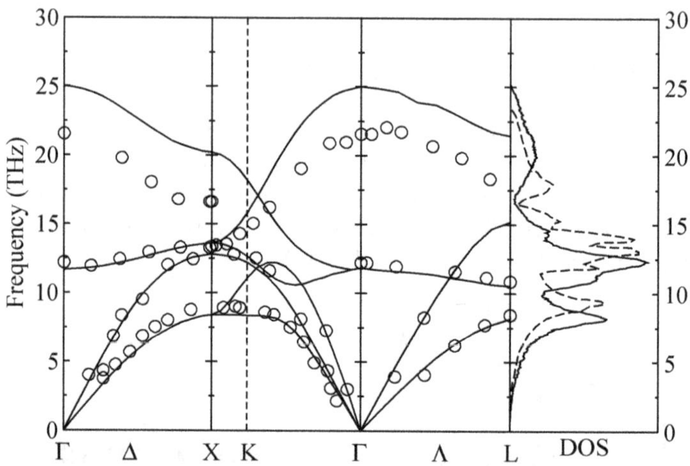

Figure 4.12. *Phonon dispersion relation computed by molecular dynamics (solid lines) compared to inelastic neutron scattering measurements (open circles) [PEC 67, SAN 70]; phonon density of states (DOS) (solid line) compared to ab initio calculations (dotted line) [SCH 94]*

4.5. Metals – isotropic many-body pair-functional potentials for metals: the modified embedded-atom method

The modified embedded-atom method (MEAM) is a modified version of the embedded-atom method designed for metals [DAW 83] to study impurities. MEAM extends this method to a number of cubic materials [BAS 91], metals and semiconductors. This method is based on the effective medium theory and considers an electronic environment interacting with the atoms. The energy of the crystal is given by:

$$E = \sum_i \left\{F_i(\rho(r_i)) + \frac{1}{2}\sum_{j \neq i} \phi_{ij}(r_{ij})\right\} \quad [4.17]$$

where ρ is the electron density and F is called the embedding energy. ϕ is a pair potential considering only the nearest neighbors that have a repulsive and an attractive component. Their analytic forms depend on the element considered and can be modified for alloys [BAS 91, KUO 03].

4.5.1. *Example: phonon dispersion relation of gold*

Using MEAM for gold crystals, the phonon dispersion relation can be computed as represented in Figure 4.13, where MEAM predictions (solid lines) and experimental data (circles and dashed line) are compared [LYN 73]. Experimental data [LYN 73] reveal that the modes at high frequencies are underestimated and that the branches are significantly flatter than the experimental ones.

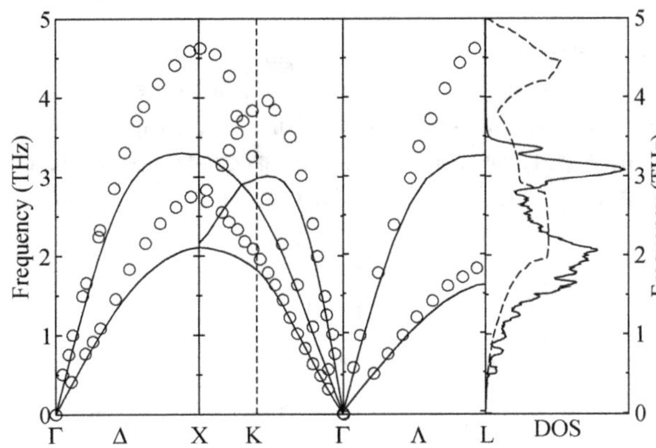

Figure 4.13. *Phonon relation dispersion of MEAM gold (solid lines) and experimental data (circles and dashed line) [LYN 73]. As shown in the standard representation, the DOS is plotted in the right-hand side inset*

4.6. Polymers and carbon-based compounds: adaptive intermolecular reactive bond order, adaptive intermolecular REBO and Dreiding potentials

4.6.1. *Adaptive intermolecular REBO potential*

REBO potential defines the system energy as the sum of the adaptive treatment of dispersion, intermolecular repulsion and torsional interactions, and is expressed as [STU 00]:

$$E = E^{REBO} + E^{LJ} + E^{tors}. \qquad [4.18]$$

As originally developed, the REBO potential exclusively includes short-range interactions. Two atoms directly interact if they are separated by a distance less than a covalent-bonding cutoff r_{ij}^{max} (e.g. 2.0 Å for C–C bonds).

Each pair of covalently bonded atoms interacts via the following potential:

$$E_{ij}^{REBO} = V_{ij}^{R}(r_{ij}) + b_{ij} V_{ij} A(r_{ij}), \qquad [4.19]$$

where V_{ij}^{R} and V_{ij}^{A} are repulsive and attractive pairwise potentials determined by atom types (carbon or hydrogen) of atoms *i* and *j*, which depend only on the distance r_{ij} between the two atoms.

In addition to the REBO potential, the adaptive intermolecular REBO (AIREBO) potential includes the mechanism for treating intermolecular interactions. These include the dispersion and short-range repulsion effects that give rise to many of the properties of liquids, polymers and thin-film hydrocarbon materials. In AIREBO potentials, the dispersion and intermolecular repulsion interactions are modeled with an L-J potential,

$$V_{ij}^{LJ}(r_{ij}) = 4\varepsilon_{ij}\left[\left(\frac{\sigma_{ij}}{r_{ij}}\right)^{12} - \left(\frac{\sigma_{ij}}{r_{ij}}\right)^{6}\right] \qquad [4.20]$$

The third component of the AIREBO potential is a term dependent on dihedral angles. The original REBO potential lacked any torsional interactions in single bonds, reflecting its original focus on network solids such as diamond and small molecular fragments relevant to the chemical vapor deposition of diamond. With no barrier to rotation in single bonds, however, the original REBO potential is poorly suited for the simulation of saturated hydrocarbons larger than methane.

The form for the torsional potential in AIREBO potential is a cosine power series in the dihedral angle ω, with a barrier height ε as shown in the following relation:

$$V^{tors}(\omega) = \varepsilon\left[\frac{256}{405}\cos^{10}\left(\frac{\omega}{2}\right) - \frac{1}{10}\right] \qquad [4.21]$$

Figure 4.14 compares the density states estimated from AIREBO and Tersoff potentials to the reference data provided by the LD predictions based on the fifth nearest neighbors interactions. This comparison clearly favors the outcomes of AIREBO potential in terms of frequency cutoff and peak amplitude over those of Tersoff. This trend is confirmed by the correspondence in the dispersion curves.

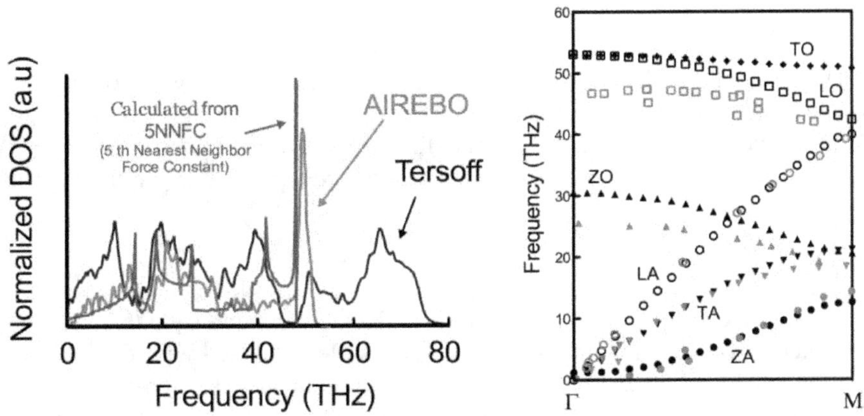

Figure 4.14. *(Left) A comparison of phonon density of states of graphene calculated from AIREBO and Tersoff potentials and from a lattice dynamics derivation using the fifth nearest neighbors force constants. (Right) Phonon dispersion relation of graphene calculated from AIREBO potential (red points) compared with experimental data points for graphite (black points) [TEW 09]. For a color version of this figure, see www.iste.co.uk/dollfus/nanodevices.zip*

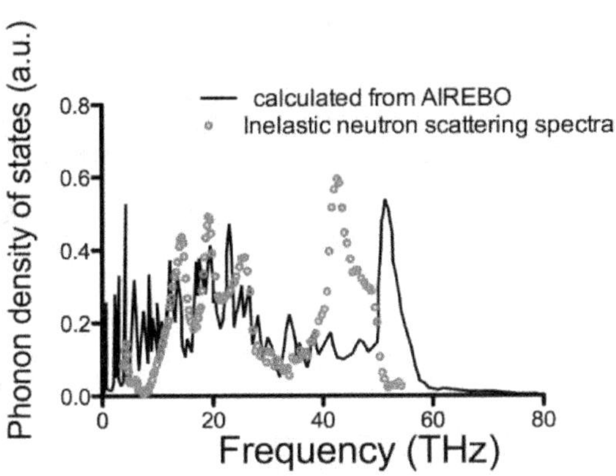

Figure 4.15. *Phonon density of states of CNT calculated from AIREBO potential (black solid line) and inelastic neutron scattering spectra of CNT measured on filter-analyzer spectrometers [ROL 00]. For a color version of this figure, see www.iste.co.uk/dollfus/nanodevices.zip*

Figure 4.15 further exemplifies the reasonable correspondence between the predictions of the AIREBO potential and the experimental data in the case of the carbon nanotube.

4.6.2. Dreiding potential

The Dreiding model is designed for the full-atomic modeling of polymers. This model has been developed by Stephen L. Mayo and was published in [MAY 90]. It includes several potentials to model the different structures found in polymers. We present those potentials applied to a specific example, i.e. a Jeffamine M600 polymer.

4.6.2.1. *Atom styles*

In the Dreiding potential, each atom is defined by an atom style that depends on its element, hybridization and geometry. Each style is labeled with five characters. The first two stand for the element symbol (Si for silicon or C_ for carbon) and the third refers to the hybridization or the geometry (1 for linear, 2 for trigonal, 3 for tetrahedral or R for resonance situation). The fourth character indicates implicit hydrogen atoms and the fifth character corresponds to alternate characteristics. Parameters are attached to the atom style, and they are used to compute the coefficients of the potentials. For Jeffamine, we have a simple structure that leads to only four atom types described in Table 4.5.

Atom	Atom style	Bond radius (Å)	Bond angle (degree)
Carbon	C_3	0.77	109.471
Oxygen	O_3	0.66	104.51
Nitrogen	N_3	0.702	106.7
Hydrogen	H_	0.33	108

Table 4.5. *Atom styles used to model Jeffamine with Dreiding model*

4.6.2.2. *Morse potentials*

To model the chemical bonds between two atoms of the polymer, Dreiding implements a Morse potential:

$$E = D[1 - e^{-\alpha(r-r_0)}]^2 \qquad [4.22]$$

The parameters depend on the atom type and the bond. The bond type sets the value of D and α. The value of r_0 depends on the atom types of the two bonded atoms. In Jeffamine, only single bonds exist; the coefficients are displayed in Table 4.7.

Bond	D (kcal/mol)	α (Å^{-1})	r_0 (Å)
C_3 – H_	35	3.16227766	1.09
C_3 – C_3	35	3.16227766	1.53
N_3 – H_	35	3.16227766	1.022
C_3 – N_3	35	3.16227766	1.462
O_3 – H_	35	3.16227766	0.98
C_3 – O_3	35	3.16227766	1.42

Table 4.6. *Bond coefficients for Morse potential*

4.6.2.3. *Angle potentials*

Dreiding proposes to use a potential to keep the angles defining the structures found in polymers, such as the alternate angles found in a carbon chain. To impose this condition, it proposes a harmonic potential:

$$E = K(\theta - \theta_0)^2 \quad [4.23]$$

where the coefficient K is equal to 100 kcal/mol and θ_0 is the bond angle of the center atom.

4.6.2.4. *Dihedral potentials*

In the Dreiding model, when the polymer contains four consecutive atoms, a potential is used to give an equilibrium angle between the two planes made by the first three atoms and the last three atoms of the set. This allows for modeling molecules included in a plane or with an out-of-plane part. The potential proposed by Dreiding is:

$$E = K(1 + d\cos(n\,\phi)) \quad [4.24]$$

These potential parameters depend only on the two center atoms. K depends on those and the number of dihedral potentials applied to those atoms. If there are two dihedral potentials applied to the two atoms, K will be half the value for this atom couple. In our polymer, the couples are always X_3 X_3, which leads to K = 1 kcal/mol but as it is applied nine times on each couple, K = 0.111111 kcal/mol. This couple also leads to $d = -1$ and $n = 3$.

4.6.2.5. *Improper potentials*

In the Dreiding potential, when an atom is bonded to exactly three other atoms, a potential is introduced to describe the planar or non-planar equilibrium position. The Dreiding model proposes the following energy definition:

$$E = K(\psi - \psi_0)^2 \quad [4.24]$$

where K depends on the hybridization of the central atom. Note that the tabulated values have to be divided by three because each improper atom has three improper potentials applied to consider all its bonded neighbors. The angle ψ is usually set to 0° except when implicit hydrogen atoms are used. Such atoms do not exist in Jeffamine.

4.6.2.6. *L-J potential*

The previous potentials only model the chemistry of the polymer, and to model the interactions between non-bonded atoms Dreiding proposes an L-J potential to model the van der Waals interaction:

$$E = 4\varepsilon\left[\left(\frac{\sigma}{r}\right)^{12} - \left(\frac{\sigma}{r}\right)^{6}\right] \quad [4.25]$$

A cutoff at 12 Å is usually considered for this interaction. The parameters depend on the interacting elements; the values of σ, the distance at which the potential cancels, and ε, the potential depth, which are the arithmetic average of the tabulated values for the two elements considered. For Jeffamine, there are only four elements that lead to 10 couples. The coefficient and couples are described in Table 4.7.

Couple	ε (kcal/mol)	σ (Å)
C – C	0.0951	3.4729905
C – O	0.0954	3.2530721
C – N	0.08625	3.3677753
C – H	0.33301652	3.1597059
O – O	0.0957	3.0331538
O – N	0.08655	3.147857
O – H	0.33301652	2.9397876
N – N	0.0774	3.2625602
N – H	0.30430248	3.0544908
H – H	0.0152	2.8464214

Table 4.7. *Coefficients for Lennard-Jones pair atoms*

4.6.2.7. *Coulomb potential*

In polymers, dipoles resulting from the bonds are modeled by partial charges on the atoms and a Coulomb potential. The partial charges q_1 and q_2 are chosen using the Gasteiger algorithm from Mulliken's definition of electronegativity [GAS 80]. This interaction is computed using the EWALD algorithm with the following Coloumb potential:

$$E = \frac{q_1 q_2}{4\pi\varepsilon_0 r} \qquad [4.26]$$

where r refers to the intercharge distance and ε_0 to the dielectric constant in vacuum.

4.6.3. *Example: DOS of the Jeffamine*

As shown by equation (7), following Coloumb potential, an MD simulation using this modeling allows for computing the vibrational DOS of Jeffamine from the atomic velocities as represented in Figure 4.16. As this prediction is the first one, validation data are not available yet. However, the expected bond frequencies can be identified in the DOS, i.e. the C–H bonds in the 87–94 THz range, the bonds between carbon atoms (simple or double) in the 18–51 THz range and the C=O bond at 52 THz.

Inaccuracies in the Dreiding potential are mainly related to the charge estimations and its relevance mainly consists of its ability to cover a broad set of systems. Finer analysis of the competition between weak bonds can be addressed by potentials such as CHARMM having the same global form as that of Dreiding but with additional complexity for describing weak bonds.

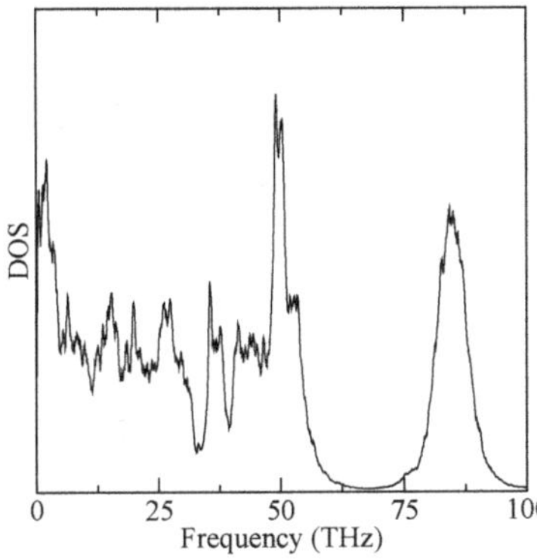

Figure 4.16. *Density of states of Jeffamine M600 polymer at 310 K from molecular dynamics simulations using the Dreiding model*

4.7. Water: TIP3P potential

The TIP3P model is a three-site water model. It includes four potentials including long-range ones such as the L-J to model the van der Waals interactions and the Coulomb potential using an EWALD algorithm and a cutoff at 12 Å, $4\pi\varepsilon_0$ being the unit scale factor. The intermolecular potentials are harmonic, they include the bond and the angle potentials to make the natural angle in water molecules that of equilibrium; r is the distance between a hydrogen and an oxygen and θ is the angle made by the two H–O bonds. R is the distance between the two atoms considered for the L-J interactions between different molecules. The charges q_1 and q_2 are the charges of the two atoms considered in the Coulomb interaction. In the TIP3P, the potential energies are defined as follows:

$$E_r = K_r\,(r - r_0)^2 \qquad [4.27]$$

$$E_\theta = K_\theta(\theta - \theta_0)^2 \qquad [4.28]$$

$$E = 4\varepsilon\left[\left(\frac{\sigma}{R}\right)^{12} - \left(\frac{\sigma}{R}\right)^{6}\right] \qquad [4.29]$$

$$E = \frac{q_1 q_2}{4\pi\varepsilon_0 R} \qquad [4.30]$$

K_r (kcal/mol)	450
r_0 (Å)	0.9572
K_θ (kcal/mol)	55
θ_0 (degree)	104.52
ε_{OH}, ε_{HH} (kcal/mol)	0
ε_{OO} (kcal/mol)	0.102
σ_{OH}, σ_{HH} (Å)	0
σ_{OO} (Å)	3.188
q_O (elementary charge)	–0.830
q_H (elementary charge)	+0.415
M_O (g/mol)	15.9994
M_H (g/mol)	1.008

Table 4.8. *Coefficients and parameters used in the TIP3P model*

4.7.1. *Example: radial density function and DOS*

The DOS of water can be computed from this model; frequencies that are absent in rigid models are included here. These calculations have been made on several water models, including TIP3P, a review on the comparison between those can be found in [JOR 83].

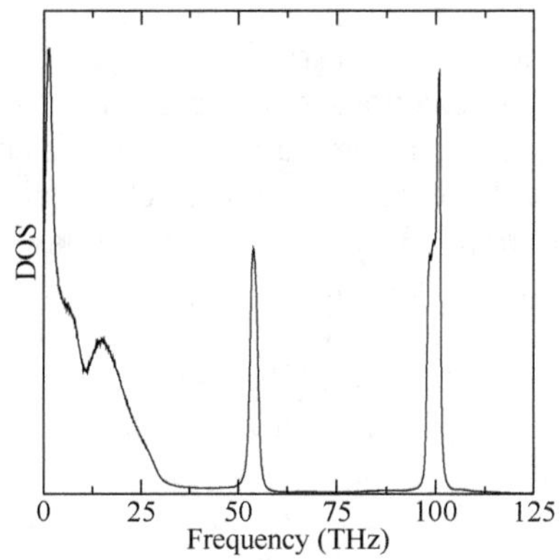

Figure 4.17. *Density of states of the TIP3P water*

The radial distribution function of the oxygen atoms can also be computed from TIP3P and compared with experimental data [JOR 83], those two outcomes are represented in Figure 4.18.

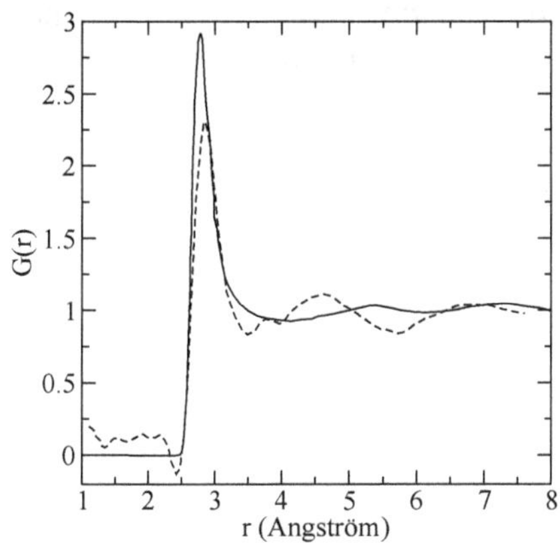

Figure 4.18. *Oxygen atoms radial distribution function in TIP3P water (solid line) compared to experimental data (dashed line)*

The main advantage of TIP3P is its ability to set interactions between one atom and a molecule other than water. The three-body potential involved is a common one that is also implemented in CHARMM for water and has a rigid and a flexible version.

4.8. Conclusion

This chapter has provided the description of a broad set of pseudopotentials applicable to generic pair interactions existing in inert gasses and also to semiconductors, oxides, metals as well as to carbon-based systems – be it in the form of solids or soft matter. The different types of interactions are dependent on positions and angles. They are either defined from basic physical laws, especially for the long-range forces – i.e. Van der Waals dipole–dipole interactions, Coulomb charge–charge interactions – or, in the case of very short ranges, from analytical functions – exponentials, power laws – yielding a global trend for strong repulsive forces. The parameters are fitted from experimental or *ab initio* data according to the expected use of the pseudopotential. Regarding the phonon properties, the dispersion curves remain a reliable check for harmonic quantities such as group velocities and elastic constant of course, but recent attempts tend to show that they also guarantee the accuracy of first anharmonic terms.

As discussed above, several graphs have reported dispersion curves computed either from LD or from MD, computed in our group or in the groups of cited authors. Very often, the agreement between pseudopotential-based computations, on the one hand, and measured or *ab initio* dispersion curves, on the other hand, remains very partial. The discrepancies usually appear at high frequencies where dispersion, i.e. higher order anharmonic terms, plays a major role. While this aspect limits the quantitative predictions of thermal properties to a typical 10–20% range of accuracy (especially at high temperatures), the qualitative relevance of pseudopotentials in terms of physical trends – temperature, pressure, size, composition, structure, etc., dependence – has been proven in countless studies.

4.9. Bibliography

[BAS 91] BASKES M.I., "Modified embedded-atom potentials for cubic materials and impurities", *Physical Review B*, vol. 46, no. 5, pp. 2727–2741, 1991.

[BIN 94] BINKS D.J., Computational modelling of zinc oxide and related oxide ceramics, PhD thesis, University of Surrey, 1994.

[DAW 84] DAW S.M., BASKES M.I., "Embedded-atom method: derivation and application to impurities, surfaces, and other defects in metals", *Physical Review B*, vol. 29 no. 12, pp. 6443–6453, 1984.

[DIC 58] DICK B.G., OVERHAUSER A.W., "Theory of the dielectric constants of alkali halide crystals", *Physical Review*, vol. 112, no. 1, pp. 90–103, 1958.

[DIN 86] DING K., ANDERSEN H.C., "Molecular-dynamics simulation of amorphous germanium", *Physical Review B*, vol. 34, no. 10, pp. 6987–6991, 1986.

[DOL 62] DOLLING G., "Inelastic scattering of neutrons in solids and liquids", *Proc. Symp. 2nd Chalk River*, p. 37, 1962.

[DOL 66] DOLLING G., COWLEY R.A., "The thermodynamic and optical properties of germanium, silicon, diamond and gallium arsenide", *Proceedings of the Physical Society*, vol. 88, no. 2, pp. 463–494, 1966.

[DOM 04] DOMINGUES G., SAULNIER J.B., VOLZ S., "Thermal relaxation times and heat conduction in β-cristobalite and α-quartz silica structures", *Superlattices and Microstructures*, vol. 35, pp. 227–237, 2004.

[EWA 21] EWALD P., "Die Berechnung optischer und elektrostatischer Gitterpotentiale", *Annals of Physics*, vol. 64, pp. 253–287, 1921.

[FEL 68] FELDMAN D.W., PARKER J.H., CHOYKE W.J. *et al.*, "Phonon dispersion curves by Raman scattering in SiC, Polytypes 3C, 4H, 6H, 15R, and 21R", *Physical Review*, vol. 173, no. 3, pp. 787–793, 1968.

[FUJ 74] FUJII Y., LUIRE N.A., PYNN R. *et al.*, "Inelastic neutron scattering from solid Ar", *Physical Review B*, vol. 10, no. 8, p. 3647, 1974.

[GAS 80] GASTEIGER J., MARSILI M., "Iterative partial equatization of orbital electronegativity – a rapid access to atomic", *Tetrahedron*, vol. 36, pp. 3219–3288, 1980.

[HEI 07] HEINO P., "Dispersion and thermal resistivity in silicon nanofilms by molecular dynamics", *The European Physical Journal B*, vol. 60, no. 2, pp. 171–179, 2007.

[HER 05] HERZBACH D., BINDER K., MÜSER M., "Comparison of model potentials for molecular dynamics simulations of silica", *Journal of Chemical Physics*, vol. 123, pp. 124711, 2005.

[HER 06] HERZBACH D., MÜSER M., "Piezoelectric coefficients by molecular dynamics simulations in the constant stress ensemble: a case study of quartz", *Computer Physics Communications*, vol. 174, no. 1, pp. 17–23, 2006.

[JOR 83] JORGENSEN W.L. *et al.*, "Comparison of simple potential functions for simulating liquid water", *Journal of Chemical Physics*, vol. 79, no. 2, pp. 926–935, 1983.

[KAB 07] KABURAKI H., LI J., YIP S. *et al.*, "Dynamical thermal conductivity of argon crystal", *Journal of Applied Physics*, vol. 102, no. 4, pp. 043514–043514-6, 2007.

[KUO 04] KUO C.L., PAULETTE C., "MEAM molecular dynamics study of a gold thin film on a silicon substrate", *Surface Science*, vol. 551, pp. 39–58, 2004.

[LEN 24] LENNARD-JONES, J.E., "On the determination of molecular fields", *Proceedings of Royal Society of London A*, vol. 106, no. 738, pp. 463–477, 1924.

[LYN 73] LYN J.W., SMITH H.G, NICKLOW R.M., "Lattice dynamics of gold", *Physical Review B*, vol. 8, no. 8, pp. 3493–3499, 1973.

[MAT 87] MATSUI M., AKAOGI M., MATSUMOTO T., "Computational model of the structural and elastic properties of the ilmenite and perovskite phases of $MgSiO_3$", *Physics and Chemistry of Minerals*, vol. 14, pp. 101–106, 1987.

[MAT 88] MATSUI M., "Molecular dynamics study of $MgSiO_3$ perovskite", *Physics and Chemistry of Minerals*, vol. 16, no. 234, pp. 234–238, 1988.

[MAT 89] MATSUI M., "Molecular dynamics study of the structural and thermodynamic properties of MgO crystal with quantum correction", *Journal of Chemical Physics*, vol. 91, no. 489, pp. 489–494, 1989.

[MAY 90] MAYO S.L., OLAFSON B.D., GODDARD III W.A., "Dreiding: a generic force field for molecular simulations", *Journal of Physical Chemistry*, vol. 93, pp. 8897–8909, 1990.

[NOR 00] NORDLUND. *et al.*, "Growth, characterization, and applications of single crystals held in Zakopane", *Computational Materials Science*, vol. 18, no. 3, pp. 283–294, 2000.

[NOW 00] NOWAK, *International Conference on Solid State Crystals*, Poland, 9–12 October 2000.

[PEC 67] PECKHAM G., "The phonon dispersion relation for magnesium oxide", *Proceedings of the Physical Society of London*, vol. 90, no. 3, pp. 657–670, 1967.

[ROL 00] ROLS S., BENES Z., ANGLARET E. *et al.*, "Phonon density of states of single-wall carbon nanotubes", *Physical Review Letters*, vol. 85, no. 24, pp. 5222–5225, 2000.

[SAN 70] SANGSTER M.J.L., PECKHAM G., SAUNDERSON D.H., "Lattice dynamics of magnesium oxide", *Journal of Physics C: Solid State Physics*, vol. 3, no. 5, pp. 1026–1036, 1970.

[SCH 94] SCHÜTT O., PAVONE P., WINDL W. *et al.*, "Ab-initio lattice dynamics and charge fluctuations in alkaline-earth oxides", *Physical Review B*, vol. 50, no. 6, pp. 3746–3753, 1994.

[SLA 31] SLATER J.C., KIRKWOOD J.G., "The Van der Waals forces in gases", *Physical Review*, vol. 37, no. 6, pp. 682–697, 1931.

[STI 85] STILLINGER F.H., WEBER T.A., "Computer simulation of local order in condensed phases of silicon", *Physical Review B*, vol. 31, no. 8, pp. 5262–5271, 1985.

[STU 00] STUART S.J., TUTEIN A.B., HARRISON J.A, "A reactive potential for hydrocarbons with intermolecular interactions", *Journal of Chemical Physics*, vol. 112, no. 14, pp. 6472–6486, 2000.

[TER 88] TERSOFF J., "New empirical approach for the structure and energy of covalent systems", *Physical Review B*, vol. 37, no. 12, pp. 6991, 1988.

[TER 89] TERSOFF J., "Modeling solid-state chemistry: interatomic potentials for multicomponent systems", *Physical Review B*, vol. 39, p. 5566, 1989.

[TEW 09] TEWARY V.K., YANG B., "Parametric interatomic potential for graphene", *Physical Review B*, vol. 79, p. 075442, 2009.

[VAN 90] VAN BEEST B.W.H., KRAMER G.J., VAN SANTEN R.A., "Force fields for silicas and aluminophosphates based on ab initio calculations ", *Physical Review Leters*, vol. 64 no. 16, pp. 1955–1958, 1990.

[VAS 07] VASHISHTA P., KALIA K., NAKANO A., "Interaction potential for silicon carbide: a molecular dynamics study of elastic constants and vibrational density of states for crystalline and amorphous silicon carbide", *Journal of Applied Physics*, vol. 101, p. 103515, 2007.

[VAS 11] VASHISHTA P., KALIA K., NAKANO A. *et al.* "Interaction potential for aluminum nitride: a molecular dynamics study of mechanical and thermal properties of crystalline and amorphous aluminum nitride", *Journal of Applied Physics*, vol. 109, p. 033514, 2011.

[VAS 90] VASHISHTA P., KALIA K., RINO J., "Interaction potential for SiO_2: a molecular-dynamics study of structural correlations", *Physical Review B*, vol. 41, no. 17, pp. 12197–12210, 1990.

[WEB 77] WEBER W., "Adiabatic bond charge model for the phonons in diamond, Si, Ge, and α-Sn", *Physical Review B*, vol. 15, no. 10, pp. 4789–4803, 1977.

5

Introduction to Quantum Transport

Brief introductions to electron transport were given in Chapters 1 and 2. The present chapter and the following chapters are fully devoted to this subject. In this chapter, we present the quantum theory of transport in the coherent regime. This means that the inelastic scattering times (electron–phonon and electron–electron interactions) are much larger than the travel times of electrons through the system. Hence, electrons can be viewed as Schrödinger waves propagating in a static potential without losing their phase coherence. Such a physical situation can occur at very low temperature (a few Kelvins) and/or in very small systems (a few nanometers). Despite these limitations, the theory of coherent transport is an important basis for all other transport formalisms. First, it describes the dynamics of electrons between collision events, which is the basis of the semiclassical theory of transport presented in Chapter 8. Second, it accurately describes elastic scattering on defects, as well as quantum tunneling. Third, the Green's function method for coherent transport, presented in this chapter, can be generalized to the non-equilibrium Green's functions (NEGF) formalism, which treats inelastic scattering and decoherence.

Several books (in particular [IMR 02, DAT 95, DAT 05]) have presented exhaustively the quantum transport theory and its relationship with experiments. This chapter only recalls the basic notions. It is mainly focused on the practical aspects of the simulation of quantum transport. It makes the link between the propagation of wavepackets and the Green's function formalism, by treating the same examples with both methods. All numerical examples from this chapter can be reproduced using the Octave/Matlab codes available at www.iste.co.uk/dollfus/nanodevices.zip.

In section 5.1, we present a detailed study of wavepacket propagation in various physical situations. This approach provides an intuitive understanding of quantum

---

Chapter written by François TRIOZON, Stephan ROCHE and Yann-Michel NIQUET.

transport and its semiclassical limit. In section 5.2, we introduce the Landauer–Büttiker formulation of transport, based on the quantum transmission of wavepackets through a conductor connected to electrodes. Finally, in section 5.3, we give a step-by-step introduction to the Green's function method, which allows calculating the quantum transmission efficiently. The link between the Green's function and wavepacket propagation is highlighted. The presentation is limited to the non-interacting case (one-electron approximation). The full NEGF formalism is addressed in the following chapter.

5.1. Quantum transport from the point of view of wavepacket propagation

5.1.1. *Calculation of wavepacket evolution*

5.1.1.1. *Time-dependent Schrödinger equation and evolution operator*

In Dirac notation, the time-dependent Schrödinger equation reads:

$$i\hbar\frac{\mathrm{d}\left|\psi(t)\right\rangle}{\mathrm{d}t} = \hat{H}(t)\left|\psi(t)\right\rangle \qquad [5.1]$$

where $\hat{H}(t)$ is the Hamiltonian operator and $|\psi(t)\rangle$ is a quantum state. The use of the Dirac notation makes equations more compact and avoids specifying the considered system and the chosen basis set for quantum states. $\hat{H}(t)$ must be Hermitian:

$$\hat{H}(t)^{\dagger} = \hat{H}(t) \qquad [5.2]$$

By linearity of the Schrödinger equation, the evolution of any state between times t_1 and t_2 is given by a linear operator $\hat{U}(t_2, t_1)$:

$$\left|\psi(t_2)\right\rangle = \hat{U}(t_2, t_1)\left|\psi(t_1)\right\rangle \qquad [5.3]$$

$\hat{U}(t_2, t_1)$ is named the evolution operator and it satisfies the Schrödinger equation:

$$i\hbar\frac{\partial\hat{U}(t_2, t_1)}{\partial t_2} = \hat{H}(t_2)\hat{U}(t_2, t_1) \qquad [5.4]$$

with the initial condition $\hat{U}(t_1, t_1) = \hat{I}$. The evolution operators between different times satisfy the relation:

$$\hat{U}(t_3, t_1) = \hat{U}(t_3, t_2)\hat{U}(t_2, t_1) \qquad [5.5]$$

In particular,

$$\hat{U}(t_1, t_2) = \hat{U}(t_2, t_1)^{-1} \quad [5.6]$$

Another important property is the unitarity of the evolution operator:

$$\hat{U}(t_2, t_1)^{\dagger} = \hat{U}(t_2, t_1)^{-1} \quad [5.7]$$

This is proven by calculating the time derivative of $\langle \psi(t) | \phi(t) \rangle$, where $|\psi(t)\rangle$ and $|\phi(t)\rangle$ are any solutions of the Schrödinger equation [5.1]. By hermiticity of $\hat{H}(t)$, this derivative equals zero. Hence, the evolution operator conserves the scalar product of any two states, which is the definition of unitarity. In particular, the norm of a state is conserved during time evolution[1]. This is necessary from physical grounds. For instance, in the one-electron approximation, a state occupied by an electron must remain normalized, yielding at any time a probability of 1 to find the electron somewhere. Moreover, considering two orthogonal states, each occupied by one electron, the states remain orthogonal during time evolution. Hence, the Pauli exclusion principle remains trivially satisfied. This is not so simple in the presence of electron–electron or electron–phonon interaction: a many-body formulation, such as the NEGF formalism presented in the following chapter, is then needed to account for the exclusion principle.

In this chapter, we will only consider time-independent Hamiltonians $\hat{H}$. Hence, the evolution operator $\hat{U}(t_2, t_1)$ only depends on the time difference $\Delta t = t_2 - t_1$. It reads:

$$\hat{U}(\Delta t) = e^{-i\hat{H}\Delta t/\hbar} \quad [5.8]$$

The exponential is most easily expressed on a basis of eigenstates of $\hat{H}$, for which we recover the evolution of stationary states:

$$\hat{U}(\Delta t) |\phi\rangle = e^{-iE\Delta t/\hbar} |\phi\rangle \quad [5.9]$$

where $|\phi\rangle$ is any eigenstate of $\hat{H}$ and E its energy.

1 Actually, in a complex vector space, the conservation of the norm implies the conservation of the scalar product. It is a necessary and sufficient condition for unitarity.

5.1.1.2. *Example: calculation of wavepacket evolution in 1D*

We consider the one-dimensional (1D) one-electron Schrödinger equation in the effective mass approximation:

$$i\hbar\frac{\partial\psi(x,t)}{\partial t} = -\frac{\hbar^2}{2m}\frac{\partial^2\psi(x,t)}{\partial x^2} + V(x)\psi(x,t) \qquad [5.10]$$

Such a 1D model can be used, for instance, to describe electron transport through multilayer systems that are homogeneous along the two other directions. The potential $V(x)$ includes the conduction band offsets between the materials (see Chapter 1) and/or the effect of an external electric field applied along x. To simplify the discussion, the effective mass is assumed to be constant in the whole system.

5.1.1.2.1. Finite difference discretization

In order to compute the time evolution of wavepackets in arbitrary potential profiles $V(x)$, we discretize the Schrödinger equation on a mesh of N sites with regular spacing a. We index the mesh points by an integer n: $x_n = na$, and we name $\psi(n,t)$ the value of the wavefunction at mesh point n. The second spatial derivative of ψ is approximated by finite differences:

$$\left.\frac{\partial^2\psi}{\partial x^2}\right|_{x=x_n} \simeq \frac{1}{a}\left[\frac{\psi(n+1,t)-\psi(n,t)}{a} - \frac{\psi(n,t)-\psi(n-1,t)}{a}\right] \qquad [5.11]$$

The Schrödinger equation becomes:

$$i\hbar\frac{\mathrm{d}\psi(t)}{dt} = H\psi(t) \qquad [5.12]$$

where $\psi(t)$ is the N-dimensional vector with components $\psi(n,t)$, and H is the matrix representation of the Hamiltonian operator on the finite difference mesh. For the present 1D model, it is a tridiagonal matrix. We define the constant:

$$\gamma = \hbar^2/2ma^2 \qquad [5.13]$$

which has the dimension of an energy, and the onsite potential:

$$V_n = V(x_n) \qquad [5.14]$$

The only non-zero matrix elements of H are:

$$\begin{aligned} H_{n,n} &= 2\gamma + V_n \\ H_{n,n+1} &= H_{n+1,n} = -\gamma \end{aligned} \quad [5.15]$$

For a uniform potential $V(x) = V_0$, the stationary solutions of the Schrödinger equation can be chosen as plane waves:

$$\psi_k(n,t) = e^{i[kna - E(k)t/\hbar]} \quad [5.16]$$

with the following dispersion relation:

$$E(k) = V_0 + 2\gamma\,[1 - \cos(ka)] \quad [5.17]$$

This can be checked by application of the Hamiltonian [5.15] to the plane waves [5.16]. $E(k)$ is represented in Figure 5.1. It approximates well the parabolic dispersion of the non-discretized Schrödinger equation for $k \ll 1/a$, i.e. for wavelengths much larger than the mesh step. However, the present model can also be interpreted as the tight-binding Hamiltonian of a 1D crystal with interatomic distance a, one orbital per atom, and nearest neighbor hopping energy γ. That is why, in Figure 5.1, $E(k)$ has been restricted to the first Brillouin zone $] - \pi/a, \pi/a]$. Such interpretation will be used in section 5.1.2.2 to illustrate interesting features occurring near a band maximum.

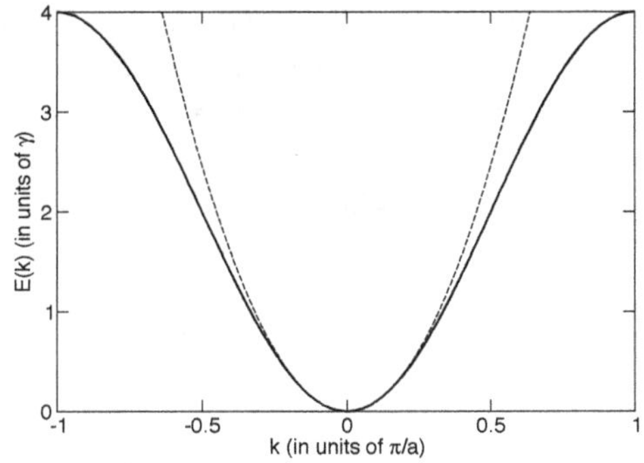

Figure 5.1. *Full line: dispersion relation of the discretized Schrödinger equation. Dashed line: parabolic dispersion ($\hbar^2 k^2/2m$) of the non-discretized equation*

5.1.1.2.2. Numerical calculation of wavepacket evolution

To compute the evolution of a given wavepacket $\psi(t)$, a natural method, used in Chapter 1, is to expand $\psi(0)$ on the basis of eigenvectors of H and to apply the phase factor [5.9] to each component. However, diagonalizing H may be computationally intensive, and even impossible for large two-dimensional (2D) and three-dimensional (3D) systems. It is numerically more efficient to compute the evolution of $\psi(t)$ by using the Taylor expansion of the evolution operator [5.8]:

$$\psi(\Delta t) = U(\Delta t)\psi(0) \simeq \sum_{p=0}^{P} \frac{1}{p!}\left(-\frac{i\Delta t}{\hbar}\right)^p H^p\psi(0) \qquad [5.18]$$

The computation of [5.18] essentially requires P iterations of H on a vector. The calculation of each product is very fast due to the sparsity of the Hamiltonian matrix: most elements are zero. Best efficiency (defined as $\Delta t/P$) is obtained by using a time step larger than the minimum phase oscillation period T_{min} of the states [LEF 91, ROC 97]. T_{min} is given by $2\pi\hbar/E_{max}$, where E_{max} is the maximum eigenvalue, in absolute value, of H. E_{max} is finite thanks to the discretization on the spatial mesh. In practice, we choose $\Delta t \simeq 20\hbar/E_{max}$, and a Taylor expansion of order $P \simeq 80$. The error is very small, comparable to machine roundoff error. The Matlab/Octave code provided in section 5.5 follows this method and allows all the 1D and 2D examples of this chapter to be reproduced within a reasonable computing time.

5.1.1.2.3. Example: group velocity of a Gaussian wavepacket

We calculate the evolution of Gaussian wavepackets in a flat potential $V_n = 0$ in order to illustrate the concept of group velocity presented in Chapter 1. We choose an effective mass $m = 0.19\ m_0$ (the transverse mass of electrons in silicon), and a mesh step $a = 0.1$ nm. The initial wavepackets are defined as:

$$\psi(n, t=0) = Ae^{ik_0 na} \times e^{-(na-x_0)^2/4\Delta_x^2} \qquad [5.19]$$

where A is chosen such that the wavepacket is normalized to unity. Two wavepackets are considered, with average wavevector $k_0 = 1\ \text{nm}^{-1}$ ($E \simeq 0.2$ eV) and $2\ \text{nm}^{-1}$ ($E \simeq 0.8$ eV), respectively. Both wavepackets have an initial width $\Delta_x = 3$ nm. Figure 5.2 shows their evolution during 10^{-13} s. As expected, the group velocity of the second wavepacket is about twice larger. The computed velocities match very well the analytical ones obtained from the dispersion relation [5.17] of the discretized Hamiltonian:

$$v(k_0) = \frac{1}{\hbar}\left.\frac{\mathrm{d}E}{\mathrm{d}k}\right|_{k=k_0} = \frac{2\gamma a}{\hbar}\sin(k_0 a) = \frac{\hbar}{ma}\sin(k_0 a) \qquad [5.20]$$

which gives $v = 6.08 \times 10^5$ m/s for $k_0 = 1$ nm^{-1} and $v = 1.21 \times 10^6$ m/s for $k_0 = 2$ nm^{-1}.

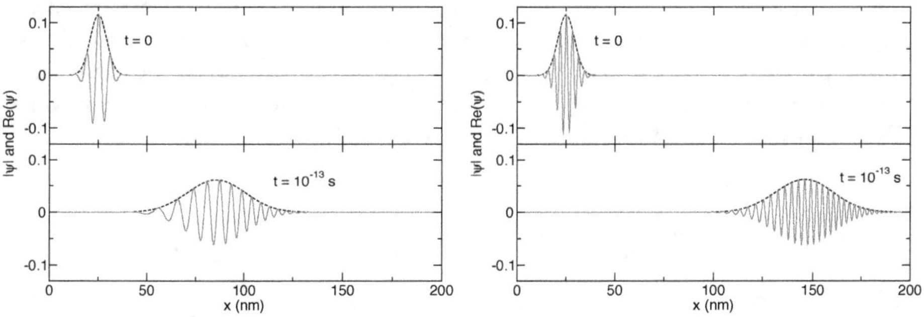

Figure 5.2. *Evolution of Gaussian wavepackets with wavevector $k_0 = 1$ nm^{-1} (left panel) and $k_0 = 2$ nm^{-1} (right panel) in a flat potential. Dashed line: modulus of the wavefunction. Solid line: real part of the wavefunction*

5.1.2. *The semi-classical limit*

In this section we illustrate by numerical simulation the semi-classical equations of motion presented in chapter 1. These equations apply to wavepackets with wavelengths much smaller than the characteristic length of variation of the potential $V(x)$. The average position $\bar{x}$ and wavevector $\bar{k}$ of a wavepacket satisfy the equations of motion:

$$\frac{\mathrm{d}\bar{x}}{\mathrm{d}t} \simeq v(\bar{k}) = \frac{1}{\hbar} \left. \frac{\mathrm{d}E}{\mathrm{d}k} \right|_{k=\bar{k}} \qquad [5.21]$$

$$\hbar \frac{\mathrm{d}\bar{k}}{\mathrm{d}t} \simeq - \left. \frac{\mathrm{d}V}{\mathrm{d}x} \right|_{x=\bar{x}} \qquad [5.22]$$

In the following we consider the motion of an electron in a uniform electric field $\mathcal{E}$ along x:

$$\frac{\mathrm{d}V}{\mathrm{d}x} = e\mathcal{E} \qquad [5.23]$$

5.1.2.1. *Acceleration near a band minimum*

Figure 5.3 shows the evolution of an electron wavepacket with initial wavevector $k_0 = 0$ in a uniform electric field $\mathcal{E} = -10^7$ V/m along x. The acceleration of the electron and the increase in the wavevector are clearly visible. The average position of the wavepacket nearly follows Newton's law with an effective mass m. More precisely, if we account for the exact group velocity [5.20] of the discretized Hamiltonian, the integration of the equations of motion [5.21] and [5.22] yields:

$$\bar{x}(t) - \bar{x}(0) = \frac{\hbar^2}{ma^2 e\mathcal{E}}\left[1 - \cos\left(\frac{ae\mathcal{E}t}{\hbar}\right)\right] \qquad [5.24]$$

which shows very good agreement with the numerical result. For $t \ll \hbar/ae\mathcal{E}$, Newton's law is indeed recovered. However, at larger t, equation [5.24] predicts a counterintuitive oscillatory motion, which is discussed in the following section.

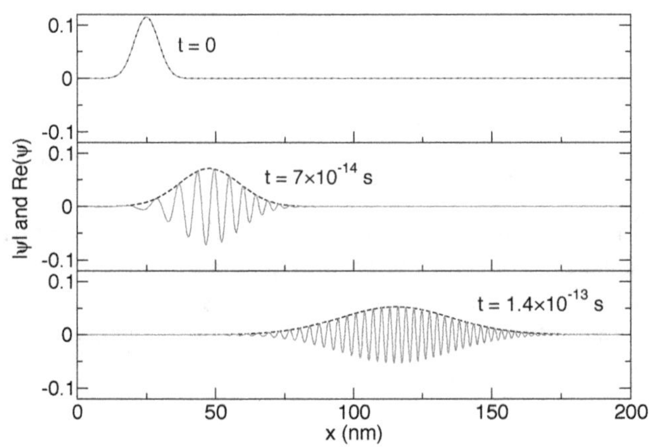

Figure 5.3. *Evolution of an electron Gaussian wavepacket with initial wavevector $k_0 = 0$ in a uniform electric field $\mathcal{E} = -10^7$ V/m*

5.1.2.2. *Motion near a band maximum*

To check the validity of equation [5.24] within a reasonable evolution (and computing) time, we increase the mesh step to $a = 0.5$ nm and keep the other parameters unchanged. Figure 5.4 shows the evolution of a wavepacket with initial wavevector $k_0 = 0$ at successive time steps $\Delta t = \pi\hbar/2ae\mathcal{E} = 2.068 \times 10^{-13}$ s. Equation [5.24] is very well verified by the simulation: the wavepacket returns to its initial position at time $t = 4\Delta t$. It is only changed by a phase factor. Figure 5.4 also shows the corresponding evolution of the wavevector and kinetic energy of the wavepacket. The electron is first accelerated, as predicted by classical mechanics.

However, when approaching the band maximum at $k = \pi/a$, the electron is slowed down and it is backscattered by the electric field for $k > \pi/a$. In the figure, k is folded back to its equivalent $k - 2\pi/a$ within the first Brillouin zone.

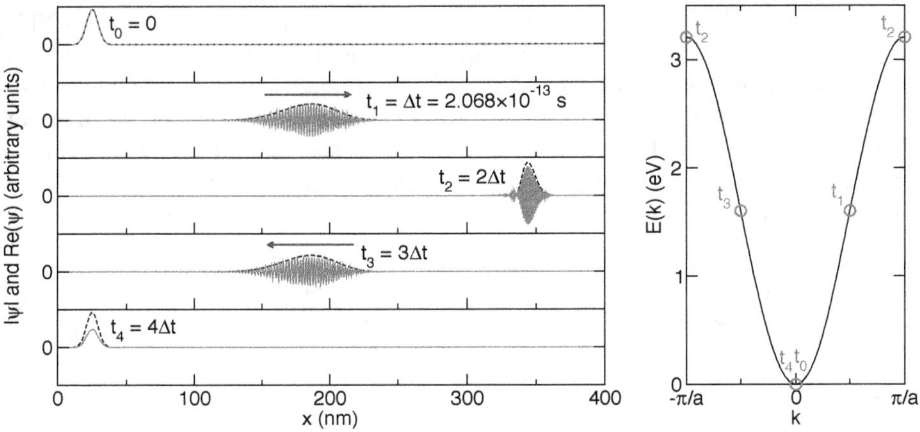

Figure 5.4. *Left panel: evolution of a Gaussian wavepacket with initial wavevector $k_0 = 0$ in a uniform electric field $\mathcal{E} = -10^7$ V/m. Mesh step $a = 0.5$ nm. Right panel: corresponding evolution of the wavevector and kinetic energy*

As previously mentioned, the discretized Hamiltonian can be viewed as a tight-binding model for a 1D crystal with lattice constant a. The backscattering of the electron at $k = \pi/a$ is due to the constructive interference of the waves backscattered by the periodic potential [ASH 76]. In real crystals subjected to an electric field, each electron does not complete full oscillations [5.24] since it is scattered many times (defects, phonons) during one oscillation period. However, if a band is completely filled, the balance between accelerated and slowed down electrons is zero. This is a way to explain why materials with completely occupied energy bands are insulating. This also explains the dynamics of holes in the valence band under an electric field.

We have shown the extreme accuracy of the semiclassical theory in the case of slowly varying potentials, which yields the success of semiclassical methods such as the Boltzmann Monte Carlo method presented in Chapter 8. Now we examine typical situations where the semiclassical theory fails.

5.1.3. *Tunneling and quantum reflection*

Keeping an effective mass $m = 0.19\ m_0$, we consider the propagation of wavepackets through a square potential barrier of height $V_b = 1$ eV and thickness

$L_b = 2$ nm. We choose a very fine mesh step $a = 0.05$ nm to have a better comparison with the analytical formula for tunneling, given in Chapter 1. Figure 5.5 shows the evolution of a wavepacket with average wavevector $k_0 = 2$ nm^{-1}, corresponding to a kinetic energy $E = 0.8$ eV. It is partially transmitted due to the tunnel effect. The evanescent wave propagating through the barrier is clearly visible in the figure. The transmission coefficient is obtained by computing the integral of $|\psi(x)|^2$ for the transmitted wave once it is well separated from the reflected one. This yields $T_{WP} = 0.071$, which is not directly comparable to the tunneling formula for $T(E)$ given in Chapter 1. This is due to the strong uncertainty on k: $\Delta_k = 1/2\Delta_x$ with $\Delta_x = 3$ nm. To obtain an exact comparison, we expand the initial wavepacket in momentum space (see Chapter 1):

$$|\psi\rangle = \int dk\tilde{\psi}(k)\,|k\rangle \qquad [5.25]$$

with

$$\tilde{\psi}(k) = \frac{\sqrt{2\Delta_x}}{(2\pi)^{1/4}} e^{-\Delta_x^2(k-k_0)^2} e^{-i(k-k_0)x_0} \qquad [5.26]$$

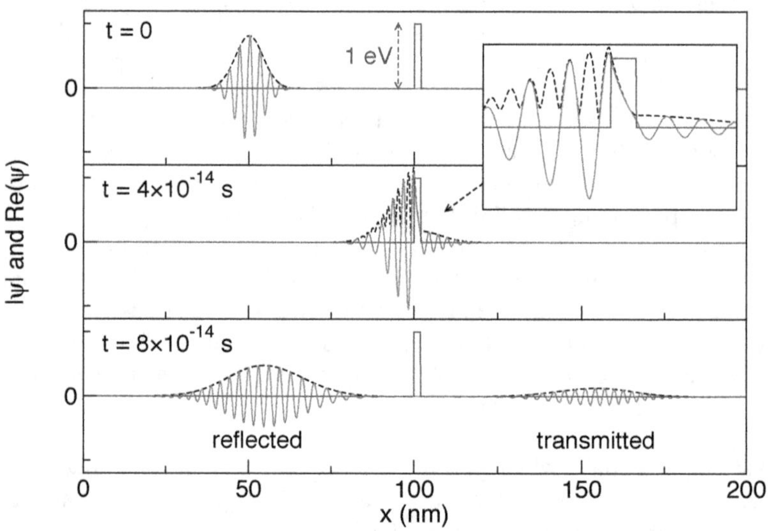

Figure 5.5. *Tunneling of a Gaussian wavepacket with initial wavevector $k_0 = 2$ nm^{-1} ($E \simeq 0.8$ eV) through a square barrier of height $V_b = 1$ eV and thickness $L_b = 2$ nm. Inset: focus on the evanescent wave inside the barrier at $t = 4 \times 10^{-14}$ s. For a color version of this figure, see www.iste.co.uk/dollfus/nanodevices.zip*

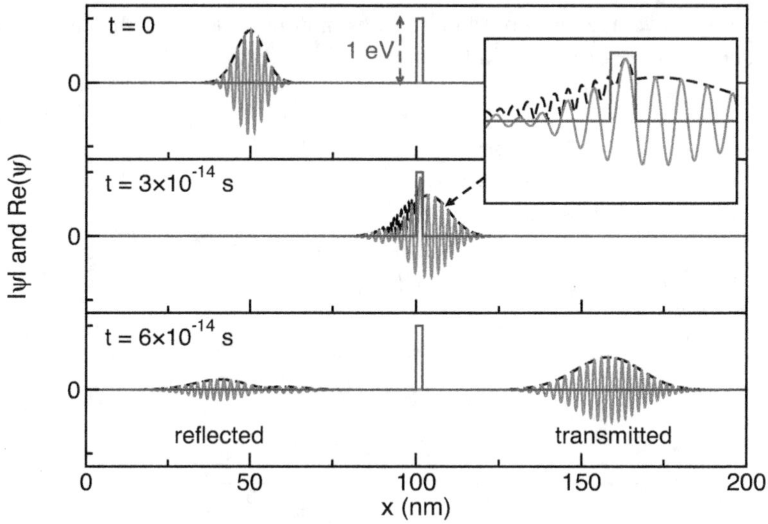

Figure 5.6. *Propagation of a Gaussian wavepacket with initial wavevector $k_0 = 3$ nm^{-1} (energy $\simeq 2$ eV) through a square barrier of height $V_b = 1$ eV and thickness $L_b = 2$ nm. Inset: focus on the propagating wave inside the barrier at $t = 3 \times 10^{-14}$ s. For a color version of this figure, see www.iste.co.uk/dollfus/nanodevices.zip*

The formula given in Chapter 1 for the transmission $T(k)$ through a square barrier is then integrated over the states $|k\rangle$ with positive velocity to obtain the transmission of the considered wavepacket:

$$T_{WP} = \int_{k>0} dk \left|\tilde{\psi}(k)\right|^2 T(k) = 0.071 \qquad [5.27]$$

in agreement with the numerical calculation. $T(k)$ and $\left|\tilde{\psi}(k)\right|^2$ are plotted in Figure 5.7.

We now increase the wavevector of the initial wavepacket to $k_0 = 3$ nm^{-1}, corresponding to a kinetic energy $E \simeq 2$ eV well above the barrier. Figure 5.6 shows the evolution of this wavepacket. It is not fully transmitted, contrary to the classical mechanics result. The calculation gives a transmission coefficient $T_{WP} = 0.924$, which is again in agreement with equation [5.27] using the new value of k_0.

Note that experimentally an electron is either totally transmitted or reflected. The transmission coefficient is the probability of transmission when performing measurements on many successive wavepackets reaching the barrier. It allows calculating the average current, but it does not provide the current fluctuations (noise). The latter are discussed in Chapter 7.

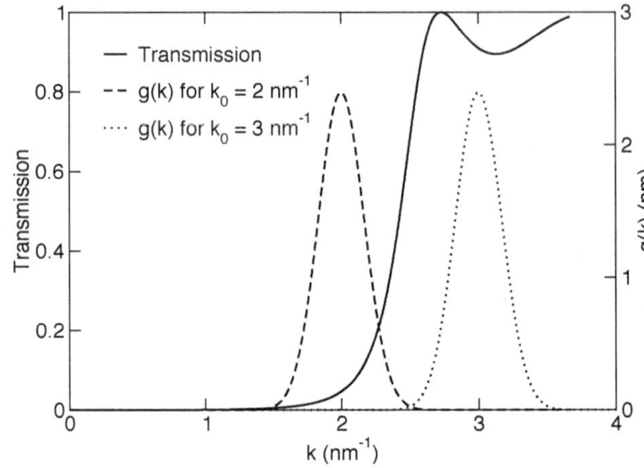

Figure 5.7. *Solid line: transmission versus wavevector of the incident wave for a square barrier of height $V_b = 1$ eV and thickness $L_b = 2$ nm. The effective mass is $m = 0.19\ m_0$. Dotted lines: spectral decomposition $g(k) = |\tilde{\psi}(k)|^2$ of the two considered wavepackets*

5.1.4. *Beyond 1D models*

5.1.4.1. *Generic example: ribbon*

We consider the Schrödinger equation in 2D, in the effective mass approximation:

$$i\hbar\frac{\partial\psi(x,y,t)}{\partial t} = -\frac{\hbar^2}{2m}\left[\frac{\partial^2\psi(x,y,t)}{\partial x^2} + \frac{\partial^2\psi(x,y,t)}{\partial y^2}\right] + V(x,y)\psi(x,y,t) \qquad [5.28]$$

and we assume that $V(x,y)$ is finite for $0 \leq y \leq W$ and $+\infty$ elsewhere. Hence, the electron is confined in a ribbon of width W along y. For a uniform potential $V(x,y) = V_0$ inside the ribbon, the stationary solutions of the Schrödinger equation can be chosen as:

$$\psi_{p,k}(x,y,t) = \sin\left(\frac{p\pi y}{W}\right) e^{i[kx - E_p(k)t/\hbar]} \qquad [5.29]$$

where $p > 0$ is an integer indexing the shape of the wavefunction along y, and k is the wavevector along x. The corresponding band structure is shown in Figure 5.8. Each value of p corresponds to a subband with the dispersion relation:

$$E_p(k) = V_0 + \frac{p^2\pi^2\hbar^2}{2mW^2} + \frac{\hbar^2k^2}{2m} \quad [5.30]$$

The confinement leads to a quantization of the wavefunction oscillation, hence of the kinetic energy along y. This simple example will be sufficient in the rest of the chapter to illustrate important aspects of quantum transport. Actually, many real systems (nanotubes, nanowires, graphene ribbons) have similar multisubband structures due to lateral confinement.

For numerical calculations, we discretize the 2D Schrödinger equation on a finite difference mesh of step a. $W = (N_y + 1)a$, where $N_y + 1$ is the number of mesh steps along y. The mesh sites are indexed by n_x and n_y, with $0 \leq n_y \leq N_y + 1$, and the wavefunctions are zero at the ribbon boundaries $n_y = 0$ and $n_y = N_y + 1$. The discretized Hamiltonian is similar to [5.15], with onsite energies $4\gamma + V(n_x, n_y)$ and coupling $-\gamma$ with the four nearest neighbors. For a uniform potential $V = V_0$, the stationary states take the form:

$$\psi_{p,k}(n_x, n_y, t) = \chi_p(n_y)e^{i[kn_xa - E_p(k)t/\hbar]} \quad [5.31]$$

with the transverse wavefunction:

$$\chi_p(n_y) = \sqrt{\frac{2}{N_y + 1}} \sin\left(\frac{p\pi n_y}{N_y + 1}\right) \quad [5.32]$$

for p integer in $[1, N_y]$. The dispersion relations read:

$$E_p(k) = V_0 + E_p^0 + 2\gamma\left[1 - \cos(ka)\right] \quad [5.33]$$

with

$$E_p^0 = 2\gamma\left[1 - \cos\left(\frac{p\pi}{N_y + 1}\right)\right] \quad [5.34]$$

The transverse wavefunctions have been normalized to unity in [5.32] for future use in this chapter.

5.1.4.2. *Intersubband scattering on a defect*

In a ribbon of width $W = 5$ nm and effective mass $m = 0.19\ m_0$, we build an initial wavepacket in the first subband, using equation [5.31] with $p = 1$, $k = 1.2\ \text{nm}^{-1}$, and a Gaussian modulation of width $\Delta_x = 10$ nm along the x axis. An impurity is placed at 1 nm from the ribbon edge, as shown in Figure 5.8 (left panel). It is modeled by a Gaussian potential of height $V_{imp} = 0.5$ eV and width $\Delta_r = 0.5$ nm:

$$V(r) = V_{imp} e^{-r^2/4\Delta_r^2} \qquad [5.35]$$

where r is the distance from the impurity. Figure 5.8 (left panel) shows the propagation of the wavepacket through the impurity potential. It splits into two transmitted and two reflected wavepackets that can be identified, from their respective modulation along y, as belonging either to the first or to the second subband. This is further supported by the analysis of the subband structure, shown in the right panel, where the initial state and the three states accessible through elastic transitions are represented. The smaller group velocity of the second subband states explains the spatial separation of the wavepackets during their propagation. This example illustrates some of the complex phenomena that occur during the scattering of an electron by an impurity or by other structural defects. The same system will be studied using the Green's function formalism at the end of this chapter.

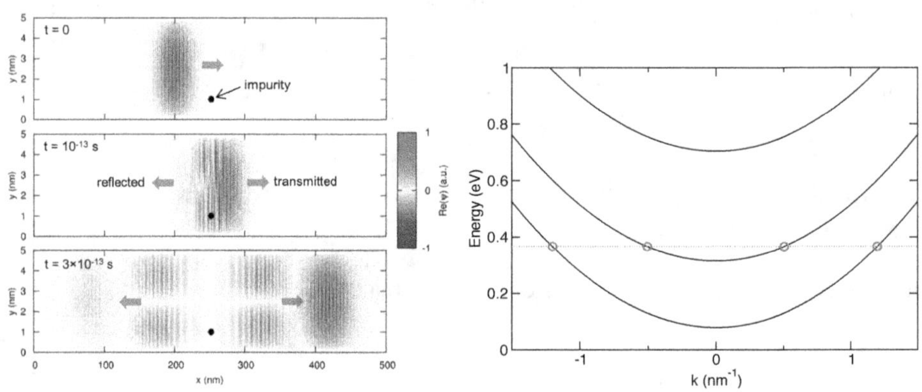

Figure 5.8. *Left panel: evolution of a wavepacket of the first subband in a 2D ribbon of width $W = 5$ nm and effective mass $m = 0.19\ m_0$. The wavepacket is scattered by an impurity and split into two transmitted and two reflected waves, identified on the subband structure shown in the right panel. Right panel: three lowest subbands of the ribbon band structure. The rightmost circle indicates the initial wavepacket and the three other circles indicate the states reached through elastic scattering. For a color version of this figure, see www.iste.co.uk/dollfus/nanodevices.zip*

5.1.5. *Simulation methods for quantum transport*

Besides its pedagogical interest, the computation of wavepacket propagation is an efficient way to study quantum transport in bulk and thin layer conductors [MAY 88, MAY 95, ROC 97, ROC 99, TRI 02]. It is less suited to inhomogeneous systems such as small conductors connected to electrodes. In the following, we present the transmission formalism for quantum transport together with the Green's function method, which is widely used for the simulation of nanodevices. The generalization to many-body Green's functions is presented in Chapter 6. Alternative quantum transport methods are presented in Chapter 7 (Bohm trajectories) and Chapter 8 (Wigner transport equation).

5.2. The transmission formalism for the conductance

In this section, we briefly present the transmission formalism, first introduced by Landauer for 1D systems [LAN 57, LAN 70] and generalized to multichannel systems (such as the ribbon of section 5.1.4.1) by Büttiker *et al.* [BUT 85]. This formalism is very well suited to the study of electron transport through a small system connected to macroscopic electrodes. Exhaustive presentations of the formalism and of its physical implications can be found in several books [IMR 02, DAT 95, DAT 05]. As mentioned at the beginning of this chapter, we limit the discussion to the coherent regime. Transport can be viewed as a succession of wavepackets injected from each electrode and partially transmitted to the other electrode without losing their phase coherence.

5.2.1. *The scattering matrix*

We consider a conductor "C" of finite length connected to a left ("L") and to a right ("R") electrode. It is assumed that the electrodes are made of ideal (crystalline) leads that connect the conductor to macroscopic electron reservoirs (made of metal or highly doped semiconductor). It is also assumed that the conductor and the ideal leads have finite transverse dimensions. Hence, we have a 1D band structure with one or several subbands as for the ribbon of section 5.1.4.1. At a given electron energy E, only the subbands $E_p(k)$ crossing this energy (see the right panel of Figure 5.8) can contribute to transport. Each crossing subband is called a conduction "channel" or "mode". The wavefunctions of the different channels are orthogonal to each other. We consider all possible incoming and outgoing Bloch waves in the two leads and define their transmission and reflection coefficients. All quantities (including the number of available channels) depend on E but we omit it in order to simplify the notations. The amplitude A_p^L of an incoming wavefunction ψ_p in channel p of the left lead is defined by:

$$\psi_p = A_p^L \psi_p^{norm} \tag{5.36}$$

where ψ_p^{norm} is a reference Bloch wave normalized to one electron per unit length. However, what is needed for studying transport is the ratio between the outgoing and incoming currents. For this purpose, we define the incoming current amplitude:

$$a_p^L = A_p^L \times \sqrt{|v_{Lp}(E)|} \quad [5.37]$$

where $v_{Lp}(E)$ is the group velocity. This definition is justified by the fact that $|a_p^L|^2$ is the number of incident electrons per unit time in channel p. Similarly, we define the incoming current amplitude a_p^R in a channel p of the right lead, and the outgoing current amplitudes in the left and right leads, denoted as, respectively, b_p^L and b_p^R. By linearity of the Schrödinger equation, the relationship between the outgoing and the incoming current amplitudes is also linear and can be put in block matrix form:

$$\begin{bmatrix} b^L \\ b^R \end{bmatrix} = S \begin{bmatrix} a^L \\ a^R \end{bmatrix} \text{ with } S = \begin{bmatrix} r^L & t^R \\ t^L & r^R \end{bmatrix} \quad [5.38]$$

where a^L and b^L are complex vectors of dimension N_L, the number of conducting channels at energy E in the left lead and similarly a^R and b^R are of dimension N_R, the number of conducting channels in the right lead. Note that the number and nature of modes may be different in each lead due to different geometry and/or electrostatic potential. S is called the scattering matrix. Its coefficients are generally complex, since the current amplitudes have arbitrary phase factors that depend on the choice for the reference normalized Bloch waves ψ_p^{norm}. From the matrix S, we obtain the reflection and transmission coefficients between the different channels. For instance, the transmission coefficient between a channel p of the left electrode and a channel q of the right electrode is:

$$T_{qp}^L = |t_{qp}^L|^2 \quad [5.39]$$

The outgoing current must be equal to the incoming current whatever the incoming amplitudes:

$$\sum_{p \in L} |b_p^L|^2 + \sum_{p \in R} |b_p^R|^2 = \sum_{p \in L} |a_p^L|^2 + \sum_{p \in R} |a_p^R|^2 \quad [5.40]$$

where we have summed the squared amplitudes, since the different channels are orthogonal to each other and do not interfere[2]. Hence, the scattering matrix S

2 This point will be discussed in more detail in section 5.3.2.2.

conserves the norm of any vector. In a complex vector space, this is equivalent to unitarity:

$$S^\dagger = S^{-1} \quad [5.41]$$

This leads to several sum rules between the transmission and reflection coefficients [DAT 95]. In particular, the unitarity of S implies:

$$\sum_{p \in L} \sum_{q \in R} T^L_{qp} = \sum_{q \in R} \sum_{p \in L} T^R_{pq} \quad [5.42]$$

Hence at any energy, the sum of the transmission coefficients from left to right equals the sum of the transmission coefficients from right to left. This is an example of Onsager symmetry, which can also be deduced by thermodynamic arguments near equilibrium [IMR 02].

Another important property of the scattering matrix is related to time-reversal symmetry [DAT 95, IMR 02]. The evolution of a wavepacket crossing the conductor between times t_1 and t_2 is given by the evolution operator:

$$\psi(t_2) = e^{-iH(t_2-t_1)/\hbar}\psi(t_1) \quad [5.43]$$

Assuming that the coefficients of the matrix H are real (which is the case in the absence of magnetic field and of spin-orbit coupling), we have:

$$\psi^*(t_2) = e^{iH(t_2-t_1)/\hbar}\psi^*(t_1) \quad [5.44]$$

Applying the evolution operator on both sides, we obtain:

$$\psi^*(t_1) = e^{-iH(t_2-t_1)/\hbar}\psi^*(t_2) \quad [5.45]$$

Leaving aside the time arguments, the state $\psi^*(t_1)$ is the evolution during a time $(t_2 - t_1)$ of an initial state $\psi^*(t_2)$. Hence by taking the conjugate of a final wavepacket $\psi(t_2)$, we obtain a wavepacket that evolves backward to the conjugate of $\psi(t_1)$. This is similar to time-reversal symmetry in classical mechanics: the reverse trajectory is valid since it obeys the same physics laws. Here, the phase of the wavefunction is reversed. This does not change the probability density but, as expected, this changes the sign of the velocities (k becomes $-k$ for a plane wave). Hence, replacing the incoming amplitudes a and outgoing amplitudes b by their conjugates, we obtain the reverse situation where the amplitudes a^* correspond to outgoing waves and the amplitudes b^* to incoming waves. Since this evolution is governed by the same Hamiltonian, the outgoing and incoming waves are related by the same matrix S:

$$a^* = Sb^* \qquad [5.46]$$

Multiplying on both sides by $S^\dagger$ and taking the conjugate, we obtain:

$$b = S^T a \qquad [5.47]$$

where S^T is the transpose of S. Since [5.47] and [5.38] are valid for any vector a, the scattering matrix is symmetric:

$$S^T = S \qquad [5.48]$$

This implies in particular the symmetry of the transmission coefficients between any mode p of lead L and any mode q of lead R:

$$T_{qp}^L = T_{pq}^R \qquad [5.49]$$

which means that the reciprocity relation [5.42] becomes valid channel by channel. In the presence of spin-orbit coupling, time-reversal symmetry still holds but the relations are more complicate. In the presence of an external magnetic field, the sign of the field must be reversed to obtain the symmetric backward evolution.

5.2.2. *Landauer–Büttiker conductance*

In order to evaluate the current through the conductor, additional assumptions are required concerning the injection of electrons from the two electrodes. It is assumed that the left and right reservoirs are at thermal equilibrium with electrochemical potentials μ_L and μ_R, respectively. It is further assumed that the occupation numbers of the wavepackets injected in each lead follow the equilibrium distribution of the reservoir from which they are injected: $f(E - \mu_L)$ for the left electrode and $f(E - \mu_R)$ for the right electrode, where E is the wavepacket energy and f is the Fermi–Dirac distribution:

$$f(E) = \frac{1}{1 + e^{E/k_B T}} \qquad [5.50]$$

Due to the Pauli exclusion principle, the leads have a limited number of electrons per unit length in each channel. Considering a lead of finite length L with periodic boundary conditions (which do not affect the physics if L is large enough), the allowed wavevectors k are:

$$k_l = l \times \frac{2\pi}{L} \qquad [5.51]$$

where l is an integer. Hence in the limit of large L, the number of allowed wavevectors between k and $k + dk$ is:

$$N_k = \frac{L}{2\pi} \times dk \qquad [5.52]$$

We consider states in channel p with energy close to E and with positive group velocity $v_p(k)$. From the previous equation, we obtain the density of such states per unit of length and of energy:

$$\rho_p^+(E) = g_s \times \frac{1}{2\pi} \times \frac{\mathrm{d}k}{\mathrm{d}E_p} = g_s \times \frac{1}{2\pi} \times \frac{1}{\hbar v_p(k)} \qquad [5.53]$$

where g_s is the spin degeneracy ($g_s = 2$ in most cases and 1 for spin-polarized transport). $\rho_p^+(E)$ is half the well-known 1D density of states, since the states $-k$ with negative velocity are not included. Let us now define the spectral density $j_{Lp}^{inj}(E)$ of the current injected in channel p of lead L. It is the injected electrical current per unit energy if the incoming states are totally occupied. It reads:

$$j_{Lp}^{inj}(E) = -e \times \rho_p^+(E) \times v_p(k) = -g_s \frac{e}{h} \qquad [5.54]$$

The important point is that $v_p(k)$ has cancelled out in the product. Hence, $j_{Lp}^{inj}(E)$ is independent of the energy and the particular shape of the dispersion relation $E_p(k)$.

The net current spectral density from channel p is the fraction of the injected current, which is transmitted to the different channels of the right electrode:

$$j_{Lp}(E) = j_{Lp}^{inj}(E) \times \sum_{q \in R} T_{qp}^L(E) = -g_s \frac{e}{h} \sum_{q \in R} T_{qp}^L(E) \qquad [5.55]$$

Let us also define the net current spectral density from all channels of the left electrode:

$$j_L(E) = \sum_{p \in L} j_{Lp}(E) = -g_s \frac{e}{h} \sum_{p \in L} \sum_{q \in R} T^L_{qp}(E) \qquad [5.56]$$

Similarly, we define the net current spectral densities from the right lead:

$$j_{Rq}(E) = -g_s \frac{e}{h} \sum_{p \in L} T^R_{pq}(E) \qquad [5.57]$$

$$j_R(E) = -g_s \frac{e}{h} \sum_{q \in R} \sum_{p \in L} T^R_{pq}(E) \qquad [5.58]$$

From [5.42], we have:

$$j_R(E) = j_L(E) \qquad [5.59]$$

Hence, at equal occupation factors, the currents coming from the left and right leads compensate at each energy E. The total current from L to R is given by the balance between the integrals of $j_L(E)$ and $j_R(E)$ weighted by their respective Fermi–Dirac occupation factors. Using [5.59] we obtain:

$$I_{LR} = \int dE j_L(E) \times [f(E - \mu_L) - f(E - \mu_R)] \qquad [5.60]$$

At equilibrium, $\mu_L = \mu_R$ and the current equals zero. Applying a bias voltage U_{LR} between the electrodes yields an electrochemical potential difference:

$$\mu_L - \mu_R = -eU_{LR} \qquad [5.61]$$

In the small bias limit, the current becomes:

$$I_{LR} = \int dE j_L(E) \times [-eU_{LR} f'(E - \mu_L)] \qquad [5.62]$$

where f' is the derivative of the Fermi–Dirac function $f(E)$, which is peaked around $E = \mu_L$ with a width of order $k_B T$. Hence, only the electrons with energies close to the Fermi level contribute to transport. The lower energy states from both electrodes are fully occupied and their contributions $j_L(E) = j_R(E)$ cancel each other. From the previous equation, we obtain the conductance in the linear regime (small bias).

$$\mathcal{G} = \mathcal{G}_0 \int dE \left[\sum_{p \in L} \sum_{q \in R} T_{qp}^L(E) \right] \times \left[-f'(E - \mu_L) \right] \qquad [5.63]$$

where

$$\mathcal{G}_0 = g_s \frac{e^2}{h} \qquad [5.64]$$

is the quantum of conductance, most often defined for a spin degenerate system:

$$\mathcal{G}_0 = \frac{2e^2}{h} = \frac{1}{12.9\ \mathrm{k}\Omega} \qquad [5.65]$$

At low temperature, $-f'(E - \mu_L)$ tends to the Dirac delta function $\delta(E - \mu_L)$. The conductance becomes:

$$\mathcal{G}(T = 0\,\mathrm{K}) = \mathcal{G}_0 \sum_{p \in L} \sum_{q \in R} T_{qp}^L(\mu_L) \qquad [5.66]$$

For an ideal conductor identical to the leads, each channel p is fully transmitted to the same channel p of the other lead. Hence, the conductance is quantized as:

$$\mathcal{G}(T = 0\,\mathrm{K}) = N_{ch}(\mu_L) \times \mathcal{G}_0 \qquad [5.67]$$

where $N_{ch}(\mu_L)$ is the number of conducting channels at the Fermi level μ_L of the system. It is possible experimentally to vary the number of channels by an external electric field, and to observe the quantized steps of conductance [WEE 88]. At higher temperature, the quantized steps are smoothed by the thermal broadening $(-f')$. However, quantized conductance has been observed at room temperature in metallic contacts of atomic size, in which the channels are well separated in energy ($\Delta E \gg k_B T = 25$ meV) [COS 95].

In theoretical and simulation studies, we often plot the linear conductance at zero temperature [5.67] as a function of energy. The conductance is expressed in units of $\mathcal{G}_0$. Thus, it reduces to the sum of the transmission coefficients. The energy represents the position of the Fermi level, hence the energy of the electrons contributing to transport. This is relevant since in nanosystems and even in more conventional field effect transistors, the position of the Fermi level with respect to the energy subbands can be largely varied under the influence of an external electric field created by a gate electrode.

From [5.60], we may also obtain the current at large bias voltage. However, such non-equilibrium condition enhances inelastic phenomena (activation of optical phonons, electron–electron scattering), which makes [5.60] invalid, except in some special cases such as thin tunnel junctions.

5.2.3. *Conductance of an ideal wire and energy dissipation*

We have seen that an ideal conductor with finite lateral dimensions (wire, ribbon) has a finite conductance proportional to the number of conducting channels. The power dissipated during the electrical measurement is:

$$P = U_{LR} I_{LR} = \frac{(I_{LR})^2}{\mathcal{G}} \quad [5.68]$$

This power is not dissipated in the ideal conductor since there are no scattering mechanisms to relax the energy of the electrons. It is actually dissipated into the macroscopic electron reservoirs, where the inelastic scattering rates are much larger, even at low temperature, due to the large available phase space for electron–electron and electron-phonon scatterings [DAT 95, IMR 02]. For instance, when $\mu_L > \mu_R$, there is an excess of electrons with energies above μ_R in the right reservoir, coming from the left reservoir. These electrons then relax their energy to restore thermal equilibrium. In the left reservoir, there is a deficit of "left moving" electrons at energies between μ_R and μ_L, since the right reservoir does not inject enough of them. Electrons of the left reservoir relax from higher energies into these empty states.

5.2.4. *Application to planar and bulk systems*

The transmission formalism is not limited to nanowires and nanoribbons. It is also applicable to planar conductors of large width and to bulk conductors. Let us consider, for instance, a thin semiconductor film. Let x be the transport direction, y is the normal to the film and z is the transverse direction. If the film is homogeneous along z, then we can model the system by a 2D slice in the (xy) plane with Born–von Karman conditions along z, associated with a wavevector k_z accounting for the transverse kinetic energy of the electrons. The 2D Hamiltonian is similar to that of the ribbon of section 5.1.4.1. The confinement along y yields a multisubband structure and the transmission coefficients are computed exactly the same way as in the 1D case, using for instance the Green's function method presented below. The difference is that these coefficients are integrated over k_z, which multiplies the number of channels proportionally to the film width. The conductance quantization therefore disappears. Another common application of the transmission formalism is a metal–insulator–metal junction made up of homogeneous thin layers of large area. The system can be modeled by a 1D Hamiltonian along the normal direction, and the transmission coefficient is integrated over the in-plane 2D wavevector. The quantum transmission formalism is particularly relevant in this case since it accurately treats tunneling through the thin insulator.

5.3. The Green's function method for quantum transmission

The Green's function formalism is one of the most convenient methods for calculating the quantum transmission through a conductor connected to electrodes. It is presented here in the one-electron approximation, but it is generalizable to the many-body theory, as detailed in the following chapter. In order to make our introduction more intuitive and related to section 5.1, we show the link with wavepacket propagation: stationary states built from the Green's functions can be viewed as wavepackets with a spatial extension tending to infinity.

5.3.1. *One-electron Green's functions: definitions and basic properties*

To preserve the generality, the Green's functions are introduced for 3D quantum systems. The use of the formalism for 1D and 2D systems is straightforward. Note that the definition of the Green's functions used in this chapter is only valid in the one-electron approximation. However, this definition coincides with the many-body definition given in Chapter 6, provided that the electrons are not interacting. Moreover, the physical interpretation of the Green's function is similar in the non-interacting and interacting cases.

5.3.1.1. *The time-dependent retarded Green's function*

For two times t and t', the "retarded" Green's operator is defined as:

$$\hat{G}^r(t,t') \equiv \begin{cases} -\frac{i}{\hbar}\hat{U}(t,t') \text{ for } t \geq t' \\ 0 \text{ for } t < t' \end{cases} \quad [5.69]$$

where $\hat{U}(t,t')$ is the evolution operator. Since the Hamiltonian is considered time independent throughout this chapter, the evolution operator and the Green's operator only depend on the time difference $t - t'$. However, we momentarily keep the two variables t and t' for more generality and symmetry in the equations. We also define the "advanced" Green's operator:

$$\hat{G}^a(t,t') \equiv \begin{cases} -\frac{i}{\hbar}\hat{U}(t,t') \text{ for } t \leq t' \\ 0 \text{ for } t > t' \end{cases} \qquad [5.70]$$

In space representation, the retarded Green's function reads:

$$G^r(\mathbf{r},t;\mathbf{r}',t') = \langle \mathbf{r}| \hat{G}^r(t,t') |\mathbf{r}'\rangle \qquad [5.71]$$

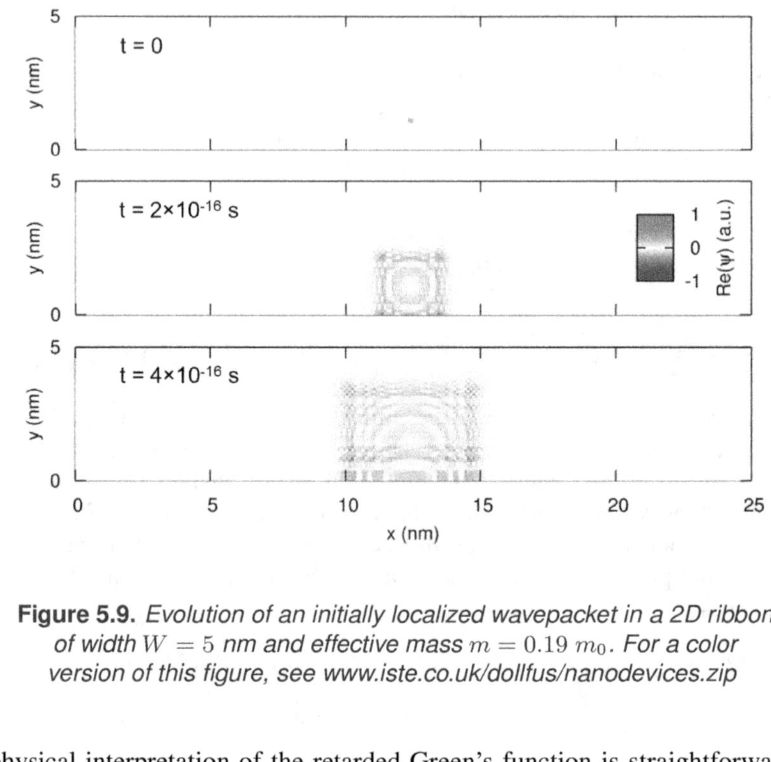

Figure 5.9. *Evolution of an initially localized wavepacket in a 2D ribbon of width $W = 5$ nm and effective mass $m = 0.19\, m_0$. For a color version of this figure, see www.iste.co.uk/dollfus/nanodevices.zip*

The physical interpretation of the retarded Green's function is straightforward. It is proportional to $\langle \mathbf{r}| \hat{U}(t,t') |\mathbf{r}'\rangle$, which is the amplitude at point $\mathbf{r}$ and at time t of a wavefunction initially localized at $\mathbf{r}'$ at time t'. Figure 5.9 shows such propagation in the 2D ribbon defined in section 5.1.4.1. It is analogous to surface waves propagating from a stone dropped in the water. The reflection of the wave on the ribbon edge is also visible. To push this interpretation further, it is interesting to determine the equation of motion for the retarded Green's function. Equation [5.4] yields:

$$i\hbar\frac{\partial \hat{G}^r(t,t')}{\partial t} = \begin{cases} \hat{H}\hat{G}^r(t,t') \text{ for } t \geq t' \\ 0 \text{ for } t < t' \end{cases} \qquad [5.72]$$

The derivative is singular when t reaches t'. $i\hbar\hat{G}^r(t,t')$ steps from 0 to $\hat{U}(t,t) = \hat{I}$, the identity operator. To take into account this singularity, equation [5.72] can be rewritten as:

$$i\hbar\frac{\partial \hat{G}^r(t,t')}{\partial t} = \hat{H}\hat{G}^r(t,t')\theta(t-t') + \hat{I}\delta(t-t') \qquad [5.73]$$

where θ is the Heaviside step function and δ is the Dirac delta function. In spatial representation:

$$i\hbar\frac{\partial}{\partial t}G^r(\mathbf{r},t;\mathbf{r}',t') = \langle\mathbf{r}|\,\hat{H}\hat{G}^r(t,t')\,|\mathbf{r}'\rangle\,\theta(t-t') + \delta^{(3)}(\mathbf{r}-\mathbf{r}')\delta(t-t') \qquad [5.74]$$

where $\delta^{(3)}$ is the 3D Dirac delta function. We now consider the 3D effective mass Hamiltonian. To avoid any mistake in the evaluation of the right-hand side, let us define $|\psi\rangle = \hat{G}^r(t,t')\,|\mathbf{r}'\rangle$. We have:

$$\langle\mathbf{r}|\,\hat{H}\,|\psi\rangle = -\frac{\hbar^2}{2m}\Delta_\mathbf{r}\psi(\mathbf{r}) + V(\mathbf{r})\psi(\mathbf{r}) \qquad [5.75]$$

where $\Delta_\mathbf{r}$ is the Laplacian with respect to the position $\mathbf{r}$. Hence, replacing $\psi(\mathbf{r})$ by its expression $\langle\mathbf{r}|\psi\rangle = G^r(\mathbf{r},t;\mathbf{r}',t')$, we obtain:

$$i\hbar\frac{\partial}{\partial t}G^r(\mathbf{r},t;\mathbf{r}',t') =$$
$$\left[-\frac{\hbar^2}{2m}\Delta_\mathbf{r} + V(\mathbf{r})\right]G^r(\mathbf{r},t;\mathbf{r}',t')\theta(t-t') + \delta^{(3)}(\mathbf{r}-\mathbf{r}')\delta(t-t') \qquad [5.76]$$

The advanced Green's function $G^a(\mathbf{r},t;\mathbf{r}',t')$ satisfies the same equation, with $\theta(t-t')$ replaced by $\theta(t'-t)$. Hence, both Green's function satisfy the time-dependent Schrödinger equation modified by a source term, which can be interpreted as an electron appearing (for G^r) or disappearing (for G^a) at position $\mathbf{r}'$ at time t'. Similar the Green's functions can be defined for other linear differential equations of physics (electrostatics, electromagnetic waves, acoustic waves, etc.).

They are sometimes called "resolvent" since, once the Green's function is known for all positions of the point source, we can build by linearity the solution of the differential equation for any boundary condition. However, an initially localized wavefunction has a maximum energy uncertainty: it contains all eigenstates of the Hamiltonian. For instance, in Figure 5.9, we can see the contribution of eigenstates with short wavelengths (i.e. high energy) comparable to the mesh step. Therefore, the time-dependent Green's functions are not convenient for computing the quantum transmission at a given energy E. An energy-resolved formalism is necessary.

5.3.1.2. *The frequency (i.e. energy) Green's functions*

From now on, we use the fact that $\hat{G}^r(t, t')$ only depends on the time difference $(t - t')$, since $\hat{H}$ is time independent. (t, t') is replaced by a single time variable t:

$$\hat{G}^r(t) = -\frac{i}{\hbar}\hat{U}(t)\theta(t) = -\frac{i}{\hbar}e^{-i\hat{H}t/\hbar}\theta(t) \qquad [5.77]$$

To obtain an energy-resolved Green's function at energy E, we filter the component of angular frequency $\omega = E/\hbar$ of $\hat{G}^r(t)$ by taking its Fourier transform:

$$\hat{G}^r(\omega) \equiv \int_{-\infty}^{+\infty} dt e^{i\omega t}\hat{G}^r(t) \qquad [5.78]$$

Since we will not use $\hat{G}^r(t)$ anymore, we have improperly chosen the same notation $\hat{G}^r$ for its Fourier transform. From equation [5.77],

$$\hat{G}^r(\omega) = -\frac{i}{\hbar}\int_{0}^{+\infty} dt e^{i\omega t}e^{-i\hat{H}t/\hbar} \qquad [5.79]$$

This integral is not convergent, as can be seen by applying the integrand to an eigenstate of $\hat{H}$. To obtain a well-defined integral, we add a small and positive imaginary part ϵ to ω:

$$\begin{aligned}\hat{G}^r_\epsilon(\omega) &\equiv -\frac{i}{\hbar}\int_{0}^{+\infty} dt e^{i\left[(\omega+i\epsilon)\hat{I}-\hat{H}/\hbar\right]t} \\ &= \left[\hbar(\omega+i\epsilon)\hat{I}-\hat{H}\right]^{-1}\end{aligned} \qquad [5.80]$$

A step-by-step derivation of the last equality can be obtained by expressing the operators on the basis of eigenstates of $\hat{H}$. The use of the small imaginary part is justified by the fact that the inverse Fourier transform of $\hat{G}^r_\epsilon(\omega)$ tends to $\hat{G}^r(t)$ when

$\epsilon \to 0^+$. This can be shown by contour integration with respect to ω in the complex plane. Another unquestionable justification for the small imaginary part is that the obtained frequency Green's function allows solving problems such as the quantum transmission, as will be shown in this chapter. Similar calculation for the advanced Green's operator [5.70] yields:

$$\hat{G}^a_\epsilon(\omega) = \left[\hbar(\omega - i\epsilon)\hat{I} - \hat{H}\right]^{-1} \quad [5.81]$$

with $\epsilon \to 0^+$. Hence, the sign of the imaginary part determines if the Green's operator is retarded or advanced. The two operators are related by:

$$\hat{G}^a_\epsilon(\omega) = \left[\hat{G}^r_\epsilon(\omega)\right]^\dagger \quad [5.82]$$

From now on, we replace the variable ω by the energy $E = \hbar\omega$, and ϵ by a small and positive $\eta = \hbar\epsilon$:

$$\hat{G}^{r,a}_\eta(E) = \left[(E \pm i\eta)\hat{I} - \hat{H}\right]^{-1} \quad [5.83]$$

Equivalently,

$$\left[(E \pm i\eta)\hat{I} - \hat{H}\right]\hat{G}^{r,a}_\eta(E) = \hat{I} \quad [5.84]$$

In the spatial representation, repeating the steps leading to equation [5.76] and taking the limit $\eta \to 0^+$, we obtain:

$$\left[E - \left(-\frac{\hbar^2}{2m}\Delta_{\mathbf{r}} + V(\mathbf{r})\right)\right] G^{r,a}(\mathbf{r}, \mathbf{r}'; E) = \delta^{(3)}(\mathbf{r} - \mathbf{r}') \quad [5.85]$$

Hence, $G^r(\mathbf{r}, \mathbf{r}'; E)$ and $G^a(\mathbf{r}, \mathbf{r}'; E)$ both satisfy the stationary Schrödinger equation at energy E, with an additional term $\delta^{(3)}(\mathbf{r} - \mathbf{r}')$ that can be interpreted as a source (or sink) of electrons at point $\mathbf{r}'$. However, we have not yet discussed the convergence of $\hat{G}^{r,a}_\eta(E)$ when $\eta \to 0^+$. This is addressed in the following section.

It is also important to mention that since $G^r(\mathbf{r}, \mathbf{r}'; E)$ is the Fourier transform of a causal function, its real and imaginary parts satisfy the Kramers–Krönig relations.

5.3.1.3. *Basic properties of the energy Green's function*

5.3.1.3.1. Finite system

$\hat{G}^r_\eta(E)$ is easily expressed in a complete basis set of eigenstates of $\hat{H}$, in which the operator $[(E + i\eta)\hat{I} - \hat{H}]$ is diagonal. Let us first consider a finite system with a complete orthonormal basis of N eigenstates $|\phi_n\rangle$ of energy E_n. Using the closure relation, we obtain:

$$\hat{G}^r_\eta(E) = \sum_n \frac{|\phi_n\rangle\langle\phi_n|}{E + i\eta - E_n} \qquad [5.86]$$

In spatial representation,

$$G^r_\eta(\mathbf{r}, \mathbf{r}'; E) = \sum_n \frac{\phi_n(\mathbf{r})\phi^*_n(\mathbf{r}')}{E + i\eta - E_n} \qquad [5.87]$$

Let us decompose the function $f_\eta(x) = 1/(x + i\eta)$ into its real and imaginary parts:

$$f_\eta(x) = \frac{x}{x^2 + \eta^2} - i\frac{\eta}{x^2 + \eta^2} \qquad [5.88]$$

The imaginary part is a Lorentzian of width η and of integral $-\pi$ between $x = -\infty$ and $x = +\infty$. Hence, it tends to $-\pi\delta(x)$ when $\eta \to 0^+$, where δ is the Dirac delta function. The real part of $f_\eta(x)$ tends to the Cauchy principal value $\mathcal{P}$ of $1/x$, which means that for any continuous function $g(x)$ that is square-integrable on an interval $[-A, +A]$, we have:

$$\lim_{\eta\to 0^+} \int_{-A}^{+A} dx \frac{x}{x^2 + \eta^2} g(x) = \lim_{a\to 0^+} \left[\int_{-A}^{-a} dx \frac{1}{x} g(x) + \int_a^A dx \frac{1}{x} g(x) \right] \qquad [5.89]$$

which removes the singularity of the integrand at $x = 0$. Finally, we can write the limit of $f_\eta(x)$ in terms of distributions:

$$\lim_{\eta\to 0^+} \frac{1}{x + i\eta} = \mathcal{P}\left(\frac{1}{x}\right) - i\pi\delta(x) \qquad [5.90]$$

From the Green's function, we obtain the local density of states per unit of volume and energy:

$$\rho(\mathbf{r}, E) = -g_s \frac{1}{\pi} \lim_{\eta\to 0^+} \mathcal{I}m\, G^r_\eta(\mathbf{r}, \mathbf{r}; E) \qquad [5.91]$$

where g_s is the spin degeneracy. Indeed, inserting [5.90] into equation [5.87] yields:

$$\rho(\mathbf{r}, E) = g_s \sum_n |\phi_n(\mathbf{r})|^2 \, \delta(E - E_n) \qquad [5.92]$$

whose integral on any energy interval yields:

$$\int_{E_1}^{E_2} dE \rho(\mathbf{r}, E) = g_s \sum_{n/E_n \in [E_1, E_2]} |\phi_n(\mathbf{r})|^2 \qquad [5.93]$$

This is the contribution of the eigenstates with energies within $[E_1, E_2]$ to the electron density at position $\mathbf{r}$, if these states are occupied.

5.3.1.3.2. Infinite system

From now on, we will instead consider an infinite system with a continuum of eigenstates. The system can be viewed as the limit of a finite box of size $L \times L \times L$ for $L \to +\infty$. Keeping η constant when L increases, the spacing between the discrete energy levels becomes much smaller than η. Thus, the Lorentzian peaks of

$$-\frac{1}{\pi} \mathcal{I}m \, G_\eta^r(\mathbf{r}, \mathbf{r}; E) = \frac{1}{\pi} \sum_n |\phi_n(\mathbf{r})|^2 \, \frac{\eta}{(E - E_n)^2 + \eta^2} \qquad [5.94]$$

merge into a smooth function. Hence, the limit $\eta \to 0^+$ is well defined for the infinite system and yields its local density of states. More rigorously, we now consider a complete basis $\{|\alpha\rangle\}$ of eigenstates of $\hat{H}$ with energies E_α, labeled by a continuous parameter[3]. The basis set is chosen orthonormal in the sense [MES 99, COH 77]:

$$\langle \alpha \, | \beta \rangle = \delta(\alpha - \beta) \qquad [5.95]$$

where δ is the Dirac delta function. The completeness relation reads:

$$\int d\alpha \, |\alpha\rangle \, \langle \alpha| = \hat{I} \qquad [5.96]$$

3 In many systems, states will be actually labeled by an integer p (e.g. the subband index) and a continuous parameter α (e.g. the wavevector). All the equations of this chapter remain valid if one replaces $\int d\alpha$ by $\sum_p \int d\alpha$.

Hence, the Green's operator can be expressed as:

$$\hat{G}^r_\eta(E) = \int d\alpha \frac{|\alpha\rangle \langle\alpha|}{E + i\eta - E_\alpha} \qquad [5.97]$$

Using the spatial representation $\phi_\alpha(\mathbf{r}) = \langle \mathbf{r}|\alpha\rangle$ and taking the limit $\eta \to 0^+$ yields:

$$\begin{aligned} G^r(\mathbf{r},\mathbf{r}';E) &= \lim_{\eta\to 0^+} G^r_\eta(\mathbf{r},\mathbf{r}';E) \qquad [5.98] \\ &= \mathcal{P}\int d\alpha \frac{\phi_\alpha(\mathbf{r})\phi^*_\alpha(\mathbf{r}')}{E - E_\alpha} - i\pi \int d\alpha \phi_\alpha(\mathbf{r})\phi^*_\alpha(\mathbf{r}')\delta(E - E_\alpha) \end{aligned}$$

Assuming that $\phi_\alpha(\mathbf{r})$ varies smoothly with α for any $\mathbf{r}$, both integrals are convergent at the values of α such that $E_\alpha = E$ since the Cauchy principal value removes the singularities. Hence, the limit $G^r(\mathbf{r},\mathbf{r}';E)$ is finite. The proof can be generalized to a continuum of eigenstates labelled by several parameters. For instance, the integration over the three components of the wavevector $\mathbf{k}$ can be reduced to 1D integrals using spherical coordinates [MES 99]. Again, the singularities are removed. An important consequence of the finiteness of the Green's function is the following. Applying $\hat{G}^r_\eta(E)$ to some state $|\psi\rangle$ localized in some finite region of space Ω, we obtain a finite limit for $\eta \to 0^+$:

$$\langle \mathbf{r}| \hat{G}^r(E) |\psi\rangle = \int_\Omega d^3r' G^r(\mathbf{r},\mathbf{r}';E)\psi(\mathbf{r}') \qquad [5.99]$$

In contrast, applying $\hat{G}^r_\eta(E)$ to an eigenstate $|\alpha\rangle$ (which is extended in the whole space) yields an infinite limit. However, we abusively define:

$$\hat{G}^{r,a}(E) = \lim_{\eta\to 0^+} \hat{G}^{r,a}_\eta(E) \qquad [5.100]$$

keeping in mind that these operators are singular. Hence, we must be careful when multiplying $\hat{G}^{r,a}(E)$ with a state or with another operator [MES 99]. As shown by equation [5.99], the product with "finite range" quantities is well defined. In case of doubt, we must go back to $\hat{G}^{r,a}_\eta(E)$ and examine carefully the convergence when $\eta \to 0^+$. From the limit of [5.82], we also have the relation:

$$\hat{G}^a(E) = \left[\hat{G}^r(E)\right]^\dagger \qquad [5.101]$$

From [5.98], we obtain the expression of the local density of states for an infinite system:

$$\rho(\mathbf{r}, E) = -g_s \frac{1}{\pi} \mathcal{I}m\, G^r(\mathbf{r}, \mathbf{r}; E) = g_s \int d\alpha\, |\phi_\alpha(\mathbf{r})|^2\, \delta(E - E_\alpha) \qquad [5.102]$$

For illustration, let us apply this formula to the 1D Hamiltonian [5.10] with $V(x) = 0$. A complete orthonormal basis set of eigenstates is provided by the plane waves $|k\rangle$:

$$\langle x|k\rangle = \frac{1}{\sqrt{2\pi}} e^{ikx} \qquad [5.103]$$

with the orthonormality relation $\langle k|k'\rangle = \delta(k - k')$. Hence:

$$\rho(x, E) = g_s \frac{1}{2\pi} \int_{-\infty}^{+\infty} dk \delta\left[E - E(k)\right] \qquad [5.104]$$

with $E(k) = \hbar^2 k^2/2m$. For $E > 0$, the equation $E(k) = E$ has two solutions $\pm\kappa = \pm\sqrt{2mE}/\hbar$. Linearizing $E(k)$ around κ and $-\kappa$, and using the identity $\delta(ax) = \delta(x)/|a|$, we obtain:

$$\rho(x, E) = g_s \frac{\sqrt{m}}{\sqrt{2}\pi\hbar} \frac{1}{\sqrt{E}} \qquad [5.105]$$

This is the density of states of a 1D system, with the Van Hove singularity in $1/\sqrt{E}$ at the band minimum. It is twice the density of states [5.53], which only included the states with positive velocity.

In systems with quantum wells (heterostructures, attractive impurity potentials), there can be localized states with a discrete energy spectrum in addition to the continuum of states. These states must be added to equation [5.97] and they appear as Dirac delta functions in the density of states, as for a finite system. The Green's function is singular at these energies and a small imaginary part η must be kept in the calculations in order to avoid numerical divergences. An example will be shown in section 5.3.4.6.

Calculating the density of states was a first example of application of the Green's function. More generally, one defines the spectral density operator:

$$\hat{A}(E) = \frac{i}{2\pi} \left[\hat{G}^r(E) - \hat{G}^r(E)^\dagger\right] = \frac{i}{2\pi} \left[\hat{G}^r(E) - \hat{G}^a(E)\right] \qquad [5.106]$$

Equation [5.97] gives:

$$\hat{A}(E) = \int d\alpha \, |\alpha\rangle \langle\alpha| \, \delta(E - E_\alpha) \tag{5.107}$$

$\hat{A}(E)$ is the "projector" on states of energy E. It allows evaluating the spectral density of any observable $\hat{B}$ at energy E, defined as:

$$b(E) = g_s \int d\alpha \, \langle\alpha| \, \hat{B} \, |\alpha\rangle \, \delta(E - E_\alpha) \tag{5.108}$$

$b(E)$ gives the contribution of states of energy E to the observable $\hat{B}$. Indeed, for any energy interval $[E_1, E_2]$,

$$\int_{E_1}^{E_2} b(E)dE = \int_{\alpha \, / \, E_\alpha \in [E_1, E_2]} d\alpha \, \langle\alpha| \, \hat{B} \, |\alpha\rangle \tag{5.109}$$

$b(E)$ is related to $\hat{A}(E)$ by:

$$b(E) = g_s \mathrm{Tr} \left[\hat{A}(E) \hat{B} \right] \tag{5.110}$$

where Tr denotes the trace operator. The equivalence with [5.108] is proven by expanding the trace on the orthonormal basis of eigenstates and by using the expression [5.107] for $\hat{A}(E)$:

$$\begin{aligned}
\mathrm{Tr} \left[\hat{A}(E) \hat{B} \right] &= \int d\beta \, \langle\beta| \left[\int d\alpha \, |\alpha\rangle \langle\alpha| \, \delta(E - E_\alpha) \right] \hat{B} \, |\beta\rangle \\
&= \int d\alpha \int d\beta \delta(\alpha - \beta) \, \langle\alpha| \, \hat{B} \, |\beta\rangle \, \delta(E - E_\alpha) \\
&= \int d\alpha \, \langle\alpha| \, \hat{B} \, |\alpha\rangle \, \delta(E - E_\alpha)
\end{aligned} \tag{5.111}$$

As a first example, with $\hat{B} = |\mathbf{r}\rangle \langle\mathbf{r}|$, equation [5.110] yields the local density of states. Equation [5.110] will be used later to calculate the spectral density of another important observable: the current operator.

5.3.1.3.3. Dyson equation

The Hamiltonian $\hat{H}$ may be expressed as the sum of a simpler Hamiltonian $\hat{H}_0$, of which we know the Green's function, and a potential:

$$\hat{H} = \hat{H}_0 + \hat{V} \qquad [5.112]$$

The Green's operator $\hat{G}^r(E)$ of the Hamiltonian $\hat{H}$ can then be obtained from the Green's operator $\hat{G}^{0r}(E)$ of $\hat{H}_0$:

$$\begin{aligned}
\hat{G}^r_\eta(E) &= \left[(E+i\eta)\hat{I} - \hat{H}\right]^{-1} \\
&= \left[(E+i\eta)\hat{I} - \hat{H}_0\right]^{-1} \left[(E+i\eta)\hat{I} - \hat{H}_0\right] \left[(E+i\eta)\hat{I} - \hat{H}\right]^{-1} \\
&= \left[(E+i\eta)\hat{I} - \hat{H}_0\right]^{-1} \left[(E+i\eta)\hat{I} - \hat{H} + \hat{V}\right] \left[(E+i\eta)\hat{I} - \hat{H}\right]^{-1} \\
&= \left[(E+i\eta)\hat{I} - \hat{H}_0\right]^{-1} \left\{\hat{I} + \hat{V}\left[(E+i\eta)\hat{I} - \hat{H}\right]^{-1}\right\} \\
&= \hat{G}^{0r}_\eta(E) + \hat{G}^{0r}_\eta(E)\hat{V}\hat{G}^r_\eta(E)
\end{aligned} \qquad [5.113]$$

At this point, the operators are well defined because of the small imaginary part η. If the potential $\hat{V}$ has finite range and if [5.113] is projected onto a state $|\psi\rangle$, which is also localized in some region of space, then the limit $\eta \to 0^+$ is well defined. This can be proven as follows. From equation [5.99], since $\psi(\mathbf{r})$ is spatially localized, the quantity $\langle\mathbf{r}|\,\hat{G}^r\,|\psi\rangle$ is finite for all $\mathbf{r}$. However, it does not decay at large $\mathbf{r}$ since, as will be discussed below, the retarded Green's functions is related to extended outgoing waves (e.g. plane waves in 1D and spherical waves in 3D). Fortunately, since $\hat{V}$ has finite range, $\hat{V}\hat{G}^r\,|\psi\rangle$ is localized in the region of influence of $\hat{V}$. Hence, the application of $\hat{G}^{0r}(E)$ to this vector yields a finite result, which proves the convergence of the last term in [5.113]. Finally, we write formally:

$$\hat{G}^r(E) = \hat{G}^{0r}(E) + \hat{G}^{0r}(E)\hat{V}\hat{G}^r(E) \qquad [5.114]$$

keeping in mind that this relation does not always make sense when it is applied to extended states. This relation is called the "Dyson equation", even if it simply relies on basic linear algebra. Dyson's genuine contribution is found in the many-body Green's functions formalism presented in the following chapter.

The only hypothesis made here is that $\hat{V}$ has finite range or at least decays sufficiently fast at long distance [MES 99]. It is important to note that equation [5.114] remains valid even if $\hat{V}$ is large. In contrast, the series

$$\hat{G}^r = \hat{G}^{0r} + \hat{G}^{0r}\hat{V}\hat{G}^{0r} + \hat{G}^{0r}\hat{V}\hat{G}^{0r}\hat{V}\hat{G}^{0r} + ... \quad [5.115]$$

obtained by iteration of [5.114] is not necessarily convergent [MES 99]. If $\hat{V}$ is small enough, we often use the first Born approximation that consists of keeping only the first term in $\hat{V}$. We will instead use the exact expression [5.114] to calculate the scattering of electrons by any localized potential, even if it is strong, without making any perturbative approximation.

A derivation similar to [5.113] leads to an alternative form of the Dyson equation, which will be used later:

$$\hat{G}^r(E) = \hat{G}^{0r}(E) + \hat{G}^r(E)\hat{V}\hat{G}^{0r}(E) \quad [5.116]$$

5.3.1.3.4. Analytical expression of the 1D Green's function

For the 1D Schrödinger equation [5.10] with uniform potential $V(x) = V_0$, it is easy to find the analytical expression of the Green's function. Equations [5.97] and [5.103] yield:

$$G^r_\eta(x, x'; E) = \frac{1}{2\pi}\int_{-\infty}^{+\infty} dk \frac{e^{ik(x-x')}}{E + i\eta - E(k)} \quad [5.117]$$

with $E(k) = V_0 + \hbar^2k^2/2m$. The integral can be calculated using a contour in complex k-plane, since the integrand is analytic in the whole plane except at the zeros of the denominator. For $E > V_0$, the integrand has two poles close to $\pm\kappa = \pm\sqrt{2m(E - V_0)}/\hbar$. Around κ, we have:

$$E(k) \simeq E + \left.\frac{\mathrm{d}E(k)}{\mathrm{d}k}\right|_{k=\kappa} (k - \kappa) = E + \hbar v(\kappa)(k - \kappa) \quad [5.118]$$

where $v(\kappa) = \hbar\kappa/m$ is the group velocity. Hence, for positive infinitesimal η, the pole lies in the upper half-plane at $k \simeq \kappa + i\eta/\hbar v(\kappa)$. Similarly, the pole close to $-\kappa$ lies in the lower half-plane at $k \simeq -\kappa + i\eta/\hbar v(-\kappa)$. For $x > x'$, the contour of integration shown in Figure 5.10 is used. At large K, the integrals on the vertical segments decrease as $1/K$, and the integral on the horizontal segment decreases as $K \times e^{-K(x-x')}$. Hence, for $K \to +\infty$, the Green's function reduces to the residue of the pole located in the upper half-plane. For $x < x'$, a similar contour in the lower half-plane is used. We obtain, for any x and x':

$$G^r(x, x'; E) = -\frac{i}{\hbar v(\kappa)} e^{i\kappa|x-x'|} = -\frac{im}{\hbar^2 \kappa} e^{i\kappa|x-x'|} \text{ for } E > V_0 \quad [5.119]$$

$$\kappa = \sqrt{2m(E - V_0)}/\hbar \quad [5.120]$$

hence $G^r(x, x'; E)$ only contains outgoing plane waves emitted from point x'. This can be inferred from the interpretation of the time-dependent Green's function [5.71] as a wavepacket outgoing from a point source (see Figure 5.9): the energy Green's function yields the component of energy E of this wavepacket.

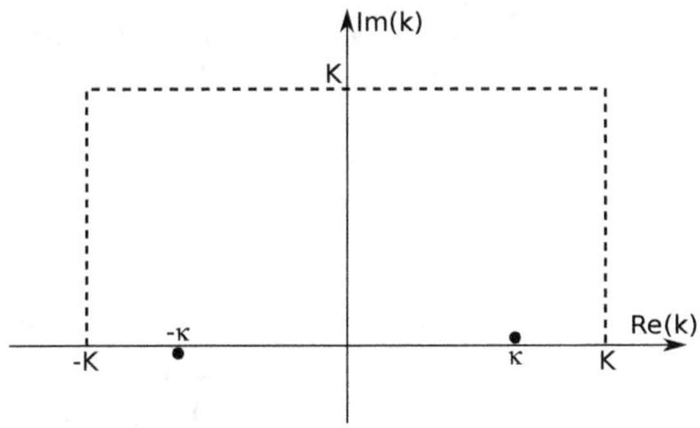

Figure 5.10. *Contour in complex k-plane used for calculating the 1D retarded Green's function for $x > x'$. The dots indicate the poles in the propagating case. The integral on the contour vanishes for $K \to \infty$ and only the residue of the pole located above the real axis remains*

For $E < V_0$, there are two poles on the imaginary axis. We obtain:

$$G^r(x, x'; E) = -\frac{m}{\hbar^2 \kappa} e^{-\kappa|x-x'|} \text{ for } E < V_0 \quad [5.121]$$

$$\kappa = \sqrt{2m(V_0 - E)}/\hbar \quad [5.122]$$

which yields evanescent waves decaying from both sides of the source at x'. Finally, we can check that the obtained expressions satisfy equation [5.85] in 1D, since the discontinuity of their derivative at $x = x'$ yields the Dirac delta function $\delta(x - x')$ in their second derivative with the correct prefactor.

5.3.1.3.5. Outgoing waves

The contour integration method used above highlights an important and general property: the retarded Green's function only contains outgoing waves from a chosen source point. For more realistic models such as tight-binding Hamiltonians, it remains possible to define the analytic continuation of the band structure and Bloch states in complex k-plane [KOH 59, CHA 82]. At a given energy, we can have poles on the real axis (propagating states) together with poles off the real axis (evanescent states). Choosing the contour in Figure 5.11[4], but with an horizontal segment at $\mathcal{I}m\,(k) = \kappa_{min}$ just below the "evanescent" poles, the integral on this segment becomes negligible for $(x - x') \gg 1/\kappa_{min}$. The Green's function is then the sum of the residues of the poles on the real axis. Equation [5.118] remains valid: for infinitesimal $\eta > 0$, the poles corresponding to positive group velocities $v(\kappa) > 0$ are in the upper half-plane and hence contribute to the Green's function for $x > x'$ only. The poles with negative group velocities contribute for $x < x'$ only. Hence, far from the source ($|x - x'| \gg 1/\kappa_{min}$), the retarded Green's function is only composed of outgoing waves. $1/\kappa_{min}$ is the largest decay length of the evanescent states at energy E. Such reasoning can be generalized in 2D and 3D. For instance, for the 3D effective mass Schrödinger equation, the retarded Green's function yields spherical waves outgoing from each source point [MES 99].

5.3.1.4. *Building the scattering states: the Lippmann–Schwinger equation*

We consider an infinite, crystalline system described by a Hamiltonian $\hat{H}_0$, modified in some finite region by a potential $\hat{V}$. The basic properties of the retarded Green's function, discussed in the previous sections, allow building eigenstates composed of a chosen incoming wave, a reflected wave and a transmitted wave. Such states, called "scattering states", have already been defined in Chapter 1 to calculate the quantum transmission through a square barrier. The Green's function allows building them for any scattering potential and also for 2D and 3D systems.

Given an incoming wave $|\varphi\rangle$, the corresponding scattering state $|\psi\rangle$ is given by the Lippmann–Schwinger equation [LIP 50]:

$$|\psi\rangle = |\varphi\rangle + \hat{G}^{0r}(E)\hat{V}\,|\psi\rangle \qquad [5.123]$$

where $\hat{G}^{0r}(E)$ is the retarded Green's operator of the unperturbed Hamiltonian $\hat{H}_0$. Equation [5.123] is not an explicit expression of $|\psi\rangle$, but it has all the required properties. First, far from the perturbed region and from eventual evanescent states localized around this region, the term $\hat{G}^{0r}(E)\hat{V}\,|\psi\rangle$ only contains outgoing waves, as proved in the previous section. The only incoming wave comes from the first term

4 Not the contour in Figure 5.10 since, for a 1D crystal, the vertical segments of the contour are conveniently chosen at the boundaries of the first Brillouin zone.

$|\varphi\rangle$. Second, the state $|\psi\rangle$ satisfying equation [5.123] is an eigenstate of energy E of the full Hamiltonian $\hat{H}$. This is shown by multiplyling the equation by $(E\hat{I} - \hat{H}_0)$. The first term of the right-hand side vanishes since $|\varphi\rangle$ is an eigenstate of $\hat{H}_0$. For the second term, we use the identity $(E\hat{I} - \hat{H}_0)\hat{G}^{0r}(E) = \hat{I}$, which makes sense here since it is projected on the localized state $\hat{V}|\psi\rangle$ (see the discussion following equation [5.100]). We finally obtain: $\hat{H}|\psi\rangle = E|\psi\rangle$.

An explicit expression of $|\psi\rangle$ is sometimes needed; [5.123] is rewritten as:

$$\left[\hat{I} - \hat{G}^{0r}(E)\hat{V}\right]|\psi\rangle = |\varphi\rangle \qquad [5.124]$$

Multiplying both sides of this equation by $\left[\hat{I} + \hat{G}^{r}(E)\hat{V}\right]$ on the left, where $\hat{G}^{r}(E)$ is the Green's operator of the full Hamiltonian, we obtain for the left-hand side:

$$\begin{aligned}(\hat{I} + \hat{G}^{r}\hat{V})(\hat{I} - \hat{G}^{0r}\hat{V})|\psi\rangle &= |\psi\rangle + \left(\hat{G}^{r} - \hat{G}^{0r} - \hat{G}^{r}\hat{V}\hat{G}^{0r}\right)\hat{V}|\psi\rangle \\ &= |\psi\rangle \end{aligned} \qquad [5.125]$$

where we have omitted the energy variable to simplify the notations, and used the Dyson equation [5.116] to cancel the last term. Finally, we obtain the explicit form for the scattering state:

$$|\psi\rangle = |\varphi\rangle + \hat{G}^{r}(E)\hat{V}|\varphi\rangle \qquad [5.126]$$

The Lippmann–Schwinger equation is very general and widely used in many domains such as the theory of collisions in particle physics or the propagation of electromagnetic fields in waveguides [LIP 50]. As with the Dyson equation, it is exact whatever the magnitude of the potential, provided that it is a short-range potential.

5.3.2. *Calculation of the transmission through a conductor connected to two electrodes*

5.3.2.1. *Decomposition of the system and contacts self-energies*

We consider a conductor "C" of finite length connected to two semi-infinite electrodes "L" (left) and "R" (right). The electrodes are assumed periodic so that we can define incoming and outgoing Bloch waves[5]. Each of the three regions is

5 Hence the contacts are assumed ideal, but this is not a true limitation of the method. To account for non-ideal contacts, we can include them in the conductor region and define the periodic electrodes beyond them. Of course, this increases the computational cost.

described by its own basis of states (which we do not specify yet). Hence, we can decompose the Hamiltonian $\hat{H}$ into operators acting in each region or coupling two different regions. We assume that the two electrodes are not directly coupled, so that $\hat{H}$ takes the block matrix form:

$$\hat{H} = \begin{bmatrix} \hat{H}_L & \hat{V}_{LC} & \hat{0} \\ \hat{V}_{CL} & \hat{H}_C & \hat{V}_{CR} \\ \hat{0} & \hat{V}_{RC} & \hat{H}_R \end{bmatrix} \quad [5.127]$$

where $\hat{H}_L$, $\hat{H}_C$ and $\hat{H}_R$ are the Hamiltonians of the isolated parts, and $\hat{V}_{CL}$ and $\hat{V}_{CR}$ are the couplings of the conductor with the left and right electrodes, respectively. By hermiticity of $\hat{H}$, we have: $\hat{V}_{LC} = \hat{V}_{CL}^{\dagger}$ and $\hat{V}_{RC} = \hat{V}_{CR}^{\dagger}$. Note that the blocks $\hat{H}_L$ and $\hat{H}_R$ are of infinite dimension, but will be tractable because of the periodicity of the electrodes.

Let us also decompose the retarded Green's function of the full system into blocks:

$$\hat{G}^r = \begin{bmatrix} \hat{G}_L^r & \hat{G}_{LC}^r & \hat{G}_{LR}^r \\ \hat{G}_{CL}^r & \hat{G}_C^r & \hat{G}_{CR}^r \\ \hat{G}_{RL}^r & \hat{G}_{RC}^r & \hat{G}_R^r \end{bmatrix} \quad [5.128]$$

In the following, we will mainly need $\hat{G}_C^r$, which is the restriction of $\hat{G}^r$ to the finite conductor. Using the Dyson equation [5.114] with

$$\hat{H}_0 = \begin{bmatrix} \hat{H}_L & \hat{0} & \hat{0} \\ \hat{0} & \hat{H}_C & \hat{0} \\ \hat{0} & \hat{0} & \hat{H}_R \end{bmatrix} \text{ and } \hat{V} = \begin{bmatrix} \hat{0} & \hat{V}_{LC} & \hat{0} \\ \hat{V}_{CL} & \hat{0} & \hat{V}_{CR} \\ \hat{0} & \hat{V}_{RC} & \hat{0} \end{bmatrix} \quad [5.129]$$

we obtain in particular:

$$\begin{aligned} \hat{G}_C^r &= \hat{G}_C^{0r} + \hat{G}_C^{0r}\hat{V}_{CL}\hat{G}_{LC}^r + \hat{G}_C^{0r}\hat{V}_{CR}\hat{G}_{RC}^r \\ \hat{G}_{LC}^r &= \hat{G}_L^{0r}\hat{V}_{LC}\hat{G}_C^r \\ \hat{G}_{RC}^r &= \hat{G}_R^{0r}\hat{V}_{RC}\hat{G}_C^r \end{aligned} \quad [5.130]$$

where $\hat{G}_C^{0r}$, $\hat{G}_L^{0r}$ and $\hat{G}_R^{0r}$ are the Green's functions of the isolated parts. Substituting the two last equations into the first one, and defining the contacts "self-energies",

$$\hat{\Sigma}_L(E) = \hat{V}_{CL}\hat{G}_L^{0r}(E)\hat{V}_{LC} \text{ and } \hat{\Sigma}_R(E) = \hat{V}_{CR}\hat{G}_R^{0r}(E)\hat{V}_{RC} \quad [5.131]$$

we obtain:

$$\hat{G}_C^r = \hat{G}_C^{0r} + \hat{G}_C^{0r}\hat{\Sigma}_L\hat{G}_C^r + \hat{G}_C^{0r}\hat{\Sigma}_R\hat{G}_C^r \qquad [5.132]$$

Multiplying both sides by $(\hat{G}_C^{0r})^{-1}$ on the left and by $(\hat{G}_C^r)^{-1}$ on the right, we finally obtain:

$$\hat{G}_C^r(E) = \left[E\hat{I} - \hat{H}_C - \hat{\Sigma}_L(E) - \hat{\Sigma}_R(E)\right]^{-1} \qquad [5.133]$$

To be more explicit, we have respecified the energy dependences. To conclude, the Green's function $\hat{G}_C^r$ is obtained by replacing the Hamiltonian $\hat{H}_C$ of the isolated conductor by $\hat{H}_C + \hat{\Sigma}_L + \hat{\Sigma}_R$, where the self-energies $\hat{\Sigma}_L$ and $\hat{\Sigma}_R$ account for the coupling with the semi-infinite contacts[6]. Hence, while the system is infinite, the calculation of $\hat{G}_C^r$ reduces to the inversion of a matrix of finite size.

5.3.2.2. *Scattering states*

Our treatment of the quantum transmission problem in terms of scattering states is largely inspired by [DAT 05]. We use the Lippmann–Schwinger equation [5.123] with the Hamiltonian $\hat{H}_0$ and the potential $\hat{V}$ of equations [5.129]. $\hat{V}$ is a peculiar potential but the Lippmann–Schwinger equation remains valid provided that $\hat{V}$ is spatially localized and that we start from an eigenstate of $\hat{H}_0$. We choose here an eigenstate $|\varphi_L\rangle$ of energy E of the isolated semi-infinite left electrode: $\hat{H}_L|\varphi_L\rangle = E|\varphi_L\rangle$. It is also eigenstate of $\hat{H}_0$ with zero components in the conductor and the right electrode. It is the sum of an incident wave and a reflected wave of the same amplitude, since the electrode boundary, when disconnected from the conductor, totally reflects the incident wave. We obtain the following scattering state $|\psi_L\rangle$ decomposed here into its components within the three regions:

$$|\psi_L\rangle = \begin{bmatrix} |\psi_{L,L}\rangle \\ |\psi_{L,C}\rangle \\ |\psi_{L,R}\rangle \end{bmatrix} = \begin{bmatrix} |\varphi_L\rangle + \hat{G}_L^{0r}(E)\hat{V}_{LC}|\psi_{L,C}\rangle \\ \hat{G}_C^r(E)\hat{V}_{CL}|\varphi_L\rangle \\ \hat{G}_R^{0r}(E)\hat{V}_{RC}|\psi_{L,C}\rangle \end{bmatrix} \qquad [5.134]$$

where, for convenience, we have used the explicit form [5.126] of the Lippmann–Schwinger equation to express $|\psi_{L,C}\rangle$, and its implicit form [5.123] to express $|\psi_{L,L}\rangle$ and $|\psi_{L,R}\rangle$. In the left electrode, the term $\hat{G}_L^{0r}(E)\hat{V}_{LC}|\psi_{L,C}\rangle$ is at an outgoing wave coming from the conductor. Its interference with the outgoing component of $|\varphi_L\rangle$ yields the reflected wave. In the right electrode, $|\psi_{L,R}\rangle = \hat{G}_R^{0r}(E)\hat{V}_{RC}|\psi_{L,C}\rangle$ is

6 The indexes "L" and "R" of the self-energies are somehow misleading, since these operator act in the subspace "C" of the finite conductor. However, we keep this standard notation.

an outgoing wave coming from the conductor, hence a transmitted wave. Hence, as expected, $|\psi_L\rangle$ is the scattering state associated with the incident part of $|\varphi_L\rangle$ coming from the left electrode. Scattering states $|\psi_R\rangle$ associated with incident waves coming from the right electrode can be constructed in the same way.

The scattering states can be defined so that they form an orthonormal basis containing all propagating states of the system. For this purpose, we start from orthonormal basis of eigenstates $\{|\varphi_{L\alpha}\rangle\}$ and $\{|\varphi_{R\alpha}\rangle\}$ for the left and right electrodes, respectively, indexed by a continuous parameter α. The orthonormality relations read[7]:

$$\begin{aligned} \langle\varphi_{L\alpha}\,|\varphi_{L\beta}\rangle &= \delta(\alpha-\beta) \\ \langle\varphi_{R\alpha}\,|\varphi_{R\beta}\rangle &= \delta(\alpha-\beta) \\ \langle\varphi_{L\alpha}\,|\varphi_{R\beta}\rangle &= 0 \end{aligned} \qquad [5.135]$$

The corresponding scattering states $|\psi_{L\alpha}\rangle$ (coming from the left) and $|\psi_{R\alpha}\rangle$ (coming from the right) satisfy the following orthonormality relations:

$$\begin{aligned} \langle\psi_{L\alpha}\,|\psi_{L\beta}\rangle &= \delta(\alpha-\beta) \\ \langle\psi_{R\alpha}\,|\psi_{R\beta}\rangle &= \delta(\alpha-\beta) \\ \langle\psi_{L\alpha}\,|\psi_{R\beta}\rangle &= 0 \end{aligned} \qquad [5.136]$$

This is a general property of the Lippmann–Schwinger equation: it preserves the orthonormality of the incident states. We will not prove it rigorously in this chapter. However, it can be inferred by considering any two normalized wavepackets, orthogonal to each other, propagating in the system [DAT 05]. As shown in section 5.1.1.1, they remain orthogonal and normalized during time evolution. The scattering states can be viewed as such orthogonal wavepackets with a spatial extension tending to infinity. A more rigorous proof can be found in [WEI 95]. This property will be important in the following. It implies in particular that two scattering states with the same energy, one coming from the left electrode and the other coming from the right electrode, are orthogonal to each other. Hence, both states can be occupied without violating the Pauli exclusion principle.

Apart from scattering states, there can also be localized states in quantum wells of the conductor, at energies below the band edges of the electrodes. These states are decoupled from the continuum of states and they do not contribute to the current. An example will be shown in section 5.3.4.6.

7 As already discussed in the footnote of section 5.3.1.3.2, all the equations remain valid in multichannel systems. The states are indexed by p, α and $\int d\alpha$ is replaced by $\sum_p \int d\alpha$. The Krönecker delta δ_{pq} appears in the orthonormality relations, since two different channels p and q are orthogonal to each other.

5.3.2.3. *Charge density*

Having defined an orthonormal basis of scattering states $|\psi_{L\alpha}\rangle$ and $|\psi_{R\alpha}\rangle$, we can calculate their contribution to the local density of states. The contribution of the "left" states at a point $\mathbf{r}$ in the conductor is:

$$\rho_L(\mathbf{r}, E) = g_s \int d\alpha \, \langle \psi_{L\alpha} | \mathbf{r} \rangle \langle \mathbf{r} | \psi_{L\alpha} \rangle \, \delta(E - E_\alpha) \qquad [5.137]$$

Replacing $|\psi_{L\alpha}\rangle$ by its component [5.134] in the conductor, we obtain:

$$\rho_L(\mathbf{r}, E) =$$
$$g_s \int d\alpha \, \langle \mathbf{r} | \hat{G}_C^r(E_\alpha) \hat{V}_{CL} | \varphi_{L\alpha} \rangle \langle \varphi_{L\alpha} | \hat{V}_{LC} \hat{G}_C^a(E_\alpha) | \mathbf{r} \rangle \, \delta(E - E_\alpha) \qquad [5.138]$$

where we have replaced the hermitic conjugate of $\hat{G}_C^r$ by the advanced Green's function $\hat{G}_C^a$. The Dirac delta function allows replacing E_α by E in the Green's functions, which are then moved out of the integral:

$$\rho_L(\mathbf{r}, E) =$$
$$g_s \langle \mathbf{r} | \hat{G}_C^r(E) \hat{V}_{CL} \left[\int d\alpha \, |\varphi_{L\alpha}\rangle \langle \varphi_{L\alpha}| \, \delta(E - E_\alpha) \right] \hat{V}_{LC} \hat{G}_C^a(E) | \mathbf{r} \rangle \qquad [5.139]$$

The term in brackets is the spectral density operator [5.107] of the isolated left electrode. Similarly to [5.106], it can be expressed as:

$$\hat{A}_L(E) = \frac{i}{2\pi} \left[\hat{G}_L^{0r}(E) - \hat{G}_L^{0a}(E) \right] \qquad [5.140]$$

After some substitutions, we finally obtain:

$$\rho_L(\mathbf{r}, E) = \frac{g_s}{2\pi} \langle \mathbf{r} | \hat{G}_C^r(E) \hat{\Gamma}_L(E) \hat{G}_C^a(E) | \mathbf{r} \rangle \qquad [5.141]$$

where $\hat{\Gamma}_L(E)$ is related to the left electrode self-energy [5.131]:

$$\hat{\Gamma}_L(E) = i \left[\hat{\Sigma}_L(E) - \hat{\Sigma}_L(E)^\dagger \right] \qquad [5.142]$$

Similarly, the contribution of the "right" states to the local density of states is:

$$\rho_R(\mathbf{r}, E) = \frac{g_s}{2\pi} \langle \mathbf{r} | \hat{G}_C^r(E) \hat{\Gamma}_R(E) \hat{G}_C^a(E) | \mathbf{r} \rangle \qquad [5.143]$$

with

$$\hat{\Gamma}_R(E) = i\left[\hat{\Sigma}_R(E) - \hat{\Sigma}_R(E)^\dagger\right] \qquad [5.144]$$

Summing equations [5.141] and [5.143], and noticing by inversion of [5.133] that

$$\hat{\Gamma}_L + \hat{\Gamma}_R = i\left[(\hat{G}^a_C)^{-1} - (\hat{G}^r_C)^{-1}\right] \qquad [5.145]$$

we obtain:

$$\begin{aligned}\rho_L(\mathbf{r}, E) + \rho_R(\mathbf{r}, E) &= \frac{ig_s}{2\pi}\langle \mathbf{r}|\,\hat{G}^r_C\left[(\hat{G}^a_C)^{-1} - (\hat{G}^r_C)^{-1}\right]\hat{G}^a_C\,|\mathbf{r}\rangle \\ &= -\frac{g_s}{\pi}\mathcal{I}m\, G^r_C(\mathbf{r}, \mathbf{r}; E)\end{aligned} \qquad [5.146]$$

We recover expression [5.102] for the density of states. However, [5.141] and [5.143] only include propagating states and do not include possible bound states in the conductor. Indeed, we will see later that $\hat{\Gamma}_L(E)$ and $\hat{\Gamma}_R(E)$ are zero at energies below the band edges of the electrodes. At the discrete energy levels of the bound states, the Green's function is singular, as discussed in section 5.3.1.3.2. Multiplying it by the zero $\hat{\Gamma}$ operators leads to inconsistent results. In order to account for the bound states, the density of states must be calculated directly from the imaginary part of $G^r_\eta(\mathbf{r}, \mathbf{r}; E)$, keeping a finite η to smooth the Dirac delta functions.

To obtain the total propagating electron density in the conductor, we sum the contributions from the left and right electrodes with the Fermi–Dirac factors. This yields:

$$n(\mathbf{r}) = \int dE\left[\rho_L(\mathbf{r}, E)f(E - \mu_L) + \rho_R(\mathbf{r}, E)f(E - \mu_R)\right] \qquad [5.147]$$

In device simulations, the carrier density $n(\mathbf{r})$ is injected into the Poisson equation, yielding a modification of the electrostatic potential, hence of the Hamiltonian. Then, the Green's function and density must be recomputed until self-consistence is achieved.

5.3.2.4. *Current operator*

Before calculating the electrical current through the conductor, it is necessary to introduce the current operator. A way to find its expression is to write the continuity equation (electron density conservation) [CAR 71]. Let us consider a system divided into two parts A and B, described by the Hamiltonian:

$$\hat{H} = \begin{bmatrix} \hat{H}_A & \hat{V}_{AB} \\ \hat{V}_{BA} & \hat{H}_B \end{bmatrix} \qquad [5.148]$$

and let us consider a wavepacket $|\psi(t)\rangle$ which we decompose into its components in the two subsystems:

$$|\psi(t)\rangle = \begin{bmatrix} |\psi_A(t)\rangle \\ |\psi_B(t)\rangle \end{bmatrix} \qquad [5.149]$$

From the time-dependent Schrödinger equation [5.1], we obtain:

$$\frac{\mathrm{d}\,\langle\psi_B(t)\;|\psi_B(t)\rangle}{\mathrm{d}t} = -\frac{\mathrm{d}\,\langle\psi_A(t)\;|\psi_A(t)\rangle}{\mathrm{d}t} = \langle\psi(t)|\,\frac{i}{\hbar}\begin{bmatrix} \hat{0} & \hat{V}_{AB} \\ -\hat{V}_{BA} & \hat{0} \end{bmatrix}|\psi(t)\rangle \qquad [5.150]$$

The derivatives define the transition rate of the probability density of the wavepacket through the interface between subsystems A and B. The current operator $\hat{J}_{AB}$ at this interface is thus identified as the term in brackets in the right-hand side, multiplied by the electron charge $-e$:

$$\hat{J}_{AB} = -\frac{ie}{\hbar}\begin{bmatrix} \hat{0} & \hat{V}_{AB} \\ -\hat{V}_{BA} & \hat{0} \end{bmatrix} \qquad [5.151]$$

First, let us check that this result is consistent with the well-known [MES 99, COH 77] current operator of the continuous Schrödinger equation [5.10]. For the discretized Hamiltonian [5.15], the coupling V_{AB} between two neighboring sites n and $n+1$ is $-\gamma = -\hbar^2/2ma^2$, where a is the mesh spacing. For any wavefunction ψ, the expectation value of the current operator [5.151] between these two sites reads:

$$\begin{aligned}\langle\psi|\,\hat{J}_{AB}\,|\psi\rangle &= \frac{ie}{\hbar}\frac{\hbar^2}{2ma^2}\left(\psi_n^*\psi_{n+1} - \psi_{n+1}^*\psi_n\right) \\ &= \frac{ie\hbar}{2ma}\left(\psi_n^*\frac{\psi_{n+1}-\psi_n}{a} - \psi_n\frac{\psi_{n+1}^*-\psi_n^*}{a}\right)\end{aligned} \qquad [5.152]$$

To go back to the continuous wavefunction, we must identify ψ_n with $\sqrt{a}\psi(x_n)$, since $|\psi_n|^2$ and $|\psi(x_n)|^2 \times a$ both give the probability to find an electron in an interval a

around x_n. Replacing the finite differences by the continous derivatives, we obtain the well-known current formula:

$$j(x) = \frac{ie\hbar}{2m}\left(\psi^*(x)\frac{\mathrm{d}\psi(x)}{\mathrm{d}x} - \psi(x)\frac{\mathrm{d}\psi^*(x)}{\mathrm{d}x}\right) \qquad [5.153]$$

It can also be checked that for an infinite periodic system, the expectation value $\langle\phi_k|\,\hat{J}_{AB}\,|\phi_k\rangle$ of the current operator on a Bloch state gives the expected result for the current density: $-e \times n \times v(k)$, where n is the probability density and $v(k)$ is the group velocity. It is worth noting that plane waves and Bloch states are not eigenvalues of the current operator. A measurement of the current through an interface during a finite time interval yields random values. In this chapter, we only address the average current through the device, given by the expectation value of the current operator. Current noise and its relation to quantum measurement is addressed in Chapter 7.

Another important property of the current operator is that for a stationary state (for instance a diffusion state coming from one electrode), the current is conserved along the system, i.e. the expectation value of the current operator does not depend on the interface chosen to define this operator.

5.3.2.5. *Electrical current*

The spectral current density [5.56] can be calculated by summing the contribution of the scattering states coming from the left electrode:

$$j_L(E) = g_s \int d\alpha\,\langle\psi_{L\alpha}|\,\hat{J}\,|\psi_{L\alpha}\rangle\,\delta\,(E - E_\alpha) \qquad [5.154]$$

where $\hat{J}$ is the current operator through any interface of the system. Here, it is more convenient to choose the current operator $\hat{J}_{CR}$ at the interface between the conductor and the right electrode. From [5.151] and [5.134] we obtain:

$$\langle\psi_{L\alpha}|\,\hat{J}_{CR}\,|\psi_{L\alpha}\rangle = \langle\varphi_{L\alpha}|\,\hat{B}(E_\alpha)\,|\varphi_{L\alpha}\rangle + \text{ (c.c.)} \qquad [5.155]$$

where (c.c.) is the complex conjugate, and with

$$\hat{B}(E_\alpha) \equiv -\frac{ie}{\hbar}\hat{V}_{LC}\hat{G}^a_C(E_\alpha)\hat{V}_{CR}\hat{G}^{0r}_R(E_\alpha)\hat{V}_{RC}\hat{G}^r_C(E_\alpha)\hat{V}_{CL} \qquad [5.156]$$

Note that the operator $\hat{B}(E_\alpha)$ only acts within the subspace of the left electrode. Inserting [5.155] into [5.154], and applying equation [5.110] with the spectral density operator [5.140] of the left electrode, we have:

$$j_L(E) = g_s \mathrm{Tr}\left[\hat{A}_L(E)\hat{B}(E)\right] + \text{(c.c.)} \quad [5.157]$$

Using a circular permutation of the operators within the trace, we obtain:

$$j_L(E) = -g_s \frac{ie}{\hbar}\frac{1}{2\pi}\mathrm{Tr}\left[\hat{\Gamma}_L(E)\hat{G}^a_C(E)\hat{\Sigma}_R(E)\hat{G}^r_C(E)\right] + \text{(c.c.)} \quad [5.158]$$

The second term (c.c.) can be grouped with the first one by another circular permutation of the operators within the trace. We finally obtain:

$$j_L(E) = -g_s \frac{e}{h}\mathrm{Tr}\left[\hat{\Gamma}_L(E)\hat{G}^a_C(E)\hat{\Gamma}_R(E)\hat{G}^r_C(E)\right] \quad [5.159]$$

Similarly, the spectral current density coming from the right electrode reads:

$$j_R(E) = -g_s \frac{e}{h}\mathrm{Tr}\left[\hat{\Gamma}_R(E)\hat{G}^a_C(E)\hat{\Gamma}_L(E)\hat{G}^r_C(E)\right] \quad [5.160]$$

As already discussed in section 5.2.2, $j_R(E) = j_L(E)$. This is checked here by replacing $\hat{\Gamma}_L$ in both traces by its expression obtained from [5.145]:

$$\hat{\Gamma}_L = i\left[(\hat{G}^a_C)^{-1} - (\hat{G}^r_C)^{-1}\right] - \hat{\Gamma}_R \quad [5.161]$$

Then, by well-chosen circular permutations of the operators in the traces, we can prove their equality.

Finally, inserting [5.159] into [5.60], the total current reads:

$$I_{LR} = -g_s \frac{e}{h} \times \int dE \mathrm{Tr}\left[\hat{\Gamma}_L(E)\hat{G}^a_C(E)\hat{\Gamma}_R(E)\hat{G}^r_C(E)\right]\left[f(E-\mu_L) - f(E-\mu_R)\right] \quad [5.162]$$

and the trace is identified with the sum of the transmissions between the different channels in [5.56]:

$$\sum_{p\in L}\sum_{q\in R} T^L_{qp}(E) = \mathrm{Tr}\left[\hat{\Gamma}_L(E)\hat{G}^a_C(E)\hat{\Gamma}_R(E)\hat{G}^r_C(E)\right] \qquad [5.163]$$

This is the "trace formula" for quantum transmission in the coherent regime [MEI 92]. This relation will be proven in a more direct manner in the specific example of section 5.3.4.3.

5.3.3. *Application to a 1D system*

We consider the 1D discretized Hamiltonian of section 5.1.1.2.1 and we use the Green's function method to calculate the quantum transmission.

5.3.3.1. *Green's function of the ideal system*

In a uniform potential ($V_n = V_0$), an orthonormal basis of eigenstates is given by plane waves $|k\rangle$:

$$\langle n|k\rangle = \sqrt{\frac{a}{2\pi}}e^{ikna} \qquad [5.164]$$

where n is a mesh site and a is the mesh spacing. k is restricted to the first Brillouin zone $]-\pi/a, \pi/a]$ and the completeness relation reads:

$$\int_{-\pi/a}^{+\pi/a} dk\, |k\rangle\, \langle k| = \hat{I} \qquad [5.165]$$

The retarded Green's function [5.97] reads:

$$G^r_\eta(n, n'; E) = \frac{a}{2\pi}\int_{-\pi/a}^{+\pi/a} dk \frac{e^{ika(n-n')}}{E + i\eta - E(k)} \qquad [5.166]$$

with $E(k) = V_0 + 2\gamma[1 - \cos(ka)]$. As in section 5.3.1.3.4, the integrand is analytic in the complex k-plane except at some singularities. For $n \geq n'$, the contour of integration shown in Figure 5.11 is used. By periodicity of the band structure, the integrals on the two vertical segments cancel each other. The integral on the horizontal segment vanishes when $K \to +\infty$. For $n \leq n'$, a similar contour in the lower half-plane is used. We obtain:

$$G^r(n, n'; E) = \begin{cases} -\frac{i}{2\gamma \sin(\kappa a)} e^{i\kappa a|n-n'|} & \text{for } V_0 < E < V_0 + 4\gamma \\ -\frac{1}{2\gamma \sinh(\kappa a)} e^{-\kappa a|n-n'|} & \text{for } E < V_0 \\ +\frac{1}{2\gamma \sinh(\kappa a)} e^{-\kappa a|n-n'|} & \text{for } E > V_0 + 4\gamma \end{cases} \quad [5.167]$$

with

$$\kappa = \begin{cases} \frac{1}{a} \arccos\left(1 - \frac{E-V_0}{2\gamma}\right) & \text{for } V_0 < E < V_0 + 4\gamma \\ \frac{1}{a} \text{acosh}\left|1 - \frac{E-V_0}{2\gamma}\right| & \text{for } E \notin [V_0, V_0 + 4\gamma] \end{cases} \quad [5.168]$$

when $a \to 0$, $G^r(n, n'; E)/a$ tends as expected from the Green's function of the continuous Hamiltonian [5.117].

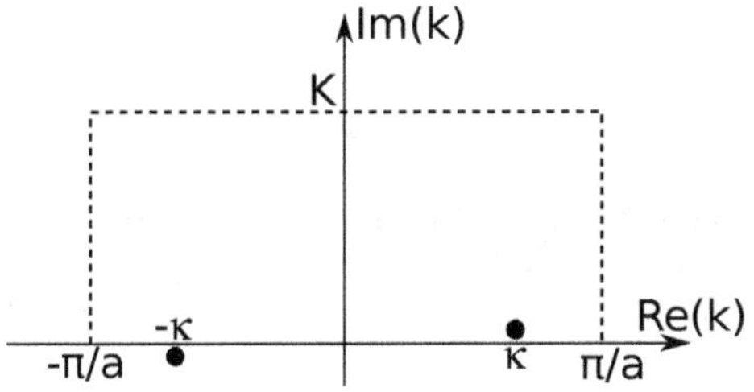

Figure 5.11. *Contour in complex k-plane used for calculating the retarded Green's function of the 1D discretized Hamiltonian for $n \geq n'$. The integral on the contour vanishes for $K \to \infty$. The dots indicate the poles in the propagating case*

5.3.3.2. *Electrodes self-energies*

A simple method to calculate the electrodes self-energies is to consider a particular decomposition of the ideal system: sites $n \leq 0$ define the left electrode, site $n = 1$ is the conductor, and sites $n \geq 2$ define the right electrode. We impose the uniform potential V_L of the right electrode in the whole system. Since the conductor has a single site, equation [5.133] becomes a scalar equation:

$$G_C^r(E) = [E - 2\gamma - V_L - \Sigma_L(E) - \Sigma_R(E)]^{-1} \quad [5.169]$$

By mirror symmetry and uniformity of the potential, $\Sigma_R(E) = \Sigma_L(E)$. Hence:

$$\Sigma_L(E) = \frac{1}{2}\left[E - 2\gamma - V_L - 1/G_C^r(E)\right] \quad [5.170]$$

$G_C^r(E)$ is the restriction to a single site of the Green's function [5.167]. We obtain:

$$\Sigma_L(E) = \begin{cases} -\gamma e^{i\kappa a} & \text{for } V_L < E < V_L + 4\gamma \\ -\gamma e^{-\kappa a} & \text{for } E < V_L \\ +\gamma e^{-\kappa a} & \text{for } E > V_L + 4\gamma \end{cases} \quad [5.171]$$

An equivalent expression for the self-energy is given in Chapter 6:

$$\Sigma_L(E) = \lim_{\eta \to 0^+} \frac{(E + i\eta - 2\gamma - V_L) + \sqrt{(E + i\eta - 2\gamma - V_L)^2 - 4\gamma^2}}{2} \quad [5.172]$$

where, whatever the energy, the complex square root is chosen with negative imaginary part. This formula is used in the code of section 5.5.

The conductor is coupled only to the first site of the left electrode via the hopping energy $-\gamma$. From [5.131], we obtain the surface Green's function of the isolated electrode:

$$\left[\hat{G}_L^{0r}(E)\right]_{00} = \frac{1}{\gamma^2}\Sigma_L(E) \quad [5.173]$$

From $\Sigma_L(E)$ one also obtains the $\Gamma_L(E)$ of equation [5.142]:

$$\Gamma_L(E) = \begin{cases} 2\gamma \sin(\kappa a) = \frac{\hbar v(\kappa)}{a} & \text{for } V_L < E < V_L + 4\gamma \\ 0 & \text{for } E \notin [V_L, V_L + 4\gamma] \end{cases} \quad [5.174]$$

All these formulas also hold for the right electrode, with possibly a different potential V_R. $\Gamma_L(E)/\hbar$ and $\Gamma_R(E)/\hbar$ can be interpreted as the escape rates per unit time of an electron from the conductor toward the left and right electrodes, respectively. This interpretation is supported by the previous equation: $\hbar/\Gamma = a/v$ is the time it takes for an electron of velocity v to go from the first site of the conductor to the neighboring site of the left electrode. It is also supported by the fact that $\Gamma_{L,R}(E)$ is related to the imaginary part of $\Sigma_{L,R}(E)$. In equation [5.133], this imaginary part is added to the conductor Hamiltonian H_C. This yields an effective non-Hermitian "Hamiltonian" with the following evolution for a state ψ of energy E inside the conductor:

$$\psi(t) = e^{-iEt/\hbar - (\Gamma_L + \Gamma_R)t/2\hbar}\psi(0) \quad [5.175]$$

Hence, $2\hbar/(\Gamma_L + \Gamma_R)$ is the lifetime of the electron inside the channel, before it escapes to one of the leads [DAT 95].

In more complex systems, there are generally no analytical formulas for the self-energies of the electrodes. However, fast and accurate numerical methods have been developed to calculate them [LOP 85, LUI 06]. They are not discussed in this chapter.

5.3.3.3. *Transmission*

We now consider a conductor with N sites $n = 1, N$ and an arbitrary potential V_n. Since the coupling is limited to first nearest neighbors, $\Sigma_L(E)$ has a single non-zero matrix element on site 1, given by [5.171]. Similarly, $\Sigma_R(E)$ has a single non-zero matrix element on site N. Hence, the trace formula [5.163] for the transmission reduces to:

$$T(E) = \Gamma_L(E) G^a_{1,N}(E) \Gamma_R(E) G^r_{N,1}(E) \quad [5.176]$$

The Γ are given by [5.174], and from [5.101], $G^a_{1,N}$ is the complex conjugate of $G^r_{N,1}$. To calculate $G^r_{N,1}$, we use the recursive Green's function technique [LAK 97]. Let us define $g^{(n)r}$ as the Green's function of the system composed of the first n sites of the conductor connected to the left electrode but disconnected from sites $n+1$ and beyond. When adding the site $n+1$ to this system, we can apply the Dyson equation as in section 5.3.2.1, V being the coupling $H_{n,n+1} = H_{n+1,n} = -\gamma$ between sites n and $n+1$. We obtain the recurrence relations:

$$\begin{aligned}
g^{(1)r}_{1,1} &= (E - H_{1,1} - \Sigma_L)^{-1} \\
g^{(n+1)r}_{n+1,n+1} &= (E - H_{n+1,n+1} - H_{n+1,n} g^{(n)r}_{n,n} H_{n,n+1})^{-1} \\
g^{(n+1)r}_{n+1,1} &= g^{(n+1)r}_{n+1,n+1} H_{n+1,n} g^{(n)r}_{n,1} \\
G^r_{N,N} &= (E - H_{N,N} - H_{N,N-1} g^{(N-1)r}_{N-1,N-1} H_{N-1,N} - \Sigma_R)^{-1} \\
G^r_{N,1} &= G^r_{N,N} H_{N,N-1} g^{(N-1)r}_{N-1,1}
\end{aligned} \quad [5.177]$$

where we have also coupled the right electrode at step N to obtain the Green's function elements $G^r_{N,N}$ and $G^r_{N,1}$ of the full system. The latter is injected in equation [5.176] for the transmission. Backward recurrence relations, also based on Dyson equations, yield additional quantities of interest such as $G^r_{n,n}$ on each layer, which gives the density of states, and $G^r_{n,1}$, which will allow diffusion states to be built explicitly. These recurrence relations read [GRO 89, LAK 97]:

$$\begin{aligned} G^r_{n,n} &= g^{(n)r}_{n,n} + g^{(n)r}_{n,n} H_{n,n+1} G^r_{n+1,n+1} H_{n+1,n} g^{(n)r}_{n,n} \\ G^r_{n,1} &= G^r_{n,n} H_{n,n-1} g^{(n-1)r}_{n-1,n-1} \end{aligned} \quad [5.178]$$

5.3.3.4. *Example: tunneling through a square barrier*

We apply this method to the square barrier considered in section 5.1.3 (height $V_b = 1$ eV and thickness $L_b = 2$ nm), with the same effective mass $m = 0.19\ m_0$ and a discretization mesh step $a = 0.05$ nm. Figure 5.12 shows the excellent agreement between the recursive Green's function calculation and the analytic formula of Chapter 1.

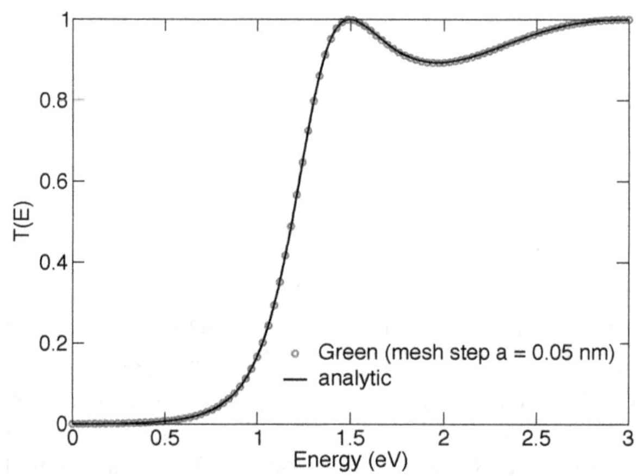

Figure 5.12. *Transmission versus energy for the square barrier of section 5.1.3. Solid line: analytic formula. Circles: recursive Green's function calculation*

5.3.4. *Application to a multichannel system*

We now consider the 2D ribbon of section 5.1.4. The system can be decomposed into layers indexed by n_x, each containing N_y sites $n_y = 1, ..., N_y$ (we exclude the boundary sites $n_y = 0$ and $n_y = N_y + 1$ where the wavefunctions cancel out). For convenience, we also index all the sites layer by layer with a single index $i = (n_x - 1)N_y + n_y$. The Hamiltonian only couples each layer to its two adjacent layers. Hence, on the basis of sites $\{|i\rangle\}$ the Hamiltonian is block tridiagonal with blocks of dimension $N_y \times N_y$. We name H_{n_x,n_x} the diagonal block of layer n_x and

H_{n_x,n_x-1}, H_{n_x,n_x+1} the coupling blocks with adjacent layers. In this specific example, H_{n_x,n_x} has diagonal elements $4\gamma + V(n_x, n_y)$ and couplings $-\gamma$ with the nearest neighbors located inside layer n_x. The interlayer couplings are as follows: $H_{n_x,n_x-1} = H_{n_x,n_x+1} = -\gamma I$ where I is the $N_y \times N_y$ identity matrix.

For a uniform potential $V(n_x, n_y) = V_0$, it is convenient to replace the basis of sites $\{|i\rangle\}$ by a basis made up of the transverse modes [5.32] of each layer. We name $|\chi_p, n_x\rangle$ the wavefunction equal to $\chi_p(n_y)$ on layer n_x and zero elsewhere. It is easy to show that:

$$\hat{H}\,|\chi_p, n_x\rangle = \left(2\gamma + V_0 + E_p^0\right)|\chi_p, n_x\rangle - \gamma\,|\chi_p, n_x - 1\rangle - \gamma\,|\chi_p, n_x + 1\rangle \quad [5.179]$$

where E_p^0 is given by equation [5.34]. Hence, for each mode p, the problem comes down to an effective 1D Hamiltonian with onsite energy $\left(2\gamma + V_0 + E_p^0\right)$ and coupling $-\gamma$ with adjacent layers. This will facilitate the calculation of the electrodes self-energies.

5.3.4.1. *Green's function of the ideal ribbon*

We go back to the initial definition [5.83] of the retarded Green's operator $\hat{G}^r(E)$. We can decompose the inversion of $[(E + i\eta)\hat{I} - \hat{H}]^{-1}$ into the inversions of the effective 1D Hamiltonians [5.179] for each mode p. From section 5.3.3.1, we know the Green's function of each 1D Hamiltonian. Switching back from the modes to the sites basis, we finally obtain:

$$G^{0r}(n_x, n_y, n'_x, n'_y; E) = \sum_{p=1}^{N_y} \chi_p(n_y)\chi_p(n'_y)G_p^{1D}(n_x, n'_x; E) \quad [5.180]$$

where G_p^{1D} is the 1D Green's function of mode p given by equations [5.167] and [5.168], with V_0 replaced by $V_0 + E_p^0$, which is the band edge of subband p. The modes such that $V_0 + E_p^0 < E < V_0 + E_p^0 + 4\gamma$ give propagating contributions to the total Green's function, with outgoing plane waves coming from layer n'_x. The other modes give evanescent contributions.

5.3.4.2. *Self-energy of the semi-infinite ribbon*

Similarly, we can decompose into modes the inversion of $[(E + i\eta)\hat{I} - \hat{H}_L^0]^{-1}$, where $\hat{H}_L^0$ is the Hamiltonian of the isolated left electrode with constant potential V_L. We obtain:

$$\Sigma_L(n_y, n'_y; E) = \sum_{p=1}^{N_y} \chi_p(n_y)\chi_p(n'_y)\Sigma_p^{1D}(E) \quad [5.181]$$

where $\Sigma_p^{1D}(E)$ is the 1D self-energy [5.171], with V_0 replaced by $V_L + E_p^0$. Note that here the self-energy has been reduced to a $N_y \times N_y$ block, since only the first layer of the conductor is coupled to the left electrode. The operator $\Gamma_L(E)$ from equation [5.142] similarly reduces to:

$$\begin{aligned}\Gamma_L(n_y, n_y'; E) &= \sum_{propagating\ modes\ p} \chi_p(n_y)\chi_p(n_y') \times 2\gamma \sin(\kappa_p a) \\ &= \sum_{propagating\ modes\ p} \chi_p(n_y)\chi_p(n_y') \times \frac{\hbar v(\kappa_p)}{a}\end{aligned} \quad [5.182]$$

The evanescent modes have cancelled out in the calculation. The wavevector κ_p of mode p is given by:

$$\kappa_p = \frac{1}{a} \arccos\left(1 - \frac{E - V_L - E_p^0}{2\gamma}\right) \quad [5.183]$$

The evanescent modes do not contribute to $\Gamma_L(E)$. Hence, as expected, they do not contribute to the transmission [5.163]. However, they must not be forgotten in $\Sigma_L(E)$ when calculating the conductor Green's function by equation [5.133]. The same equations hold for Σ_R and Γ_R, with the potential V_L replaced by V_R.

5.3.4.3. *Transmission*

We consider a conductor with N_x layers $n_x = 1, ..., N_x$. The calculation of the Green's function can be done using the recursive algorithm of section 5.3.3.3, where scalar quantities become matrix blocks of size $N_y \times N_y$. Σ_L is added to the first diagonal block of the Hamiltonian (layer 1) and Σ_R is added to the last diagonal block (layer N_x). We obtain the Green's function block $G_{N_x,1}$ between the first and the last layer of the conductor.

Then, there are two options. One is to apply the trace formula [5.163], which reduces to:

$$\sum_{p \in L} \sum_{q \in R} T_{qp}^L = \mathrm{Tr}\left[\Gamma_L(E) G_{1,N_x}^a(E) \Gamma_R(E) G_{N_x,1}^r(E)\right] \quad [5.184]$$

with $G_{1,N_x}^a(E) = \left[G_{N_x,1}^r(E)\right]^\dagger$.

The other option is to use the Fisher–Lee formula [FIS 81], which gives better physical insight since it provides the transmission coefficients T_{qp}^L between the different modes. It is based on the direct analysis of the scattering states in order to obtain the scattering matrix elements t_{qp}^{L} defined in section 5.2.1. Let us consider an incident wave in mode p with wavevector κ_p in the left electrode. One can define the corresponding eigenstate of the isolated electrode, which is the sum of the incident wave and the reflected wave at the surface $n_x = 0$:

$$\varphi_L(n_x, n_y) = \chi_p(n_y)\left[e^{i\kappa_p(n_x-1)a} - e^{-i\kappa_p(n_x-1)a}\right] \text{ for } n_x \leq 0 \qquad [5.185]$$

Here, the incident wave has been normalized to one electron per mesh step a. We name $\varphi_{L,0}$ the restriction of φ_L to layer $n_x = 0$:

$$\varphi_{L,0} = -2i\sin(\kappa_p a)\,\chi_p \qquad [5.186]$$

Equation [5.134] gives the amplitude of the corresponding scattering state on layer N_x:

$$\begin{aligned}\psi_{L,N_x} &= G^r_{N_x,1}H_{1,0}\varphi_{L,0}\\ &= 2i\gamma\sin(\kappa_p a)\,G^r_{N_x,1}\chi_p\\ &= i\frac{\hbar v_L(\kappa_p)}{a}G^r_{N_x,1}\chi_p\end{aligned} \qquad [5.187]$$

where $v_L(\kappa_p)$ is the velocity of the incident wave. Assuming that the last layer N_x of the conductor is identical to the layers of the right electrode, the scattering state on this layer does not contain any reflected contribution. It is a linear combination of transmitted and evanescent waves. Hence to obtain the amplitude of the transmitted mode q, we simply project the scattering state on this mode on layer N_x. We then obtain the ratio between the transmitted amplitude in mode q and the incident amplitude in mode p:

$$\frac{B_q^R}{A_p^L} = i\frac{\hbar v_L(\kappa_p)}{a}\chi_q^\dagger G^r_{N_x,1}\chi_p \qquad [5.188]$$

We must renormalize the amplitudes with the group velocities, as in equation [5.37], to obtain the coefficients of the scattering matrix. We finally obtain the Fisher–Lee formula [FIS 81]:

$$t_{qp}^L = i\frac{\hbar}{a}\sqrt{v_L(\kappa_p)v_R(\kappa_q)}\chi_q^\dagger G^r_{N_x,1}\chi_p \qquad [5.189]$$

The equivalence with the trace formula [5.184] can be shown by expressing Γ_L and Γ_R on the basis of modes, in which they are diagonal with matrix elements $\hbar v/a$ for propagating modes and 0 for evanescent modes (see equation [5.182]). For each couple of propagating modes p and q, we recover the square modulus of equation [5.189].

We can check the time-reversal symmetry [5.48] of the scattering matrix. We have:

$$t_{pq}^R = i\frac{\hbar}{a}\sqrt{v_L(\kappa_p)v_R(\kappa_q)}\chi_p^\dagger G_{1,N_x}^r \chi_q \quad [5.190]$$

Since the Hamiltonian is Hermitian and has real coefficients, it is symmetric. Hence, $G^r(E) = [(E + i\eta)I - H]^{-1}$ is also symmetric. Using the fact that the modes χ_p are real, we can show that:

$$\chi_p^\dagger G_{1,N_x}^r \chi_q = \chi_q^\dagger G_{N_x,1}^r \chi_p \quad [5.191]$$

Hence, we obtain the expected result: $t_{pq}^R = t_{qp}^L$.

The Fisher–Lee formula is straightforwardly generalized to the case where the right electrode is different from the left electrode (e.g. different width), hence has a different basis of modes.

5.3.4.4. *Generalization to more complex Hamiltonians*

The trace formula [5.163] is applicable to any Hamiltonian that does not couple directly the left and right electrodes. The recursive Green's function algorithm presented in section 5.3.3.3 is applicable to any Hamiltonian expressed on a localized basis set with finite range coupling (e.g. tight-binding), since we can always decompose the system into layers such that the Hamiltonian takes a block tridiagonal form.

The Fisher–Lee formula has also been generalized to tight-binding Hamiltonians [AND 91, KHO 05]. This requires special care for computing the propagating and evanescent modes of the electrodes [BOY 96], calculating the self-energies [RIV 03, LUI 06], and properly coupling the modes with the conductor Green's function [KHO 05].

5.3.4.5. *Calculation of local quantities*

The recursive Green's functions algorithm allows calculating the density of states [5.102] and integrated charge density [5.147] at each site of the conductor by performing the backward recurrence [5.178]. It also allows calculating other local quantities such as the current density [CRE 04]. In the following section, we will show the shape of a diffusion state in the conductor.

5.3.4.6. *Numerical example: inter-subband scattering on an impurity*

We consider exactly the same system as in section 5.1.4.2: a ribbon of width $W =$ 5 nm and a repulsive impurity potential located at 1 nm from the ribbon edge. The recursive Green's function method is used together with the Fisher–Lee formula to calculate the transmission coefficients between the different channels. The results are shown in Figure 5.13. The linear conductance at $T = 0$ K is strongly reduced close to the band edges, with respect to the ideal case. This is because the band edge states are more sensitive to impurity scattering. This is due to their small velocity v and correspondingly large density of states (proportional to $1/v$), leading to a large number of final states for backscattering. The transmission coefficients between the two first subbands are shown in the figure. We see, for instance, that the first subband is partially scattered toward the second subband in its energy range. This was already observed on wavepacket propagation in section 5.1.4.2.

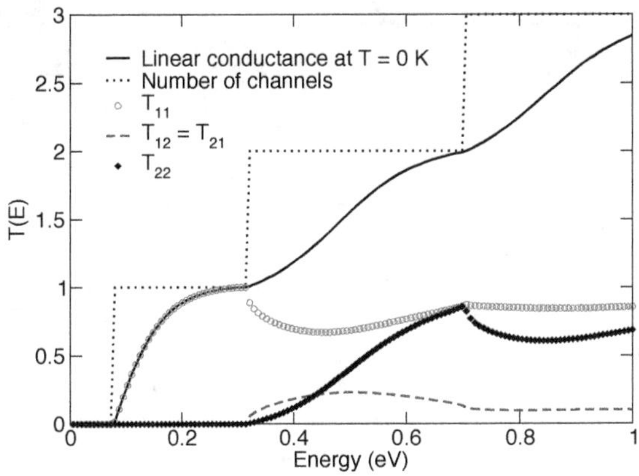

Figure 5.13. *Transmission coefficients versus energy in a ribbon of width $W = 5$ nm with a repulsive impurity potential. Solid line: linear conductance at $T = 0$ K, in units of $\mathcal{G}_0$. Dotted line: number of conducting channels. Others: transmission coefficients between the two first channels*

It is very instructive to change the sign of the impurity potential, which then becomes attractive. The transmission coefficients are shown in Figure 5.14. They differ strongly from the repulsive case in Figure 5.13. For instance, the first subband has a zero transmission dip at a particular energy. The density of states of the conductor is plotted on the same energy axis. It exhibits peaks around $E = 0, 0.25$ and 0.65 eV. The last two coincide with the transmission dips. The first peak is a localized state inside the quantum well, decoupled from the leads since its energy is below the first conducting channel. Its contribution to the density of states is a Dirac delta function but we have added a finite imaginary part in the calculation of the Green's function in order to smooth this peak. The second peak is a state of the second subband that is trapped in the quantum well. It is called a quasi-bound state since it is not strictly localized. It is instead hybridized with the conducting states of the first subband. This phenomenon is called Fano resonance [FAN 61] and leads to this peculiar shape of the transmission spectrum. Similarly, the third peak at 0.657 eV is a quasi-bound state of the third subband. This is illustrated in Figure 5.15, where a scattering state at this energy is plotted. Such resonances have been predicted in various systems, for instance in graphene nanoribbons [BIE 09].

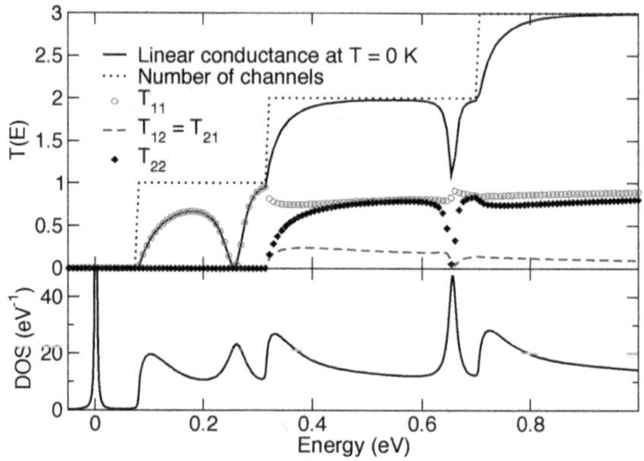

Figure 5.14. *Top panel: transmission coefficients versus energy in a ribbon of width $W = 5$ nm with an attractive impurity potential. Same legends as in Figure 5.13. Bottom panel: total density of states of the conductor (length $L = 10$ nm), showing peaks corresponding to the bound state ($E \simeq 0$) and to quasi-bound states of subbands 2 and 3 ($E \simeq 0.25$ and 0.65 eV). An imaginary part $\eta = 10^{-3}$ eV was added to the energy in the calculation of the Green's function in order to smooth the Dirac peak of the bound state*

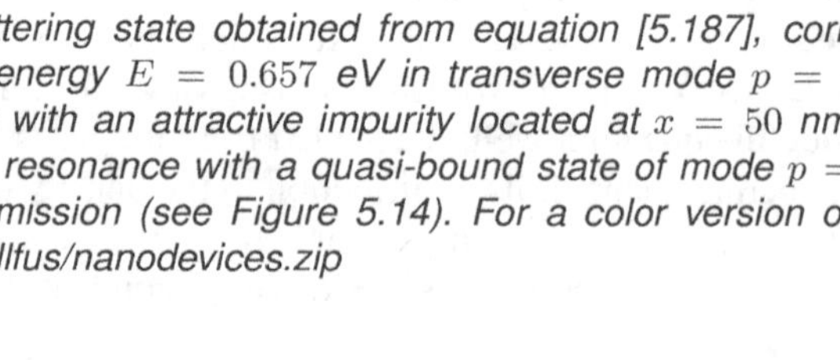

Figure 5.15. *Scattering state obtained from equation [5.187], corresponding to an incident wave of energy $E = 0.657$ eV in transverse mode $p = 2$, in a ribbon of width $W = 5$ nm with an attractive impurity located at $x = 50$ nm and $y = 1$ nm. There is a strong resonance with a quasi-bound state of mode $p = 3$, which nearly cancels the transmission (see Figure 5.14). For a color version of this figure, see www.iste.co.uk/dollfus/nanodevices.zip*

This is an example where the first-order perturbation theory of scattering, based on Fermi's golden rule (see Chapter 1), fails. Indeed, the squared matrix element of the scattering potential, appearing in the first-order approximation, does not depend on the sign of the potential. Hence, it predicts the same scattering rate for a repulsive and an attractive impurity. The Green's function calculation shows that it is not the case. This highlights the relevance of quantum transport formalisms for treating more accurately scattering on impurities and structural defects.

5.4. Conclusion

In this chapter, we have detailed the notions introduced in Chapter 1 about the quantum theory of electron transport by calculating wavepacket propagation in various cases. We have addressed the transmission formalism, which treats transport through a small conductor connected to electrodes. We have also introduced the Green's functions formalism, an efficient way to calculate quantum transmission. We have shown examples where the semiclassical theory of transport is not sufficient, or where elastic scattering by defects is not described by the first-order perturbation theory. However, many important aspects were not addressed at all.

Electron–phonon scattering plays an important role at room temperature or at large bias voltage. The quantum theory of transport accounting for electron–phonon coupling is addressed in the following chapter. It is based on the many-body NEGF formalism which, in the non-interacting case, reduces to the formalism presented in this chapter. The Boltzmann and Wigner Monte Carlo methods addressed in Chapter 8 also include electron–phonon scattering, together with other scattering mechanisms.

The present chapter was limited to transport in the stationary regime. Simulating transient regimes and noise is of great importance for technological applications. The Monte Carlo methods presented in Chapter 8 can address these issues. Chapter 7 presents an alternative quantum formalism that is also well suited for this purpose.

5.5. Matlab/Octave codes

The codes available at www.iste.co.uk/dollfus/nanodevices.zip are compatible with Matlab and Octave. The main code "wavepacket_evolution.m" generates the example from section 5.1.4.2. The main code "transmission_green.m" generates the example from section 5.3.4.6 (Figure 5.14). With simple modifications of the parameters, all the 1D and 2D examples of this chapter can be reproduced.

5.6. Bibliography

[AND 91] ANDO T., "Quantum point contacts in magnetic fields", *Physical Review B*, vol. 44, no. 15, p. 8017, 1991.

[ASH 76] ASHCROFT N.W., MERMIN N.D., *Solid State Physics*, Brooks/Cole, Salt Lake City, 1976.

[BIE 09] BIEL B., BLASE X., TRIOZON F. *et al.*, "Anomalous doping effects on charge transport in graphene nanoribbons", *Physical Review Letters*, vol. 102, no. 9, p. 096803, 2009.

[BOY 96] BOYKIN T.B., "Generalized eigenproblem method for surface and interface states: the complex bands of GaAs and AlAs", *Physical Review B*, vol. 54, no. 11, p. 8107, 1996.

[BUT 85] BUTTIKER M., IMRY Y., LANDAUER R. *et al.*, "Generalized many-channel conductance formula with application to small rings", *Physical Review B*, vol. 31, no. 10, p. 6207, 1985.

[CAR 71] CAROLI C., COMBESCOT R., NOZIÈRES P. *et al.*, "Direct calculation of the tunneling current", *Journal of Physics C: Solid State Physics*, vol. 4, p. 916, 1971.

[CHA 82] CHANG Y.-C., SCHULMAN J.N., "Complex band structures of crystalline solids: An eigenvalue method", *Physical Review B*, vol. 25, no. 6, p. 3975, 1982.

[COH 77] COHEN TANNOUDJI C., DIU B., LALOË F., *Quantum Mechanics*, Wiley, New York, 1977.

[COS 95] COSTA-KRAMER J., GARCIA N., GARCIA-MOCHALES P. *et al.*, "Nanowire formation in macroscopic metallic contacts: quantum mechanical conductance tapping a table top", *Surface Science*, vol. 342, nos.1–3, p. L1144, 1995.

[CRE 04] CRESTI A., GROSSO G., PASTORI PARRAVICINI G., "Chiral symmetry of microscopic currents in the quantum Hall effect", *Physical Review B*, vol. 69, no. 23, p. 233313, 2004.

[DAT 95] DATTA S., *Electronic Transport in Mesoscopic Systems*, Cambridge University Press, Cambridge, 1995.

[DAT 05] DATTA S., *Quantum Transport: Atom to Transistor*, Cambridge University Press, Cambridge, 2005.

[FAN 61] FANO U., "Effects of configuration Interaction on Intensities and Phase Shifts", *Physical Review*, vol. 124, no. 6, p. 1866, 1961.

[FIS 81] FISHER D.S., LEE P.A., "Relation between conductivity and transmission matrix", *Physical Review B*, vol. 23, no. 12, p. 6851, 1981.

[GRO 89] GROSSO G., MORONI S., PASTORI PARRAVICINI G., "Electronic structure of the InAs-GaSb superlattice studied by the renormalization method", *Physical Review B*, vol. 40, no. 18, p. 12328, 1989.

[IMR 02] IMRY Y., *Introduction to Mesoscopic Physics*, Second Edition, Oxford University Press, Oxford, 2002.

[KHO 05] KHOMYAKOV P.A., BROCKS G., KARPAN V. *et al.*, "Conductance calculations for quantum wires and interfaces: Mode matching and Green's functions", *Physical Review B*, vol. 72, no. 3, p. 035450, 2005.

[KOH 59] KOHN W., "Analytic properties of Bloch waves and wannier functions", *Physical Review*, vol. 115, no. 4, p. 809, 1959.

[LAK 97] LAKE R., KLIMECK G., BOWEN R.C. *et al.*, "Single and multiband modeling of quantum electron transport through layered semiconductor devices", *Journal of Applied Physics*, vol. 81, no. 12, p. 7845, 1997.

[LAN 57] LANDAUER R., "Spatial variation of currents and fields due to localized scatterers in metallic conduction", *IBM Journal of Research and Development*, vol. 1, no. 3, p. 223, 1957.

[LAN 70] LANDAUER R., "Electrical resistance of disordered one-dimensional lattices", *Philosophical Magazine*, vol. 21, no. 172, p. 863, 1970.

[LEF 91] LEFORESTIER C., BISSELING R.H., CERJAN C. *et al.*, "A comparison of different propagation schemes for the time dependent Schrödinger equation", *Journal of Computational Physics*, vol. 94, no. 1, pp. 59-80, 1991.

[LIP 50] LIPPMANN B.A., SCHWINGER J., "Variational principles for scattering processes. I", *Physical Review*, vol. 79, no. 3, p. 469, 1950.

[LOP 85] LOPEZ SANCHO M.P., LOPEZ SANCHO J.M., RUBIO J., "Highly convergent schemes for the calculation of bulk and surface Green functions", *Journal of Physics F: Metal Physics*, vol. 15, p. 851, 1985.

[LUI 06] LUISIER M., SCHENK A., FICHTNER W., "Atomistic simulation of nanowires in the sp3d5s* tight-binding formalism:From boundary conditions to strain calculations", *Physical Review B*, vol. 74, p. 205323, 2006.

[MAY 88] MAYOU D., "Calculation of the conductivity in the short-mean-free-path regime", *Europhysics Letters*, vol. 6, no. 6, p. 549, 1988.

[MAY 95] MAYOU D., KHANNA S.N., "A real-space approach to electronic transport", *Journal de Physique I France*, vol. 5, no. 9, p. 1199, 1995.

[MEI 92] MEIR Y., WINGREEN N.S., "Landauer formula for the current through an interacting electron region", *Physical Review Letters*, vol. 68, no. 16, p. 2512, 1992.

[MES 99] MESSIAH A., *Quantum Mechanics*, Dover, New York, 1999.

[RIV 03] RIVAS C., LAKE R., "Non-equilibrium Green function implementation of boundary conditions for full band simulations of substrate-nanowire structures", *Physica Status Solidi (b)*, vol. 239, no. 1, p. 94, 2003.

[ROC 97] ROCHE S., MAYOU D., "Conductivity of quasiperiodic systems: a numerical study", *Physical Review Letters*, vol. 79, no. 13, p. 2518, 1997.

[ROC 99] ROCHE S., MAYOU D., "Formalism for the computation of the RKKY interaction in aperiodic systems", *Physical Review B*, vol. 60, no. 1, p. 322, 1999.

[TRI 02] TRIOZON F., VIDAL J., MOSSERI R. *et al.*, "Quantum dynamics in two- and three-dimensional quasiperiodic tilings", *Physical Review B*, vol. 65, no. 22, p. 220202(R), 2002.

[WEE 88] VAN WEES B.J., VAN HOUTEN H., BEENAKKER C.W.J. *et al.*, "Quantized conductance of point contacts in a two-dimensional electron gas", *Physical Review Letters*, vol. 60, no. 9, p. 848, 1988.

[WEI 95] WEINBERG S., *The Quantum Theory of Fields, Volume 1: Foundations*, Cambridge University Press, Cambridge, 1995.

6

Non-Equilibrium Green's Function Formalism

This chapter describes the main lines of the non-equilibrium Green's function formalism.

6.1. Second quantization and time evolution pictures

In this section, we briefly introduce second quantization that is best adapted to the definition and use of Green's functions. We start from a complete set of single particle wave functions $u_k(\mathbf{r})$ (where $\mathbf{r}$ is the set of coordinates) and build from them all possible many-particle states with varying total number of particles. These correspond to symmetrized products of the $u_k(\mathbf{r})$ for bosons and to antisymmetrized products (Slater determinants) for fermions. In the Hilbert space spanned by these many-particle states, we define operators c_k that annihilate a particle in state k and their counterpart c_k^+, that create a particle in this state. From the symmetry properties and the orthonormality of the many-particle states, we get the following commutation relations:

$$\left[c_k, c_{k'}^+\right]_\pm = \delta_{k,k'} \qquad \left[c_k, c_{k'}\right]_\pm = \left[c_k^+, c_{k'}^+\right]_\pm = 0, \tag{6.1}$$

where the $+$ sign (anticommutator) denotes fermions and the $-$ sign (commutator) denotes bosons. With this formulation, any Hamiltonian h_0, which is the sum of identical independent particle contributions, can be written as:

$$H_0 = \sum_{k,k'} (h_0)_{k'k} c_{k'}^+ c_k, \tag{6.2}$$

Chapter written by Michel LANNOO and Marc BESCOND.

where $(h_0)_{k'k}$ is the conventional single particle matrix element. In the case of fermions with two-particle interactions $V(\mathbf{r}, \mathbf{r}')$ well-known rules [SLA 63, MAT 67, HAU 96] for the matrix elements between Slater determinants give an interaction Hamiltonian:

$$V = 1/2 \sum_{k_1,k_2,k_3,k_4} V_{k_1,k_2,k_3,k_4} c^+_{k_1} c^+_{k_2} c_{k_3} c_{k_4}, \qquad [6.3]$$

with

$$V_{k_1,k_2,k_3,k_4} = \int d^3\mathbf{r} d^3\mathbf{r}' u^*_{k_1}(\mathbf{r}) u^*_{k_2}(\mathbf{r}') V(\mathbf{r}, \mathbf{r}') u_{k_3}(\mathbf{r}') u_{k_4}(\mathbf{r}). \qquad [6.4]$$

We can also express h_0 and V in terms of the field operators $\Psi(\mathbf{r}) = \sum_k u_k(\mathbf{r}) c_k$, which gives:

$$\begin{aligned} H_0 &= \int d^3\mathbf{r} \Psi^+(\mathbf{r}) h_0(\mathbf{r}) \Psi(\mathbf{r}) \\ V &= 1/2 \int d^3\mathbf{r} d^3\mathbf{r}' \Psi^+(\mathbf{r}) \Psi^+(\mathbf{r}') V(\mathbf{r}, \mathbf{r}') \Psi(\mathbf{r}') \Psi(\mathbf{r}) \\ \left[\Psi(\mathbf{r}), \Psi^+(\mathbf{r}')\right]_\pm &= \delta_{\mathbf{r}-\mathbf{r}'}. \end{aligned} \qquad [6.5]$$

Time evolution pictures [HAU 96, MAH 00]: In the following, we shall be interested by the time dependence of different quantities. If we consider time-independent operators O, there are three possible pictures (for simplicity we express H in units of $\hbar$ in all expressions involving the time dependence of operators and reintroduce $\hbar$ in the calculation of physical quantities):

– the Schrödinger picture in which the wave function $\phi_S(t) = e^{-iHt}\phi(0)$ is time dependent and the operators O remain constant;

– the Heisenberg picture where the operators are defined by $O_H(t) = e^{iHt} O e^{-iHt}$ and act on constant wave functions;

– the "interaction" picture appropriate to the case $H = H_0 + V$ where usually H_0 is an independent particle Hamiltonian and V is the complex interaction part. In this case, the wave function $\widehat{\phi}(t)$ develops in time under the influence of V, i.e.:

$$\widehat{\phi}(t) = e^{iH_0t}\phi_S(t) = e^{iH_0t}e^{-iHt}\phi(0), \qquad [6.6]$$

while the operators are given by

$$\widehat{O}(t) = e^{iH_0t} O e^{-iH_0t}. \qquad [6.7]$$

As we shall see, the interaction picture is particularly adapted to a perturbation expansion in powers of $V = H - H_0$, which is the basis of the so-called diagrammatic technique. It will be also useful for what follows to define the $S(t, t')$ matrix that connects the wave functions at times *t* and *t*' in the interaction picture. From equation [6.6], we get:

$$\begin{aligned}\widehat{\phi}(t) &= e^{iH_0 t} e^{-iHt} e^{iHt'} e^{-iH_0 t'} \widehat{\phi}(t') \\ &= S(t, t')\widehat{\phi}(t').\end{aligned} \qquad [6.8]$$

It is shown in different textbooks [MAT 67, HAU 96, MAH 00, RAM 86] that equation [6.8] can be expressed as a time-ordered product:

$$\begin{aligned}S(t, t') &= Texp\{-i \int_{t'}^{t} dt_1 \widehat{V}(t_1)\} \\ &= \sum_{n=0}^{\infty} \frac{(-i)^n}{n!} \int_{t'}^{t} dt_1 \cdots dt_n T\{\widehat{V}(t_1)\widehat{V}(t_2) \cdots \widehat{V}(t_n)\},\end{aligned} \qquad [6.9]$$

where T is the time ordering operator that orders times from right (earlier) to left (later). Expression [6.9] is the basis of the expansion of the Green's functions in power of the interactions between particles, which corresponds to the diagrammatic technique.

6.2. General definition of the Green's functions, their physical meaning and their perturbation expansion

In the following, we adopt a shorthand notation in which (1) stands for $(\mathbf{r_1}, t_1)$ where $\mathbf{r_1}$ represents the spatial coordinates and t_1 is the time. The general definition of the Green's function for fermions is:

$$G(1, 2) = -i\langle T\left[\Psi_H(1)\Psi_H{}^+(2)\right]\rangle, \qquad [6.10]$$

where the operators Ψ_H and $\Psi_H{}^+$ are expressed in the Heisenberg picture and where $\langle\rangle$ describes an averaging appropriate to the physical situation under consideration. For $t_1 > t_2$, the Green's function describes the probability amplitude that an electron injected at time t_2 and point $\mathbf{r_2}$ will be found at $t_1, \mathbf{r_1}$. For $t_2 > t_1$, it describes the symmetrical propagation of a hole. It is also of interest to define other Green's functions again for fermions:

$$\begin{aligned}
&- \text{Lesser than: } G^{<}(1,2) &&= i\langle \Psi_H{}^{+}(2)\Psi_H(1)\rangle. \\
&- \text{Greater than: } G^{>}(1,2) &&= -i\langle \Psi_H(1)\Psi_H{}^{+}(2)\rangle. \\
&- \text{Retarded: } G^{r}(1,2) &&= \theta(t_1 - t_2)[G^{>}(1,2) - G^{<}(1,2)]. \\
&- \text{Advanced: } G^{a}(1,2) &&= -\theta(t_2 - t_1)[G^{>}(1,2) - G^{<}(1,2)],
\end{aligned} \qquad [6.11]$$

with the obvious relation:

$$G^r - G^a = G^> - G^<. \qquad [6.12]$$

These various Green's functions correspond to important physical properties. First of all, $G^{\gtrless}$ are directly related to observables. Indeed, the electron density is directly expressed as:

$$n(\mathbf{r}_1, t_1) = -iG^{<}(1,1) \qquad [6.13]$$

In the same way, the quantum formulation of the particle current density is written as [HAU 96]:

$$\vec{j} = \frac{\hbar}{2m} \lim_{2\to 1} (\vec{\nabla_2} - \vec{\nabla_1}) G^{<}(1,2). \qquad [6.14]$$

As we shall see G^r and G^a contain spectroscopic information while the full G can be expanded systematically in perturbation theory.

We now consider the most important problem that is related to the perturbation expansion of the Green's function. Indeed, as mentioned before, the Hamiltonian is the sum of a part H_0 (relatively simple to solve and usually connected to an independent particle picture) and a more complex interaction part V, which is the main reason for the problem. This is usually tackled via a perturbation expansion, which we present first at equilibrium and then generalize to the non-equilibrium case.

6.2.1. *Equilibrium perturbation expansion at T = 0 K*

The general expression [6.10] for the Green's function then becomes:

$$G(1,2) = -i \frac{< \Phi_0 | T \left[\Psi_H(1) \Psi_H^{+}(2) \right] | \Phi_0 >}{< \Phi_0 | \Phi_0 >}, \qquad [6.15]$$

where $|\Phi_0>$ is the exact unknown ground state. To tackle this problem, we need to express G in terms of the non-interacting ground state $|\phi_0>$. The simplest way is to consider that the interaction is turned on adiabatically from time $t=-\infty$ to $t=0$ and then is suppressed in the same way up to $t=+\infty$ in order to return to its initial state. From the Gell-Mann and Low theorem [GEL 51], we can write, using the time evolution operator [6.9]:

$$|\Phi_0>=S(0,-\infty)|\phi_0>, \qquad <\Phi_0|=<\phi_0|S(+\infty,0), \tag{6.16}$$

To transform equation [6.15], we use equation [6.16] and the relation $\Psi_H(t) = S(0,t)\widehat{\Psi}(t)S(t,0)$ obtained from the definition [6.8] of $S(t,t')$ and use the fact that $S(t_i,t_k)S(t_k,t_j)=S(t_i,t_j)$ to get the final expression for G:

$$G(1,2) = -i\frac{<\phi_0|T\left[S(+\infty,-\infty)\widehat{\Psi}(1)\widehat{\Psi}^+(2)\right]|\phi_0>}{<\phi_0|S(\infty,-\infty)|\phi_0>}, \tag{6.17}$$

which is the desired expression in terms of the unperturbed ground state. Replacing S in expression [6.17] by its series expansion [6.9] gives the general perturbation expansion of G, which we now write in terms of the matrix elements of G between $u_{k_1}(\mathbf{r_1})$ and $u_{k_2}(\mathbf{r_2})$:

$$G_{k_1,k_2} = \frac{\sum_{n=0}^{\infty}\frac{(-i)^n}{n!}\int_{-\infty}^{+\infty}dt'_1\cdots dt'_n <\phi_0|T\left[\widehat{V}(t'_1)\cdots\widehat{V}(t'_n)\widehat{c}_{k_1}(t_1)\widehat{c}^+_{k_2}(t_2)\right]|\phi_0>}{\sum_{n=0}^{\infty}\frac{(-i)^n}{n!}\int_{-\infty}^{+\infty}dt'_1\cdots dt'_n <\phi_0|T\left[\widehat{V}(t'_1)\cdots\widehat{V}(t'_n)\right]|\phi_0>}. \tag{6.18}$$

As discussed later on with specific examples, the possibility of effectively using this perturbation expansion relies on the fact that H_0 of equation [6.2] is quadratic in the creation and destruction operators, which allows to make use of Wick's theorem [MAT 67, HAU 96, MAH 00].

6.2.2. *Non-equilibrium perturbation expansion*

The difficulty here is that one deals with irreversible systems and that one cannot expect that they will recover the initial state at $t\to+\infty$. A general method to tackle this problem has been proposed by several authors [HAU 96, RAM 86, CRA 68, KEL 65]. It consists of calculating the Green's function on the time contour shown in Figure 6.1 where we can recover a final situation identical to the initial situation. This makes it possible to obtain the same structure of the equations as in the zero temperature case, giving the same perturbation expansion.

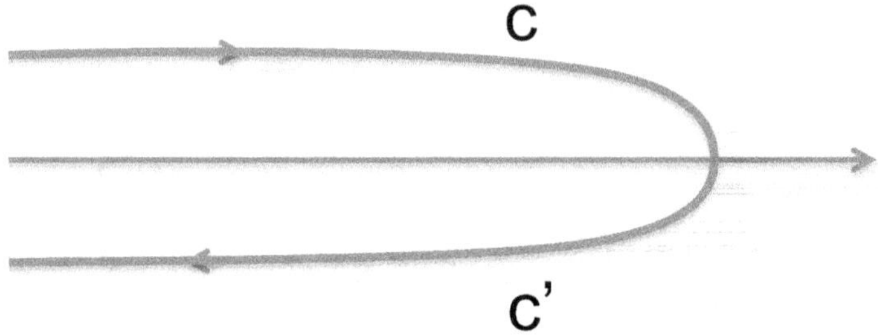

Figure 6.1. *The Keldysh contour in the complex time plane*

However, the price to pay is that there are more components of Green's function. We obviously have:

$$G(1,2) = \begin{cases} G_c \text{ if } t_1, t_2 \in c, \text{ usual time-ordered expression [6.10] on one branch} \\ -i < \Psi_H(1)\Psi_H{}^+(2) >= G^>(1,2) \text{ if } t_1 \in c', t_2 \in c \\ i < \Psi_H^+(2)\Psi_H(1) >= G^<(1,2) \text{ if } t_1 \in c, t_2 \in c' \\ G_{\overline{c}} \text{ if } t_1, t_2 \in c', \text{ idem as } G_c \text{with anti-time ordering.} \end{cases} \quad [6.19]$$

where $G^>$ $G^<$ have the same definition as in equation [6.11]. In most papers, we replace G_c and $G_{\overline{c}}$ by the retarded and advanced Green's functions also defined by equation [6.11] since we have:

$$\begin{aligned} G^r &= G_c - G^< = G^> - G_{\overline{c}} \\ G^a &= G_c - G^> = G^< - G_{\overline{c}}. \end{aligned} \quad [6.20]$$

Here, we adopt this approach and will then work with the four Green's functions defined by [6.11].

It is now necessary to give precisions about the initial state of the system. We take the commonly adopted point of view that the system is initially composed of three disconnected parts: source, active region and drain [Figure 6.2]. The source and drain are assumed to be describable in an independent electron picture and are taken to be initially at equilibrium with their own electrochemical potential. Of course, as discussed in several textbooks [HAU 96, MEI 92], this approach is more adapted to the study of the steady state and should fail for the description of some transient phenomena.

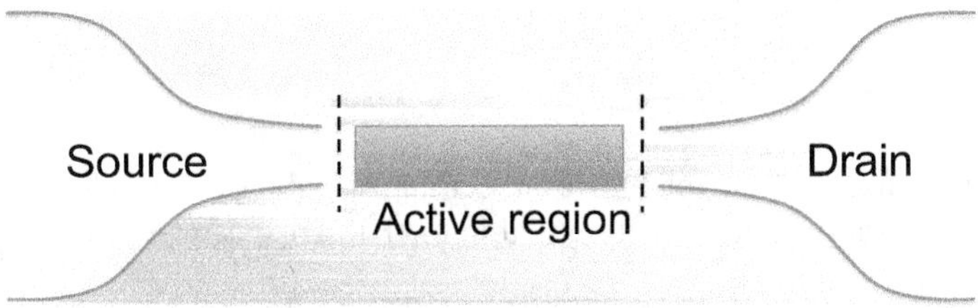

Figure 6.2. *Standard geometry of the system composed of an active region connected to two reservoirs at equilibrium: the source and drain*

6.3. Stationary Green's functions and fluctuation-dissipation theorem

For the steady-state systems, the Green's function $G(1,2)$ only depends on the difference $\tau = t_1 - t_2$. It is then often interesting to take the Fourier transform:

$$G(\mathbf{r_1}, \mathbf{r_2}, \omega) = \int d\tau e^{i\omega\tau} G(\mathbf{r_1}, \mathbf{r_2}, \tau). \quad [6.21]$$

We shall now derive some general relations between these frequency-dependent Green's functions and end this section by demonstrating the fluctuation-dissipation theorem valid for any system at equilibrium.

6.3.1. *General relations*

6.3.1.1. $G^<$

We start from the general expression:

$$G^<_{k,k'}(\tau) = i < e^{iHt_2} c_k^+ e^{iH(t_1-t_2)} c_{k'} e^{-iHt_1} > . \quad [6.22]$$

We then express the previous equation in terms of the eigenstates $|m>$ of H of energy E_m. This gives:

$$G^<_{k,k'}(\tau) = i \sum_{n,m} f_n e^{-i(E_n - E_m)\tau} < n|c_k^+|m >< m|c_{k'}|n >, \quad [6.23]$$

where f_n is the occupation factor of state $|n>$. From this expression, we can easily show that:

$$G^{<}_{k,k'}(\tau) = -\left[G^{<}_{k',k}(-\tau)\right]^*, \qquad [6.24]$$

which, when Fourier transformed, gives under matrix form (in the following, we adopt the notation $\mathbf{O}$ for any operator acting within the Hilbert space spanned by the single-particle states):

$$\mathbf{G}^{<}(\omega) = \left[\mathbf{G}^{<}(\omega)\right]^+, \qquad [6.25]$$

with a similar relation for $\mathbf{G}^{>}$.

6.3.1.2. *Relation between* $\mathbf{G}^{\mathrm{r}}$ *and* $\mathbf{G}^{\mathrm{a}}$

We start from the definitions [6.11], take their Fourier transform and use equation [6.25] to obtain:

$$\mathbf{G}^{\mathbf{a}}(\omega) = -\left[\mathbf{G}^{\mathbf{r}}(\omega)\right]^+. \qquad [6.26]$$

6.3.1.3. *Expression of* $\mathbf{G}^{\mathrm{r}}(\omega)$ *in terms of* $\mathbf{G}^{>}(\omega) - \mathbf{G}^{<}(\omega)$

Here, we express the Fourier transform of $\mathbf{G}^{\mathrm{r}}$ given by equation [6.11] under the form:

$$\mathbf{G}^{\mathbf{r}}(\omega) = \int_0^{\infty} d\tau \left[\mathbf{G}^{>}(\tau) - \mathbf{G}^{<}(\tau)\right] e^{(\omega+i\eta)\tau}, \qquad [6.27]$$

where the small imaginary part $i\eta$ is added to ensure convergence. Expressing $\mathbf{G}^{\gtrless}(\tau)$ in terms of their Fourier transform $\mathbf{G}^{\gtrless}(\omega)$, we easily get:

$$\begin{aligned}\mathbf{G}^{\mathbf{r}}(\omega) &= \frac{i}{2\pi}\int_{-\infty}^{\infty} d\omega' \frac{\mathbf{G}^{>}(\omega') - \mathbf{G}^{<}(\omega')}{\omega - \omega' + i\eta} \\ &= \frac{1}{2}\left[\mathbf{G}^{>}(\omega) - \mathbf{G}^{<}(\omega)\right] + \frac{i}{2\pi}\mathcal{P}\int_{-\infty}^{\infty} d\omega' \frac{\mathbf{G}^{>}(\omega') - \mathbf{G}^{<}(\omega')}{\omega - \omega'}.\end{aligned} \qquad [6.28]$$

This expression will prove very useful in practical implementations.

6.3.2. *Equilibrium: the fluctuation-dissipation theorem*

This shows the relations that exist between the different Green's functions at equilibrium and their expression in terms of the spectral function and the Fermi function. For this, we start from the Fourier transform of $G^<$, which can be written at equilibrium:

$$G^<(\mathbf{r_1},\mathbf{r_2},\omega) = i\int d\tau e^{i\omega\tau}\mathrm{Tr}\frac{e^{-\beta(H-\mu N)}}{Z}\Psi^+(\mathbf{r_2})e^{iH\tau}\Psi(\mathbf{r_1})e^{-iH\tau}, \quad [6.29]$$

where Z is the partition function, $\beta = \hbar/kT$ and Tr stands for the trace. Inserting again the eigenstates $|n>$ of H for a given number of electrons, we can write:

$$G^<(\mathbf{r_1},\mathbf{r_2},\omega) = \frac{2\pi i}{Z}\sum_{n,m} e^{-\beta(E_n-\mu N_n)} < n|\Psi^+|m >< m|\Psi|n > \\ \delta(\omega + E_m - E_n). \quad [6.30]$$

We can express $G^>$ in the same way and obtain:

$$G^>(\mathbf{r_1},\mathbf{r_2},\omega) = -\frac{2\pi i}{Z}\sum_{n,m} e^{-\beta(E_m-\mu N_m)} < n|\Psi^+|m >< m|\Psi|n > \\ \times\delta(\omega + E_m - E_n). \quad [6.31]$$

Since $E_n - E_m = \omega$ and $N_m - N_n = 1$, we can write:

$$\mathbf{G}^>(\omega) = -e^{-\beta(\omega-\mu)}\mathbf{G}^<(\omega). \quad [6.32]$$

If we now define the spectral function $A(\mathbf{r_1},\mathbf{r_2},\omega)$ as:

$$\mathbf{A}(\omega) = i\left[\mathbf{G}^>(\omega) - \mathbf{G}^<(\omega)\right] = i\left[\mathbf{G^r}(\omega) - \mathbf{G^a}(\omega)\right], \quad [6.33]$$

we get:

$$\mathbf{A}(\omega) = -i\left[1 + e^{\beta(\omega-\mu)}\right]\mathbf{G}^<(\omega), \quad [6.34]$$

which means that, under matrix form:

$$\mathbf{G}^{<}(\omega) = if(\omega)\mathbf{A}(\omega)$$
$$\mathbf{G}^{>}(\omega) = -i(1 - f(\omega))\mathbf{A}(\omega), \quad [6.35]$$

where $f(\omega)$ is the Fermi function.

6.4. Dyson's equation and self-energy: general formulation

We shall now show one can derive Dyson's equation (which is the basis of all Green's function calculations) in the case of interacting particles by introducing the self-energy concept. All this will be done directly in the non-equilibrium case. We start by introducing the equation of motion for non-interacting electrons.

6.4.1. *Equation of motion for non-interacting electrons*

We start from a non-interacting Hamiltonian H_0 of the form equation [6.2] or [6.5] and consider the equation of evolution of the field operator in the Heisenberg picture, $\Psi_{H_0}(\mathbf{r}, t) = e^{iH_0 t}\Psi(\mathbf{r})e^{-iH_0 t}$:

$$i\frac{\partial \Psi_{H_0}(\mathbf{r}, t)}{\partial t} = \left[\Psi_{H_0}(\mathbf{r}, t), H_0\right]. \quad [6.36]$$

From equation [6.5], this can be written as:

$$i\frac{\partial \Psi_{H_0}(\mathbf{r}, t)}{\partial t} = e^{iH_0 t}\Big\{\int d\mathbf{r}' \left[\Psi(\mathbf{r})\Psi^{+}(\mathbf{r}')h_0(\mathbf{r}')\Psi(\mathbf{r}')\right.$$
$$\left. -\Psi^{+}(\mathbf{r}')h_0(\mathbf{r}')\Psi(\mathbf{r}')\Psi(\mathbf{r})\right]\Big\}e^{-iH_0 t}. \quad [6.37]$$

We transform the second term by permuting $\Psi(\mathbf{r}')$ and $\Psi(\mathbf{r})$ with a change in sign and then using equation [6.5]. This gives:

$$i\frac{\partial \Psi_{H_0}(\mathbf{r}, t)}{\partial t} = h_0(\mathbf{r})\Psi_{H_0}(\mathbf{r}, t). \quad [6.38]$$

This can be used to take the time derivative of the general expression of the Green's function given by equation [6.10], which can be written as:

$$i\frac{\partial}{\partial t_1}G_0(1,2) = \langle T\left[\frac{\partial \Psi_{H_0}(1)}{\partial t_1}\Psi^{+}_{H_0}(2)\right]\rangle + \langle(\frac{\partial T}{\partial t_1})\left[\Psi_{H_0}(1)\Psi^{+}_{H_0}(2)\right]\rangle, \quad [6.39]$$

where $\frac{\partial T}{\partial t_1}$ formally represents the contribution of the time ordering operator to the derivative. From equation [6.38], the first term is just $h_0 G_0(1,2)$. From equation [6.11], the second term is zero for $G_0^{\gtrless}$, which are not time-ordered. On the other hand, when t_1 and t_2 belong to the same branch of the contour c or c' in Figure 6.1, then this second term becomes $\delta(t_1 - t_2)\left[\Psi_{H_0}(1)\Psi_{H_0}^+(2) + \Psi_{H_0}^+(2)\Psi_{H_0}(1)\right]$, which from equation [6.5] gives $\delta(t_1 - t_2)\delta(\mathbf{r_1} - \mathbf{r_2})$. We can thus rewrite equation [6.39]:

$$\left[i\frac{\partial}{\partial t_1} - h_0(\mathbf{r_1})\right] G_0(1,2) = \widehat{\delta}(1,2), \qquad [6.40]$$

where in the 2×2 representation given by $G^{\gtrless}$ or $G^{r,a}$ of equation [6.11] is:

$$\widehat{\delta}^{<,>} = 0 \qquad \widehat{\delta}^{r,a} = \delta(t_1 - t_2)\delta(\mathbf{r_1} - \mathbf{r_2}). \qquad [6.41]$$

6.4.2. *Dyson's equation and its justification for interacting electrons*

The calculation of the Green's function becomes quite complex if we want to take into account the interactions between particles, either direct (i.e. coulombic) or mediated by phonons or photons. We might first want to use the equation of motion method discussed in the previous section. As shown in [DIV 08], this makes it possible to express the one-particle Green's function $G(1,2)$ in terms of the two-particle one $G(1,2,3,4)$ and so on. This leads to a hierarchy of equations that are in general impossible to solve.

To tackle this problem, the idea is thus to start from the perturbation expansion given by equation [6.18] extended to the non-equilibrium case by replacing the mean values $< \phi_0|...|\phi_0 >$ by the appropriate averages $< ... >$. Non-vanishing terms must contain an equal number of creation and destruction operators. Furthermore, to each $\hat{c}_k$ must be associated a $\hat{c}_k^+$. Thus, we have to look for all possible contributions where pairs $\hat{c}_k\hat{c}_k^+$ can be associated with the form of a so-called contraction $\langle T\hat{c}_k(t_1)\hat{c}_k^+(t_2)\rangle = iG_0(1,2)$. Details of such an approach are given in [MAT 67]. The result is that the non-vanishing terms can be represented graphically in Figure 6.4 for the electron-phonon system to be discussed later. In such diagrams, thin lines with arrows correspond to free Green's functions obtained from a contraction, the thick line with arrows is the final Green's function and the dotted line corresponds to the phonon contribution, the direction of the arrows representing increasing times. All these contributions can be either calculated by brute force or by using the following general rules [MAT 67, MAH 00]:

– Wick's theorem stating that the result is the sum of all pairwise contractions [MAT 67, MAH 00];

– unconnected diagrams cancel between the numerator and denominator of equation [6.18];

– for the term of order n in the expansion, there exists $n!$ equivalent contributions due to different time orderings that cancel the $1/n!$.

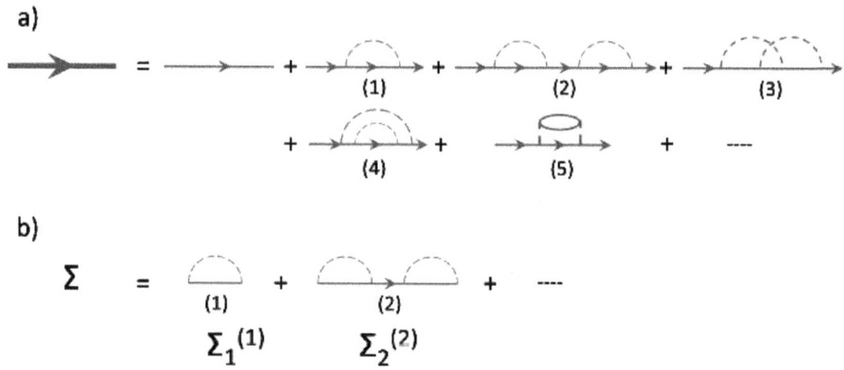

Figure 6.3. *Diagrammatic representation of a) the interacting Green's function and b) the self-energy of the electron–phonon interaction*

In Figure 6.3, we can note that each term contains one incoming and one outgoing thin line corresponding to G_0. Thus, this figure can be written under matrix form:

$$\mathbf{G} = \mathbf{G_0} + \mathbf{G_0}[\mathbf{\Sigma_1}^{(1)} + \mathbf{\Sigma_2}^{(2)} + \mathbf{\Sigma_3}^{(2)} + \mathbf{\Sigma_4}^{(2)} + \mathbf{\Sigma_5}^{(2)} + ...]\mathbf{G_0}, \qquad [6.42]$$

where each $\mathbf{\Sigma_i^{(n)}}$ is the part contained between the two lines featuring $\mathbf{G_0}$ representing the ith diagram of order n in the perturbation.

The problem of equation [6.42] is that, in general, it is impossible to calculate all diagrams (and perform their sum which might diverge [HAU 96]) and perform their sum. Usually, we calculate the lowest order like $\mathbf{\Sigma_1}^{(1)}$ and performs a partial infinite sum corresponding to powers of $\mathbf{\Sigma_1}^{(1)}$. For this let us look at $\mathbf{\Sigma_2}^{(2)}$. Its contribution to $\mathbf{G}$ can be written as $\mathbf{G_0\Sigma_1}^{(1)}\mathbf{G_0\Sigma_1}^{(1)}\mathbf{G_0}$, i.e. either $\mathbf{G_0}(\mathbf{\Sigma_1}^{(1)}\mathbf{G_0})^2$ or $(\mathbf{G_0\Sigma_1}^{(1)})^2\mathbf{G_0}$. This is typical of a "reducible" diagram that can be expressed as a product of two lower order "irreducible" diagrams. This makes it possible to perform partial infinite summations based on a given low order $\mathbf{\Sigma_i}^{(n)}$. Considering as a first example all reducible diagrams based on $\mathbf{\Sigma_1}^{(1)}$ alone, we can write two equivalent equations:

$$\mathbf{G} = \mathbf{G_0}\sum_{k=0}^{\infty}[\mathbf{\Sigma_1}^{(1)}\mathbf{G_0}]^k = \mathbf{G_0}[\mathbf{I} - \mathbf{\Sigma_1}^{(1)}\mathbf{G_0}]^{-1}, \qquad [6.43]$$

$$= \sum_{k=0}^{\infty} [\mathbf{G_0 \Sigma_1}^{(1)}]^n = [\mathbf{I} - \mathbf{G_0 \Sigma_1}^{(1)}]^{-1} \mathbf{G_0} \qquad [6.44]$$

Equation [6.44] is one example of Dyson's equation, which we shall rewrite in different forms later, before we generalize the argument leading to equation [6.44]. We can define a "self-energy" of interaction $\mathbf{\Sigma_I}$ as a finite sum of lower order $\mathbf{\Sigma_i}^{(n)}$ and sum over all higher order "reducible" diagrams built from $\mathbf{\Sigma_I}$. Then, the net result for $\mathbf{G}$ is again given by equation [6.44] with the replacement of $\mathbf{\Sigma_1}^{(1)}$ by $\mathbf{\Sigma_I}$. This can be rewritten under the usual form of Dyson's equation:

$$\mathbf{G} = \mathbf{G_0} + \mathbf{G_0 \Sigma_I G}, \qquad [6.45]$$

$$= \mathbf{G_0} + \mathbf{G \Sigma_I G_0}, \qquad [6.46]$$

or in integral form as:

$$G(1,2) = G_0(1,2) + \int G_0(1,3)\Sigma_I(3,4)G(4,2)d3d4, \qquad [6.47]$$

$$= G_0(1,2) + \int G(1,3)\Sigma_I(3,4)G_0(4,2)d3d4, \qquad [6.48]$$

the integrations being performed on the contour of Figure 6.1. We now introduce another formulation of Dyson's equation that will prove of interest in the following. For this, we take the derivative $i\frac{\partial}{\partial t_1}G(1,2)$ and make use of equation [6.40]. This leads to:

$$\left[i\frac{\partial}{\partial t_1} - h_0(\mathbf{r_1})\right] G(1,2) = \widehat{\delta}(1,2) + \int \Sigma_I(1,3)G(3,2)d3. \qquad [6.49]$$

It is also possible to derive a symmetrical result by taking the derivative $i\frac{\partial}{\partial t_2}G(1,2)$ in the second equation [6.48]. We also use for this the symmetrical of equation [6.40], which is easily shown to be obtained by changing $i\frac{\partial}{\partial t_1}$ into $-i\frac{\partial}{\partial t_2}$. This gives:

$$\left[-i\frac{\partial}{\partial t_2} - h_0(\mathbf{r_2})\right] G(1,2) = \widehat{\delta}(1,2) + \int G(1,3)\Sigma_I(3,2)d3. \qquad [6.50]$$

The power of Dyson's equation is that it can provide the solution of the problem once the self-energy Σ_I is known. However, this is misleading since the problem is then reported on the determination of Σ_I. The validity of the method thus relies on the fact that we can identify a dominant family of diagrams and neglect all others. Another interesting point is that, as it was defined, the self-energy is a combination of diagrams containing thin lines, i.e. depends on G_0 and can be written $\Sigma_I(G_0)$. This can lead in many cases to divergent series [MAT 67]. However, it is possible to renormalize each G_0 by replacing it by the result of the expansion of G, i.e. "bare" Green's functions G_0 replaced by "dressed" Green's functions. This leads to the replacement of $\Sigma_I(G_0)$ by $\Sigma_I(G)$. The advantage of this way of proceeding is that it preserves conservation laws as shown in [BAY 61, BAY 62]. The difficulty is that Dyson's equation now has to be solved self-consistently since equation [6.46] now contains $\Sigma_I(G)$. Despite the required increase in computational power, this is the most commonly adopted method, at least for the electron–phonon coupling, together with the use of the lowest order term $\Sigma_1^{(1)}$ for $\Sigma_I(G)$. However, care must be taken when adapting these weak coupling methods to intermediate or strong coupling situations where they can lead to incorrect results [WHI 92, LEV 93]. In such situations, they can be thought to provide analytic continuations, which have to be compared to the results of strong coupling models if these can be solved.

6.4.3. *Detailed form of Dyson's equation for non-equilibrium*

The expressions [6.49] and [6.50] require the calculation of an integral of the form:

$$\mathbf{A}(t_1,t_2) = \int_{\mathcal{C}} \mathbf{B}(t_1,t_3)\mathbf{C}(t_3,t_2)dt_3, \qquad [6.51]$$

over the contour $\mathcal{C}$ where each operator $\mathbf{O}$ can take the four determinations $\mathbf{O}^{\gtrless}$ or $\mathbf{O}^{r,a}$. As specified by Langreth theorem and shown in [LAN 76, HAU 96], this can be done by a deformation of the contour in order to get two contours (Figure 6.1). From this, we get:

$$\mathbf{A}^<(t_1,t_2) = \int_{-\infty}^{+\infty} dt_3 \left[\mathbf{B}^r(t_1,t_3)\mathbf{C}^<(t_3,t_2) + \mathbf{B}^<(t_1,t_3)\mathbf{C}^a(t_3,t_2)\right], \qquad [6.52]$$

$$\mathbf{A}^r(t_1,t_2) = \int_{-\infty}^{+\infty} dt_3\mathbf{B}^r(t_1,t_3)\mathbf{C}^r(t_3,t_2). \qquad [6.53]$$

When applied to $\mathbf{G}^<$ and $\mathbf{G}^r$ in equations [6.49] this leads to:

$$\begin{aligned}\left[i\tfrac{\partial}{\partial t_1} - h_0(\mathbf{r}_1)\right] G^<(1,2) &= \textstyle\int \left[\Sigma_I^r(1,3)G^<(3,2) + \Sigma_I^<(1,3)G^a(3,2)\right] d3,\\ \left[i\tfrac{\partial}{\partial t_1} - h_0(\mathbf{r}_1)\right] G^r(1,2) &= \delta(1,2) + \textstyle\int \Sigma_I^r(1,3)G^r(3,2)d3.\end{aligned} \qquad [6.54]$$

The Fourier transform of these equations can be condensed under operator form:

$$\begin{aligned} [\omega - \mathbf{H_0} - \mathbf{\Sigma}^r(\omega)]\, \mathbf{G}^<(\omega) &= \mathbf{\Sigma}^r(\omega)\mathbf{G}^a(\omega), \\ [\omega - \mathbf{H_0} - \mathbf{\Sigma}^r(\omega)]\, \mathbf{G}^r(\omega) &= \mathbf{I}, \end{aligned} \quad [6.55]$$

where **I** is the unit operator. From this, we easily obtain:

$$\begin{aligned} \mathbf{G}^< &= \mathbf{G}^r\mathbf{\Sigma}^< G^a, \\ \mathbf{G}^r &= (\omega - \mathbf{H_0} - \mathbf{\Sigma}^r(\omega))^{-1}. \end{aligned} \quad [6.56]$$

As discussed in [THY 08], the last expression for $\mathbf{G}^<$ is generally valid except when true localized states arise.

6.5. Some examples

For illustration, we first calculate the retarded Green's function of a one-dimensional (1D) free electron system. We then do the same for a 1D tight-binding chain by repeated use of Dyson's equation and show their equivalence for energies near the band minimum. Finally, we derive the general form of the contact self-energy resulting from the coupling of the active region with the source and drain in Figure 6.2. We also express the Green's functions in the energy domain where $\epsilon = \hbar\omega$ instead of the frequency domain obtained by direct Fourier transform of the time-dependent equation.

6.5.1. *1D free electron retarded Green's function*

From definition [6.55] with $\Sigma^r = 0$ we get, using a plane wave basis normalized over the length L, with eigenvalues ϵ_k:

$$\begin{aligned} g_0^r(x, x', \epsilon) = \tfrac{1}{\hbar} g_0^r(x, x', \omega) &= \tfrac{1}{L} \textstyle\sum_{k'} \frac{e^{ik'(x-x')}}{\epsilon - \epsilon_{k'}}, \\ &= -\tfrac{1}{2\pi} \cdot \tfrac{2m}{\hbar^2} \int \frac{dk' e^{ik(x-x')}}{k'^2 - k^2}, \end{aligned} \quad [6.57]$$

with $k = \sqrt{\frac{2m\epsilon}{\hbar^2}}$. For $x - x' > 0$, this can be integrated by residues in the upper half plane and for $x - x' < 0$ in the lower half plane. The net result is, for $\epsilon > 0$:

$$g_0^r(x, x', \epsilon) = -i \frac{m}{\hbar^2} \frac{e^{ik|x-x'|}}{k}. \quad [6.58]$$

6.5.2. *1D tight-binding chain*

We want to calculate the on-site Green's function of a homogeneous tight-binding chain with nearest neighbor's interaction $-\beta$ and zero on-site potential (Figure 6.4(a)). We shall do this as an illustration of a repeated use of Dyson's equation but also as an introduction to the concept of contact self-energy discussed below. We consider only retarded Green's functions g^r. We first calculate the surface Green's function g_s^r of the surface atoms ± 1 of the semi-infinite chain of Figure 6.4(b). We start from a situation where the bond between, say, atom 1 and 2 is broken. Then, in this situation $(g_0^r)_{11} = 1/E$ (where E is the energy) and $(g_0^r)_{22} = g_s^r$. We then reestablish the bond between atoms 1 and 2 giving $(g^r)_{11} = g_s^r$. The Dyson equation that relates these two situations is:

$$\begin{aligned} g_{11}^r &= (g_0^r)_{11} - (g_0^r)_{11}\beta g_{21}^r, \\ g_{21}^r &= \quad (g_0^r)_{22}(-\beta)g_{11}^r. \end{aligned} \qquad [6.59]$$

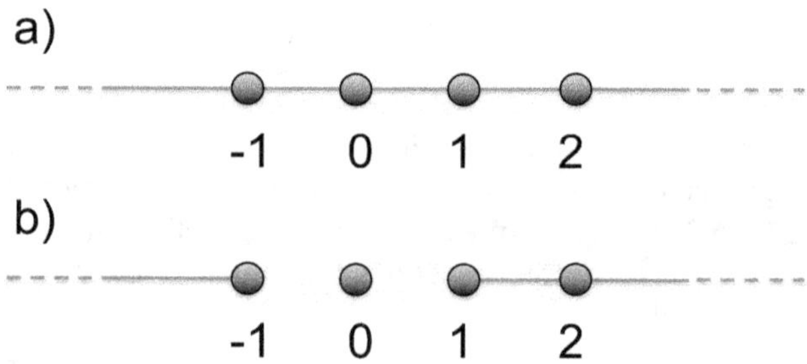

Figure 6.4. *Linear tight-binding chain considered in this part with numbered atoms a). b) Similar linear chain but isolating atom 0*

Finally, this gives a quadratic equation for g_s^r:

$$g_s^r = \frac{1}{E}\left[1 + \beta^2 (g_s^r)^2\right], \qquad [6.60]$$

whose solutions are:

$$g_s^r = \frac{E}{2\beta^2} \pm \left(\frac{E^2}{4\beta^4} - \frac{1}{\beta^2}\right)^{1/2}, \qquad [6.61]$$

for which we have to take the proper determination for the range of energies of interest.

To calculate g_{00} in Figure 6.4(a), we then reestablish the bonds between atom 0 and 1,-1, which can again be done through the use of Dyson's equation as in equations [6.59] and [6.60] to get:

$$G^r_{00} = g^r_{00} + g^r_{00} 2\beta^2 g^r_s G^r_{00}. \quad [6.62]$$

With $g^r_{00} = \frac{1}{E}$ and g^r_s from equation [6.61] we get for G^r_{00} the final result:

$$G^r_{00} = \begin{cases} -\ i(4\beta^2 - E^2)^{-1/2} \text{ for } -2\beta \leq E \leq 2\beta, \\ -\quad (E^2 - 4\beta^2)^{-1/2} \text{ for } E \leq -2\beta. \end{cases} \quad [6.63]$$

If we consider the free electron limit where $E = -2\beta + \epsilon$ and $\epsilon \ll 4\beta$, then we obtain within the continuum:

$$G^r_{00} \simeq -\frac{i}{\sqrt{4\beta\epsilon}}. \quad [6.64]$$

If we recall that the tight binding dispersion law is $E = -2\beta coska \simeq -2\beta + \beta a^2 k^2 \cdots$, then β is related to the effective mass by $\beta = \frac{\hbar^2}{2ma^2}$ so that $\beta\epsilon = (\frac{\hbar^2}{2m})^2 \frac{k^2}{a^2}$. Substituting in equation [6.64], this shows that:

$$G^r_{00}(\epsilon) = a g^r_0(\epsilon), \quad [6.65]$$

in free electron limit. This is also an illustration of the fact that the effective mass picture can be discretized by a tight-binding Hamiltonian provided that the Green's functions are related by equation [6.65], where a is the size of the mesh.

6.5.3. *Contact self-energy*

It is now useful to apply Dyson's equation to the configuration in Figure 6.2. We start from the decoupled central part of Green's function $\mathbf{g}_{cc}$ and assume that the interaction self-energy $\mathbf{\Sigma}_I$ only occurs in this active region, which is connected to the source and drain by coupling matrices $\mathbf{V}_{c\alpha}$ with $\alpha = L$(left) for the source and $\alpha = R$(right) for the drain. Then Dyson's equation reads, if there is no direct coupling between the contacts:

$$\begin{aligned} \mathbf{G}_{cc} &= \mathbf{g}_{cc} + \mathbf{g}_{cc} \textstyle\sum_\alpha \mathbf{V}_{c\alpha}\mathbf{G}_{\alpha c} + \mathbf{g}_{cc}\mathbf{\Sigma}_I \mathbf{G}_{cc}, \\ \mathbf{G}_{\alpha c} &= \mathbf{g}_{\alpha\alpha}\mathbf{V}_{\alpha c}\mathbf{G}_{cc}, \end{aligned} \quad [6.66]$$

which gives:

$$\begin{aligned} \mathbf{G}_{cc} &= \mathbf{g}_{cc} + \mathbf{g}_{cc}(\mathbf{\Sigma}_C + \mathbf{\Sigma}_I)\mathbf{G}_{cc}, \\ \mathbf{\Sigma}_c &= \sum\nolimits_{\alpha=L,R} \mathbf{V}_{c\alpha}\mathbf{g}_{\alpha\alpha}\mathbf{V}_{\alpha c} = \mathbf{\Sigma}_{cL} + \mathbf{\Sigma}_{cR}. \end{aligned} \quad [6.67]$$

As discussed before, the source and drain are assumed to be described by independent particles and to be at thermal equilibrium with their own Fermi level μ_L or μ_R or Fermi distribution f_L or f_R. This means that $\Sigma_{c\alpha=L\text{or}R}$ obeys the fluctuation-dissipation theorem. From equation [6.67], the retarded component is given by:

$$\mathbf{\Sigma}^r_{c\alpha} = \mathbf{V}_{c\alpha}\mathbf{g}^r_{\alpha\alpha}\mathbf{V}_{\alpha c}, \quad [6.68]$$

while from equations [6.34] and [6.35]:

$$\mathbf{\Sigma}^<_{c\alpha} = 2\pi i \mathbf{V}_{c\alpha}\delta(\omega - \mathbf{H}_\alpha)\mathbf{V}_{\alpha c} f_\alpha(\omega). \quad [6.69]$$

All these relations are under matrix form and imply summations over the eigenstates of the source and drain.

6.6. The ballistic regime

We shall first derive the famous Landauer formula [LAN 57, LAN 89, BÜT 85] from elementary considerations. We then apply the NEGF theory to the ballistic case and show how this generalizes the Landauer formula. We finally make the connection to scattering theory, which is essentially equivalent.

6.6.1. *The Landauer formula*

We consider the device in Figure 6.2 consisting of a conducting central part between two perfect contacts. As discussed before, this means that the left and right reservoirs inject electrons into the active part with their own thermal equilibrium distribution $f_L(\epsilon)$ and $f_R(\epsilon)$. Each contact is assumed to allow for perfect injection and no reflection. The active part is taken to be a perfect nanowire with constant potential and independent 1D propagation modes due to lateral confinement. Each 1D mode leads to a current, which is the sum of the left and right injection:

$$I = \sum_k \rho_k v_k \left[f_L(\epsilon_k) - f_R(\epsilon_k)\right], \quad [6.70]$$

where the ϵ_k of this subsection are the energy eigenvalues ($\epsilon_k = \hbar\omega_k$) and where the electron density $\rho_k = -e/L$ and their velocity $v_k = \frac{1}{\hbar}\frac{d\epsilon_k}{dk}$. Changing the discrete sum to an integral, we readily obtain:

$$I = -\frac{2e}{h}\int d\epsilon\,[f_L(\epsilon) - f_R(\epsilon)], \qquad [6.71]$$

where the factor 2 corresponds to spin degeneracy. If the applied potential V_{RL} between drain (R) and source (S) is small, then $\mu_L - \mu_R = eV_{RL}$ is also small. As the Fermi function is $f(\epsilon - \mu)$, we can expand the integrand in equation [6.71] to get:

$$I = \frac{2e}{h}\int d\epsilon\frac{df}{d\epsilon}(\mu_L - \mu_R)\mathrm{d}\epsilon, \qquad [6.72]$$

which can also be written:

$$I = \frac{2e^2}{h}V_{RL}. \qquad [6.73]$$

The quantity $G_0 = 2e^2/h$ is the quantum of conductance corresponding to one individual mode. The conductance thus increases by steps equal to G_0 when V_{RL} is increased corresponding to the contribution of new 1D modes. We can extend the previous arguments by considering a density of modes $M(\epsilon)$ as well as the possibility that the active region contains elastic scatterers so that the transmission coefficient is $T(\epsilon)$ instead of 1. This obviously leads to the generalized Landauer formula [MEI 92, DAT 95] extending equation [6.71]:

$$I = -\frac{2e}{h}\int M(\epsilon)T(\epsilon)\,[f_L(\epsilon) - f_R(\epsilon)]\,d\epsilon. \qquad [6.74]$$

6.6.2. *The NEGF derivation of the Landauer formula*

We now want to calculate $\frac{\partial n(\mathbf{r},t)}{\partial t}$ from equation [6.13] in terms of the time derivatives of $G^<(1,2)$. The first one is given by equation [6.54] deduced from equation [6.49]. The second one can be deduced in the same way from equation [6.50] giving:

$$\left[i\frac{\partial}{\partial t_2} - h_0(\mathbf{r_2})\right] G^<(1,2) = \int \left[G^r(1,3)\Sigma_I^<(3,2) + G^<(1,3)\Sigma_I^a(3,2)\right] d3. \quad [6.75]$$

The derivative of $n(\mathbf{r}, t)$ of equation [6.13] with respect to time can thus be deduced from that of the $\mathbf{G}^<$ matrix, which we express as:

$$i\frac{\partial}{\partial t}\mathbf{G}^<(t,t) = i \lim_{t_1,t_2 \to t}(\frac{\partial}{\partial t_1} + \frac{\partial}{\partial t_2})\mathbf{G}^<(t_1,t_2). \qquad [6.76]$$

We calculate this by taking the difference of equations [6.54] and [6.75]. For simplicity, we express the result under matrix form noting that the matrix elements of $h_0(\mathbf{r_1})G(1,2)$ are those of $\mathbf{GH_0}$ while those of $h_0(\mathbf{r_2})G(1,2)$ correspond to $\mathbf{H_0G}$. We also use the fact that all quantities depend on the time difference $t_1 - t_3 = \tau$ and obtain:

$$i\frac{\partial}{\partial t}\mathbf{G}^<(t,t) = [\mathbf{G}^<(t,t), \mathbf{H}_0] + \int [\mathbf{\Sigma}_I^r(\tau)\mathbf{G}^<(-\tau) + \mathbf{\Sigma}_I^<(\tau)\mathbf{G}^a(-\tau)$$
$$-\mathbf{G}^r(\tau)\mathbf{\Sigma}_I^<(-\tau) - \mathbf{G}^<(\tau)\mathbf{\Sigma}_I^a(-\tau)]d\tau. \qquad [6.77]$$

Using the definitions of equation [6.11] for the retarded and advanced quantities (also valid for $\mathbf{\Sigma}^r$ and $\mathbf{\Sigma}^a$), we can transform equation [6.77] into:

$$i\frac{\partial}{\partial t}\mathbf{G}^<(t,t) = [\mathbf{G}^<(t,t), \mathbf{H}_0] + \int_{-\infty}^{0} d\tau[\mathbf{\Sigma}_I^>(\tau)\mathbf{G}^<(-\tau) - \mathbf{\Sigma}_I^<(\tau)\mathbf{G}^>(-\tau)$$
$$-\mathbf{G}^>(\tau)\mathbf{\Sigma}_I^<(-\tau) + \mathbf{G}^<(\tau)\mathbf{\Sigma}_I^>(-\tau)]. \qquad [6.78]$$

If we take the trace of this expression over a suitable basis set, then from trace invariance and changing τ to $-\tau$ in the last two terms, we obtain:

$$i\mathrm{Tr}\frac{\partial \mathbf{G}^<}{\partial t} = \mathrm{Tr}[\mathbf{G}^<, \mathbf{H}_0]$$
$$+\mathrm{Tr}\int_{-\infty}^{+\infty} d\tau[\mathbf{\Sigma}_I^>(\tau)\mathbf{G}^<(-\tau) - \mathbf{\Sigma}_I^<(\tau)\mathbf{G}^>(-\tau)]. \qquad [6.79]$$

It is this general formula that we shall exploit in several situations and, first of all, to generalize the Landauer formula.

We thus consider a ballistic system corresponding to the situation where the left part is disconnected and then reconnected (see Figure 6.5). Then, the current is the charge increase per unit time on the right side, $ie\mathrm{Tr}_R\frac{\partial \mathbf{G}^<(t,t)}{\partial t}$ where e is the absolute value of the electron charge. To apply equation [6.79], we use the fact, as shown in section 6.5.3, that one can treat the problem in a basis set corresponding to the right-hand side only with the condition of adding a contact self-energy $\mathbf{\Sigma}_L$. In this situation, $\mathbf{H}_0$ corresponds to the disconnected situation and $\mathrm{Tr}_R[\mathbf{H}_0, \mathbf{G}^<] = 0$. Furthermore, $\mathbf{\Sigma}_I = 0$ for a ballistic system but has to be replaced by $\mathbf{\Sigma}_L$. We thus get:

$$\begin{aligned} J_L = & \quad ie\mathrm{Tr}_R \tfrac{\partial \mathbf{G}^<(t,t)}{\partial t} \\ = & \; e\mathrm{Tr}_R \textstyle\int_{-\infty}^{+\infty} d\tau [\mathbf{\Sigma}_L^>(\tau)\mathbf{G}^<(-\tau) - \mathbf{\Sigma}_L^<(\tau)\mathbf{G}^>(-\tau)], \end{aligned} \qquad [6.80]$$

which can be Fourier transformed and written in the energy domain ($\epsilon = \hbar\omega$). For this, we note that in our previous general treatment all quantities that have the dimension of an energy were in units of $\hbar$ while the Green's functions were expressed in inverse units. Thus:

$$J_L = -\frac{e}{\hbar}\mathrm{Tr}_R \int_{-\infty}^{+\infty} \frac{d\epsilon}{2\pi}[\mathbf{G}^>(\epsilon)\mathbf{\Sigma}_L^<(\epsilon) - \mathbf{G}^<(\epsilon)\mathbf{\Sigma}_L^>(\epsilon)], \qquad [6.81]$$

since the product $\mathbf{G}(\epsilon)\mathbf{\Sigma}(\epsilon) = \mathbf{G}(\omega)\mathbf{\Sigma}(\omega)$.

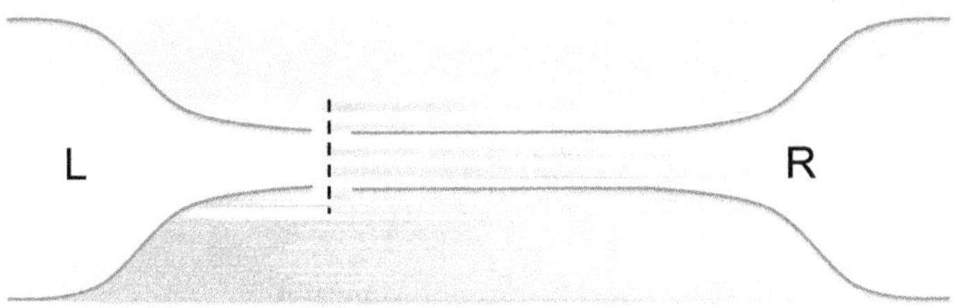

Figure 6.5. *Calculation of the current: the system is now virtually divided in two parts, left and right*

We restrict ourselves to the central active region by using the concept of contact self-energy $\mathbf{\Sigma}_C$. With $\mathbf{G}^\gtrless = \mathbf{G}^r\mathbf{\Sigma}_C^\gtrless\mathbf{G}^a$, we get:

$$\begin{aligned} J_L = & -\frac{e}{\hbar}\mathrm{Tr}_C \int_{-\infty}^{+\infty} \frac{d\epsilon}{2\pi}[\mathbf{G}^r(\epsilon)\mathbf{\Sigma}_C^>(\epsilon)\mathbf{G}^a(\epsilon)\mathbf{\Sigma}_L^<(\epsilon) \\ & -\mathbf{G}^r(\epsilon)\mathbf{\Sigma}_C^<(\epsilon)\mathbf{G}^a(\epsilon)\mathbf{\Sigma}_L^>(\epsilon)]. \end{aligned} \qquad [6.82]$$

We can now use the fact shown by equations [6.69] and [6.35]:

$$\mathbf{\Sigma}_{C\alpha}^<(\epsilon) = i\mathbf{\Gamma}_\alpha(\epsilon)f_\alpha(\epsilon)$$

$$\Sigma^{>}_{C\alpha}(\epsilon) = -i\Gamma_\alpha(\epsilon)[1 - f_\alpha(\epsilon)] \qquad [6.83]$$
$$\Gamma_\alpha(\epsilon) = 2\pi \mathbf{V}_{C\alpha}\delta(\epsilon - \mathbf{H}_\alpha)\mathbf{V}_{\alpha C},$$

where all quantities are in the energy domain. As $\Sigma_C = \Sigma_L + \Sigma_R$, the terms in Σ_L disappear and we get:

$$J_L = -\frac{e}{\hbar}\int_{-\infty}^{+\infty} \frac{d\epsilon}{2\pi}\mathrm{Tr}_C[\mathbf{G}^r(\epsilon)\Gamma_R(\epsilon)\mathbf{G}^a(\epsilon)\Gamma_L(\epsilon)](f_L(\epsilon) - f_R(\epsilon)). \qquad [6.84]$$

A similar derivation applies to J_R with the same result, showing current conservation. This is the generalization of the Landauer formula (equation [6.74]) giving the appropriate method for calculating $M(\epsilon)T(\epsilon)$ for any ballistic system [KEL 65]. Let us note that this approach allows us to handle a Hartree or Schrödinger–Poisson self-consistent procedure since this still represents a single particle theory. For this, we need the particle density, which is given by equation [6.13]. We extend equation [6.13] to define the particle density operator $\mathbf{n} = -i\mathbf{G}^<(t,t)$, which, from equations [6.56] and [6.82] and performing a Fourier transform, gives:

$$\mathbf{n} = \int \frac{d\epsilon}{2\pi}\mathbf{G}^r(\epsilon)[\Gamma_L(\epsilon)f_L(\epsilon) + \Gamma_R(\epsilon)f_R(\epsilon)]\mathbf{G}^a(\epsilon). \qquad [6.85]$$

From this, we can obtain the local density as $n(\mathbf{r}) =< \mathbf{r}|\mathbf{n}|\mathbf{r} >$ in the continuous case or $n_i = Tr_i < i\alpha|\mathbf{n}|i\alpha >$ in the tight binding case where $|i\alpha >$ are the atomic states of site i.

6.6.3. *Relation with scattering theory*

Some calculations on these systems have made use of a scattering approach. This is characterized by the fact that incident waves from the left (source), populated with the Fermi function f_L, are scattered by the active region, giving rise to a transmitted wave in the drain. A similar situation occurs for states incident from the right that are populated according to f_R.

As shown in [DEL 04], the approach proceeds by calculating the retarded Green's function starting from the decoupled situation in Figure 6.2 and then reconnecting the three parts. The assumptions are thus essentially the same as in the NEGF ballistic theory and give identical results such as equations [6.84] and [6.85].

6.7. The electron–phonon interaction

We derive here, as concisely as possible, the full expression of the electron–phonon self-energy in the Born approximation. We first define the phonon states, then the electron–phonon coupling constants and finally evaluate the lowest order significant term in the expansion of equation [6.18] of the Green's function. We end up with simplified expressions commonly applied to semiconductors.

6.7.1. *The phonon Hamiltonian*

We call $u_{i\alpha}$ the α component of the displacement of atom i. To take into account the case of different atomic masses, we define the renormalized atomic displacement:

$$v_{i\alpha} = \frac{u_{i\alpha}}{\sqrt{M_i}}, \qquad [6.86]$$

where M_i is the mass of atom i. The phonon Hamiltonian is, thus, in the harmonic approximation:

$$H_{ph} = -\frac{\hbar^2}{2}\sum_{i\alpha}\frac{\partial^2}{\partial v_{i\alpha}^2} + 1/2\sum_{i\alpha,j\beta} v_{i\alpha}D_{i\alpha,j\beta}v_{j\beta}. \qquad [6.87]$$

Calling v_q the eigenstates of the matrix D and taking into account that they can be complex numbers (e.g. Bloch waves in crystals) we transform equation [6.87] into:

$$H_{ph} = \sum_q[-\frac{\hbar^2}{2}\frac{\partial^2}{\partial v_q^*\partial v_q} + \frac{1}{2}\omega_q^2 v_q^* v_q]. \qquad [6.88]$$

where q is, for the moment, a general phonon index and is not specified as a wave vector. We now define the creation a_q^+ and destruction a_q operators:

$$a_q^+ = \frac{1}{\sqrt{2\hbar\omega_q}}(-\hbar\frac{\partial}{\partial v_q} + \omega_q v_q^*)$$

$$a_q = \frac{1}{\sqrt{2\hbar\omega_q}}(\hbar\frac{\partial}{\partial v_q^*} + \omega_q v_q), \qquad [6.89]$$

from which the phonon Hamiltonian can be written under second quantized form:

$$H_{ph} = \sum_q \hbar\omega_q(a_q^+ a_q + 1/2). \qquad [6.90]$$

6.7.2. *Electron–phonon interaction*

Its first order matrix elements between two electronic states are given by:

$$V_{kl} = \sum_q (\frac{\partial V_{kl}}{\partial v_q}) \cdot v_q. \qquad [6.91]$$

From equation [6.89] and expressing V in second quantization $\sum_{kl} V_{kl} c_k^+ c_l$, we get:

$$V = \sum_{k,l,q} M_{kl}(q) c_k^+ c_l A_q, \qquad [6.92]$$

with

$$M_{kl}(q) = \sqrt{\frac{\hbar}{2\omega_q} \frac{\partial V_{kl}}{\partial v_q}}; \qquad A_q = a_q + a_q^+. \qquad [6.93]$$

This is the form suitable for calculating the Green's function expansion equation [6.18].

6.7.3. *Electron–phonon self-energy*

We start to evaluate the lowest order terms in the expansion of equation [6.18]. The first term $n = 1$ vanishes since it is linear in A_q and that the average of a_q or a_q^+ vanishes. The first significant term is thus of second order and can be written as:

$$(-i)^2 \sum_{kl,k'l'} M_{kl}(q) M_{k'l'}(q^*) \; \langle \; T \hat{c}_k^+(t_1') \hat{c}_l(t_1') \hat{c}_{k'}^+(t_2') \hat{c}_{l'}(t_2') \hat{c}_{k_1}(t_1) \hat{c}_{k_2}^+(t_2) \rangle$$
$$\langle T A_q(t_1') A_{q^*}(t_2') \rangle. \qquad [6.94]$$

In each of the two averages there are as many creation as destruction operators, which is fortunate since otherwise the result would be zero. Furthermore to each $\hat{c}_k$ must be associated a $\hat{c}_k^+$ in order to give a non-zero result. Thus, we have to look for all possible contributions where pairs $\hat{c}_k^+ \hat{c}_k$ can be associated to form a "contraction". The non-vanishing results can be represented graphically in terms of diagrams [MAT 67, HAU 96, MAH 00] of which we show an example later. All these contributions can be calculated either by brute force or by using the following general rules: (1) the cancellation of the unconnected diagrams between numerator and denominator in [MAT 67], (2) existence of $n!$ equivalent contributions due to different time orderings canceling the $1/n!$ and (3) Wick's theorem for contractions [MAT 67]. Once this is done, the important contribution between k_1 and k_2 is given by:

$$-\sum_{l,q} M_{k_1 l}(q) M_{lk_2}(q^*) iG^0_{k_1 k_2}(t_1, t_1') iG^0_{ll}(t_1' t_2') iG^0_{k_2 k_2}(t_2' t_2) iD_{qq^*}(t_1' t_2'), \qquad [6.95]$$

with

$$iG^0_{\alpha\alpha}(t_i, t_j') = \langle T\hat{c}_\alpha(t_i)\hat{c}^+_\alpha(t_j')\rangle$$
$$iD^0_{qq^*}(t_1' t_2') = \langle T\hat{A}_q(t_1')\hat{A}_{q^*}(t_2')\rangle. \qquad [6.96]$$

The corresponding diagram is shown in Figure 6.6 where the straight line represents the Green's function and the dashed line represents the interaction. This diagram is of the Hartree–Fock type. There also exists a Hartree one which is usually neglected but is essential for polaronic systems [HYL 94].

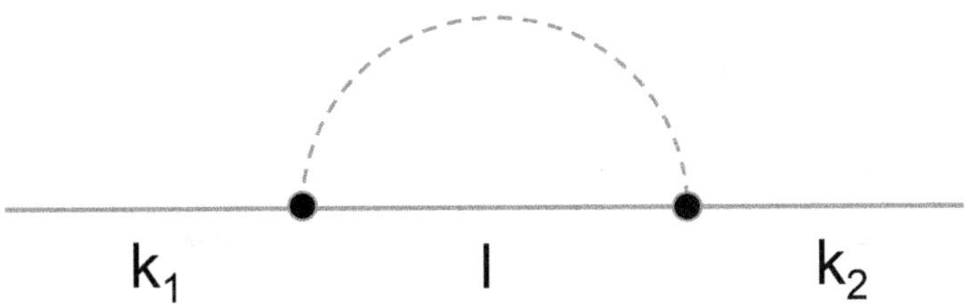

Figure 6.6. *Diagram corresponding to the electron–phonon self-energy*

Now we want to express equation [6.95] as the first term of the expansion of Dyson's equation $\mathbf{G} = \mathbf{G}_0 + \mathbf{G}_0\boldsymbol{\Sigma}(G_0)\mathbf{G}_0 + \cdots$. As equation [6.95] is the expansion of iG, we get the following expression for the electron–phonon self-energy:

$$\boldsymbol{\Sigma}_{e-ph}(t_1', t_2') = \sum_q \mathbf{M}(q)\mathbf{G}^0(t_1', t_2') D^0_{qq^*}(t_1', t_2')\mathbf{M}^+(q^*). \qquad [6.97]$$

We have now to calculate the phonon's Green's function. For applications to the non-equilibrium case, it will be enough to determine $D^{0(\gtrless)}$ for which there is no time ordering. For this we note that, from the equation of evolution like equations [6.36]–[6.38] $\hat{a}_q(t) = a_q e^{-i\omega_q t}$, that $D^{0<}(t_1', t_2')$ is given by the normal time ordering, i.e. equation [6.96] without T and $D_0^{>}(t_1', t_2')$ by the reverse ordering. This gives:

$$iD^{0\gtrless}(t_1', t_2') = (n_q + 1)e^{\pm i\omega_q(t_1 - t_2)} + n_q e^{\mp i\omega_q(t_1 - t_2)}, \qquad [6.98]$$

where n_q is the phonon occupation number assumed to be the equilibrium $n_q = [e^{\frac{\hbar\omega}{k_B T}} - 1]^{-1}$. We can finally determine $\Sigma^{\gtrless}$ by noticing that products such as $G^0 D^0$ in equation [6.97] with the same time dependence obviously cause no problems on the time contour in Figure [6.1]. Thus:

$$\mathbf{\Sigma}^{\gtrless}_{e-ph}(t_1, t_2) = i\sum_q \mathbf{M}(q)\mathbf{G}^{0\gtrless}(t_1, t_2)D^{0\gtrless}_{qq^*}(t_1, t_2)\mathbf{M}^+(q^*). \qquad [6.99]$$

Fourier transforming equation [6.99] and writing the result in the energy domain $\epsilon = \hbar\omega$ give the well-known expression [MAT 67, MAH 00]:

$$\mathbf{\Sigma}^{\gtrless}_{e-ph}(\epsilon) = \sum_q \mathbf{M}(q)\Big\{(n_q + 1)\mathbf{G}^{0\gtrless}(\epsilon \mp \hbar\omega_q) + n_q\mathbf{G}^{0\gtrless}(\epsilon \pm \hbar\omega_q)\Big\}\mathbf{M}^+(q^*). \qquad [6.100]$$

6.7.4. *Explicit expressions for $\Sigma^{\gtrless}_{e-ph}$ in the effective mass approximation*

Usually bulk calculations for semiconductors rely on the effective mass approximation (EMA) together with the concept of deformation potentials. We present here the application of the same notions to the case of confined structures. The justification of this approach is based on the fact that several local physical quantities, e.g. the local dielectric constant [DEL 04, DEL 03], reach their bulk values within one interatomic distance from the interfaces. This argument will be applied to the electron–phonon self-energy. Of course, this will neglect possible fine structure due to phonon confinement but should not affect average quantities.

6.7.4.1. *Longitudinal acoustic phonons*

Let us first discuss the effect of longitudinal acoustic phonons. The electronic states near a conduction band minimum are, in the EMA, taken as $\phi(\mathbf{r})e^{i\mathbf{kr}}$ where $\phi(\mathbf{r})$ is periodic. In the long wavelength limit, a similar approximation holds for acoustic phonons for which there is no relative atomic motion in the unit cell. Thus, here $u_{i\alpha} = u_\alpha(\mathbf{q})e^{i\mathbf{qR_i}}$ where $u_\alpha(\mathbf{q})$ represents a common translation of the atoms of the unit cell, the mass thus being the total mass in the cell. Using the fact that the normalized eigenmodes are $u_{i\alpha} = \frac{u(\mathbf{q})}{\sqrt{N}} e^{i\mathbf{q}\mathbf{R_i}}$, where N is the number of cells in the crystal of volume V, we can write $M_{k,k'}$ between two Bloch states as:

$$M_{\mathbf{k},\mathbf{k}'}(\mathbf{q}) = \sqrt{\frac{\hbar}{2M\omega_q}} \frac{1}{V\sqrt{N}} \int e^{i(\mathbf{k}'-\mathbf{k}+\mathbf{q})\mathbf{r}} [|\phi(\mathbf{r})|^2 \sum_i \frac{\partial V}{\partial u_{i\alpha}} e^{i\mathbf{q}(\mathbf{R_i}-\mathbf{r})}] d^3\mathbf{r}. \quad [6.101]$$

The second term in the integrand is periodic and we replace it by its average over one cell, neglecting the so-called Umklapp processes. This leaves us with:

$$M_{\mathbf{k},\mathbf{k}'} = \sqrt{\frac{\hbar}{2M\omega_q}} \frac{1}{\sqrt{N}} \delta_{\mathbf{k}',\mathbf{k}-\mathbf{q}} \frac{1}{\Omega} \int_\Omega |\phi(\mathbf{r})|^2 \sum_i \frac{\partial V}{\partial u_{i\alpha}} e^{i\mathbf{q}(\mathbf{R_i}-\mathbf{r})} d^3\mathbf{r}, \quad [6.102]$$

where the integral is now restricted to one unit cell of volume Ω. Contributions to this integral are limited in space and in the long wavelength limit $\mathbf{q} \to 0$ we can expand the exponential. The term $\mathbf{q} = \mathbf{0}$ corresponds to a translation and vanishes. One is then left with the first-order term and we can rewrite equation [6.102] as:

$$M_{\mathbf{k},\mathbf{k}'} = i\mathbf{q}\sqrt{\frac{\hbar}{2M\omega_q}} \frac{1}{\sqrt{N}} \delta_{\mathbf{k}',\mathbf{k}-\mathbf{q}} \frac{1}{\Omega} \int_\Omega |\phi(\mathbf{r})|^2 \sum_i \frac{\partial V}{\partial u_{i\alpha}} \mathbf{n}.(\mathbf{R_i} - \mathbf{r}) d^3\mathbf{r}, \quad [6.103]$$

where $\mathbf{n}$ is the unit vector of $\mathbf{q}$. For longitudinal modes, the last term is a constant with dimension of an energy called the deformation potential. Thus, we can write:

$$M_{\mathbf{k},\mathbf{k}-\mathbf{q}} = M_\mathbf{q} = i\mathbf{q}\sqrt{\frac{\hbar}{2M\omega_q}} \frac{\Delta}{\sqrt{N}}. \quad [6.104]$$

Finally, substituting equation [6.104] into [6.100] gives, for acoustic phonons:

$$\Sigma_{ac}^{\gtrless}(\mathbf{r_1}, \mathbf{r_2}, \epsilon) = \frac{\Delta^2}{N} \sum_q \frac{\hbar}{2M} \frac{q^2}{\omega_q} [(n_q + 1) G^{0\gtrless}(\mathbf{r_1}, \mathbf{r_2}, \epsilon \mp \hbar\omega_q)$$
$$+ n_q G^{0\gtrless}(\mathbf{r_1}, \mathbf{r_2}, \epsilon \pm \hbar\omega_q] e^{i\mathbf{q}(\mathbf{r_1}-\mathbf{r_2})}. \quad [6.105]$$

This is a basic formula for current applications. We can still simplify it [JIN 06] by noticing that, for long-range acoustic phonons, $\hbar\omega_q \ll k_B T$, $n_q \simeq n_q + 1 \simeq \frac{k_B T}{\hbar\omega}$. If one also neglects the ω_q dependences of $G^{0\gtrless}$, we get:

$$\Sigma_{ac}^{\gtrless}(\mathbf{r_1}, \mathbf{r_2}, \epsilon) = \frac{\Delta^2 k_B T}{M v_l^2} G^{0\gtrless}(\mathbf{r_1}, \mathbf{r_2}, \epsilon) \frac{1}{N} \sum_{\mathbf{q}} e^{i\mathbf{q}(\mathbf{r_1}-\mathbf{r_2})}, \qquad [6.106]$$

where $v_l = \frac{\omega_q}{\|\mathbf{q}\|}$ is the velocity of sound waves. A final approximation is to write $\frac{1}{N}\sum_{\mathbf{q}} e^{i\mathbf{q}(\mathbf{r_1}-\mathbf{r_2})} = \Omega\delta(\mathbf{r_1} - \mathbf{r_2})$, where Ω is the volume of the unit cell. This leads to the widely used expression [JIN 06]:

$$\Sigma_{ac}^{\gtrless}(\mathbf{r_1}, \mathbf{r_2}, \epsilon) = \frac{\Delta^2 k_B T}{\rho v_l^2} G^{0\gtrless}(\mathbf{r_1}, \mathbf{r_2}, \epsilon)\delta(\mathbf{r_1} - \mathbf{r_2}), \qquad [6.107]$$

where ρ is the density. There are several difficulties with this expression. The main one is that the $\mathbf{q}$ wave vectors of the phonons are limited to the first Brillouin zone. Thus, at least, $\delta(\mathbf{r_1} - \mathbf{r_2})$ should be a broadened δ function, roughly constant and equal to $1/\Omega$ over the unit cell (where Ω is the volume of the unit cell) and zero outside. This means that when discretizing the continuous equations, care should be taken about the size a of the cubic mesh for $\delta(\mathbf{r_1} - \mathbf{r_2})$ to remain diagonal, i.e. treated as an on-site quantity. One has thus to perform a delicate compromise for the choice of a, i.e. small enough to provide an accurate approximation to the EMA continuous equations but large enough for $\Sigma^{\gtrless}$ to be treated as an on-site quantity. A value of $a \sim 2$Å for silicon seems appropriate since the size of the unit cell is $(5, 43\text{Å})^{33}$. Nevertheless, it should be of interest to test this point and also solve equation [6.106] in an appropriate way.

6.7.4.2. *Optical phonons*

For indirect gap semiconductors there are several conduction band minima and there can be intervalley scattering. We can still apply the previous formalism by expanding the phonons q around $\mathbf{Q} = \mathbf{k_j} - \mathbf{k_i}$, the difference of position of the two minima in reciprocal space. As the optical phonons are flat, we can take an average value of ω_Q. The self-energy can then be expressed in the same form as equation [6.105] giving [MAH 87]:

$$\Sigma_{op}^{\gtrless}(\mathbf{r_1}, \mathbf{r_2}, \epsilon) = \frac{\hbar(\Delta_0 Q)^2}{2\omega_Q \rho}[(n_{\mathbf{Q}} + 1)G^{0\gtrless}(\mathbf{r_1}, \mathbf{r_2}, \epsilon \mp \hbar\omega_{\mathbf{Q}})$$
$$+ n_{\mathbf{Q}} G^{0\gtrless}(\mathbf{r_1}, \mathbf{r_2}, \epsilon \pm \hbar\omega_{\mathbf{Q}}]\delta(\mathbf{r_1} - \mathbf{r_2}). \qquad [6.108]$$

6.7.5. *Conserving properties: continuity equations*

An important criterion for the quality of approximate expressions of the interaction self-energy is that it respects general conservation laws. For transport properties, this corresponds to the continuity equations. We discuss this subject here with some emphasis on the electron–phonon case. Finally, we investigate the lowest order approximation (LOA) and its possible analytic continuation. The continuity equation is expressed by the fact that at a point $\mathbf{r}$:

$$div.\overrightarrow{j} + \frac{\partial n(\mathbf{r})}{\partial t} = 0, \qquad [6.109]$$

where $n(\mathbf{r})$ and $\overrightarrow{j}$ are the particle and current densities, respectively. From equation [6.13], we can express $\frac{\partial n}{\partial t}$ as:

$$\frac{\partial n}{\partial t} = -i\langle \mathbf{r}|\frac{\partial}{\partial t}\mathbf{G}^{<}|\mathbf{r}\rangle, \qquad [6.110]$$

given by equation [6.77]. On the other hand, we can use the definition of equation [6.14] for the current density to write:

$$\overrightarrow{j} = -\frac{\hbar i}{2m}\langle \Psi^{+}(\mathbf{r},t)(\nabla\Psi(\mathbf{r},t)) - (\nabla\Psi^{+}(\mathbf{r},t))\Psi(\mathbf{r},t)\rangle, \qquad [6.111]$$

so that $\overrightarrow{\nabla}.\overrightarrow{j}$ can be written as:

$$\nabla.\overrightarrow{j} = -\frac{\hbar i}{2m}\langle \Psi^{+}(\mathbf{r},t)(\nabla^2\Psi(\mathbf{r},t)) - (\nabla^2\Psi^{+}(\mathbf{r},t))\Psi(\mathbf{r},t)\rangle. \qquad [6.112]$$

Expressing $h_0(\mathbf{r}) - V(\mathbf{r})$ in units of $\hbar$ as $-\frac{\hbar^2}{2m}\Delta^2$, we easily get:

$$div.\overrightarrow{j} = i\langle \mathbf{r}|[\mathbf{G}^{<}, \mathbf{H}_0]|\mathbf{r}\rangle. \qquad [6.113]$$

Combining equations [6.110], [6.113], and [6.77] readily gives at point $\mathbf{r}$:

$$div.\overrightarrow{j}(\mathbf{r}) + \frac{\partial n(\mathbf{r})}{\partial t} = \int_{-\infty}^{0} d\tau \langle \mathbf{r}|\mathbf{\Sigma}_I^{<}(\tau)\mathbf{G}^{>}(-\tau) + \mathbf{G}^{>}(\tau)\mathbf{\Sigma}_I^{<}(-\tau)$$
$$-\mathbf{\Sigma}_I^{>}(\tau)\mathbf{G}^{<}(-\tau) - \mathbf{G}^{<}(\tau)\mathbf{\Sigma}^{>}(-\tau)|\mathbf{r}\rangle. \qquad [6.114]$$

6.7.5.1. *Current conservation*

To be coherent, Σ_I in the active region must be chosen in such a way that current conservation expressed by equation [6.109] is satisfied. This requires that the right-hand side of the equation [6.114] vanishes. Following [MAH 87], let us then write this condition more explicitly:

$$\int_{-\infty}^{0} d\mathbf{r}' d\tau \langle \mathbf{r} | \mathbf{\Sigma}_I^{<}(\mathbf{r}, \mathbf{r}', \tau) \mathbf{G}^{>}(\mathbf{r}', \mathbf{r}, -\tau) + \mathbf{G}^{>}(\mathbf{r}, \mathbf{r}', \tau) \mathbf{\Sigma}_I^{<}(\mathbf{r}', \mathbf{r}, -\tau)$$
$$- \mathbf{\Sigma}_I^{>}(\mathbf{r}, \mathbf{r}', \tau) \mathbf{G}^{<}(\mathbf{r}', \mathbf{r}, -\tau) - \mathbf{G}^{<}(\mathbf{r}, \mathbf{r}', \tau) \mathbf{\Sigma}_I^{>}(\mathbf{r}', \mathbf{r}, -\tau) | \mathbf{r} \rangle = 0 \qquad [6.115]$$

We first try to see if this is satisfied for the electron–phonon interaction in the LOA. The self-energy in real space can be obtained from equation [6.99] as:

$$\Sigma_{e-ph}^{\gtrless}(1,2) =$$
$$\sum_{\mathbf{k},\mathbf{q}} \langle \mathbf{r_1} | \mathbf{q} \rangle M_{\mathbf{k},\mathbf{k}-\mathbf{q}} G^{0\gtrless}_{\mathbf{k}-\mathbf{q},\mathbf{k}-\mathbf{q}}(t_1, t_2) M^*_{\mathbf{k},\mathbf{k}-\mathbf{q}} \langle \mathbf{k} | \mathbf{r_2} \rangle D^{0\gtrless}_{\mathbf{q}}(t_1, t_2), \qquad [6.116]$$

where again 1 stands for $\mathbf{r}_1, t_1$. Using equation [6.104] and changing the summation from $\mathbf{k}$ to $\mathbf{k} - \mathbf{q}$, we can rewrite this since $\langle \mathbf{r} | \mathbf{k} \rangle \propto e^{i\mathbf{r}\mathbf{k}}$

$$\Sigma_{e-ph}^{\gtrless}(1,2) = g^{\gtrless}(1,2) G^{0\gtrless}(1,2), \qquad [6.117]$$

with:

$$g^{\gtrless}(1,2) = \sum_{\mathbf{q}} |M_{\mathbf{q}}|^2 D^{0\gtrless}_{\mathbf{q}}(t_1, t_2) e^{i\mathbf{q}(\mathbf{r_1} - \mathbf{r_2})}. \qquad [6.118]$$

Now we can rewrite equation [6.115] with these notations:

$$\int d\tau d\mathbf{r}' [G^{0>}(\mathbf{r}, \mathbf{r}', \tau) G^{0<}(\mathbf{r}', \mathbf{r}, -\tau) [g^{<}(\mathbf{r}', \mathbf{r}, -\tau) - g^{>}(\mathbf{r}, \mathbf{r}', \tau)]$$
$$+ [G^{0<}(\mathbf{r}, \mathbf{r}', \tau) G^{0>}(\mathbf{r}', \mathbf{r}, -\tau)(g^{<}(\mathbf{r}, \mathbf{r}', \tau) - g^{>}(\mathbf{r}', \mathbf{r}, -\tau))] = 0. \qquad [6.119]$$

From the definition of equation [6.118] of $g^{\gtrless}(1,2)$ and the expression [6.98] of D^0_q, it is clear that $g^{>}(1,2) = g^{<}(2,1)$ so that the conservation condition of equation [6.119] is automatically satisfied. The conclusion of this section is that the LOA to the expansion of the Green's function or to the Dyson equation, as given by equations [6.95] and [6.97], i.e.:

$$\mathbf{G} = \mathbf{G}^0 + \mathbf{G}^0\boldsymbol{\Sigma}(\mathbf{G}^0)\mathbf{G}^0, \qquad [6.120]$$

gives rise to a current conserving approximation. This result can be extended to Dyson's equation under the form:

$$\mathbf{G} = \mathbf{G}^0 + \mathbf{G}^0\boldsymbol{\Sigma}(\mathbf{G})\mathbf{G}, \qquad [6.121]$$

with the same mathematical expression [6.117] for $\boldsymbol{\Sigma}$ but where $\boldsymbol{\Sigma}(\mathbf{G}^0)$ is replaced by $\boldsymbol{\Sigma}(\mathbf{G})$. With this the conservation condition, [6.119] is also fulfilled, replacing $\mathbf{G}^{0\gtrless}$ by $\mathbf{G}^{\gtrless}$. This is the so-called "SCBA", for self-consistent Born approximation. However, the resolution of equation [6.121] requires an iteration procedure, which is time consuming and only contains part of the higher order diagrams.

The previous arguments can be applied directly to the electron–photon interaction. They can also be extended to a large class of self-energies, the so-called Φ derivable self-energies that are defined as functionals $\Sigma = \delta\Phi[G]/\delta G$ built according to Luttinger and Ward [LUT 60b, LUT 60a]. These functionals have been proven quite generally [BAY 61, BAY 62] to provide conserving approximations. However, they are usually limited to simple approximations such as those discussed previously, and also the GW approximation [THY 08] for electron–electron interactions.

In fact, all such self-consistent approximations of $\boldsymbol{\Sigma}(\mathbf{G})$ are based on analytic continuation techniques for the Green's function itself. By construction, they ignore vertex corrections that might be of importance. It was shown that in some cases they lead to worse descriptions than the LOA itself [WHI 92, LEV 93]. In fact, it is possible to apply analytic continuation techniques not on $\mathbf{G}$ but on the LOA expression of the current itself. This leads to accurate results at a lesser cost of computational power as is discussed in detail in [MER 12, MER 13].

6.7.5.2. *Other expression for the current*

We now derive a general expression for the electrical current valid at any point in the system. We divide the system into a left and right part and write the current as the total charge derivative:

$$J = ie\mathrm{Tr}_R\frac{\partial \mathbf{G}^<}{\partial t}(t,t), \qquad [6.122]$$

i.e. the charge per unit time flowing from the left to the right. Making use of the general expression [6.78], we assume that a conservative Σ_I has been chosen that eliminates the last term on the right. This leaves us with:

$$J = e\mathrm{Tr}_R[\mathbf{G}^<(t,t), \mathbf{H}_0], \quad [6.123]$$

where $\mathbf{H}_0$ is the single particle Hamiltonian. We recall that in the time domain $\mathbf{H}_0$ is in units of $\hbar$. We then Fourier transform and express the result in the energy domain. We also note that if we express $\mathbf{H}_0$ as the sum of the Hamiltonians of the disconnected parts plus the part which connects them, which can be expressed as $\mathbf{H}_{0,LR} + \mathbf{H}_{0,RL}$, then we get :

$$J = \frac{e}{h}\mathrm{Tr}_R \int d\epsilon (\mathbf{G}^<_{RL}(\epsilon)\mathbf{H}_{0,LR} - \mathbf{H}_{0,RL}\mathbf{G}^<_{LR}(\epsilon)). \quad [6.124]$$

This results from the fact that the trace of the commutators is zero for the disconnected parts. We should also note that in equation [6.124] $\mathbf{H}_0$ is expressed in normal energy units. It is also possible to calculate J_L and J_R directly at the contacts by a generalization of equation [6.84] proposed by different authors, but equation [6.124] is simpler for applications. In cases where there is doubt that $\mathbf{\Sigma_I}$ is conservative, then we can use equation [6.124] all along the system to identify to what extent it remains constant.

6.7.5.3. *Band to band recombination*

Time-dependent perturbations with large enough frequencies can induce band to band generation and recombination. This is obviously true when treating the electron–photon interaction. We summarize here the elements presented in [AEB 11, AEB 12] allowing to treat this important situation, especially for the modeling of photovoltaic systems.

Everything starts from the continuity equation [6.114] with $\frac{\partial n}{\partial t} = 0$ in the steady state. In a semiconductor and for the whole set of bands, the right-hand side of equation [6.114] then gives zero if $\mathbf{\Sigma_I}$ is properly chosen to conserve the total current. However, this is not so for each subband, conduction or valence, taken separately. In general, we are interested in the total contribution of one of these bands, which represents the total particle rate R into this band. As discussed in [AEB 11, AEB 12], we can thus take the trace of equation [6.114] over band B, then use the transformation from equations [6.78]–[6.79] due to the cyclic property of the trace to finally obtain, for frequencies corresponding to a band whose limit can depend on position $\mathbf{r}$:

$$R = \frac{1}{h}\int d^3\mathbf{r} \int_{B(\mathbf{r})} d\epsilon \langle \mathbf{r}|\mathbf{\Sigma}^<_I(\epsilon)\mathbf{G}^>(\epsilon) - \mathbf{\Sigma}^>_I(\epsilon)\mathbf{G}^<(\epsilon)|\mathbf{r}\rangle. \quad [6.125]$$

The expression of $\mathbf{\Sigma}_I^{\gtrless}$ for photons has the same functional form as in equation [6.100]. We can thus evaluate R by integrating over the conduction band alone, ignoring the intraband transitions, i.e. the terms in $\mathbf{G}^{\gtrless}(\epsilon + \hbar\omega_q)$. This leaves us with:

$$R = \frac{1}{h}\int d^3\mathbf{r}\int_{B(\mathbf{r})} d\epsilon \sum_q \langle \mathbf{r}|\mathbf{M}(\mathbf{q})\Big\{ n_q \mathbf{G}_v^{0<}(\epsilon - \hbar\omega_q)\mathbf{G}_c^{0>}(\epsilon)$$

$$-(n_q+1)\mathbf{G}_v^{0>}(\epsilon + \hbar\omega_q)\mathbf{G}_c^{0<}(\epsilon)\Big\}\mathbf{M}^+(\mathbf{q})|\mathbf{r}\rangle, \qquad [6.126]$$

where M is the matrix coupling valence and conduction states and $\mathbf{G}_{v,c}^{0\gtrless}$ are the submatrices in the corresponding subspace. As pointed out in [AEB 11, AEB 12], the first positive term in equation [6.126] corresponds to optical absorption, while the last negative term, respectively, corresponds to stimulated (in n_q) and spontaneous emission.

This is the result deduced from the LOA but we might also want to use the analytic continuation consisting of the replacement of G^0 by G (SCBA). This is often done but at the expense of heavy self-consistent numerical calculations, which might not always give an accurate result for reasons outlined before.

Finally, this approach combined with the inclusion of different relaxation processes (like electron–phonon) provides the basis of the treatment of photovoltaic effects in these systems.

6.7.6. *Summary of the important formulae*

Most expressions derived in the previous sections, e.g. $G(\mathbf{r}_1, t_1, \mathbf{r}_2, t_2)$, have been expressed in a continuous space and time representation. Usually, for steady state, we prefer to work in the energy domain obtained by Fourier transform followed by the replacement of ω by $\epsilon/\hbar$ giving $G(\mathbf{r}_1, \mathbf{r}_2, \epsilon)$ and the true Hamiltonian matrix elements instead of expressing them in units of $\hbar$. Furthermore, for practical calculations we need to reduce the problem to a finite set of discrete equations leading to a matrix formulation. This can be done either by expressing all quantities in a finite basis set or by discretizing the equations or even by using a mixed method. Furthermore, the equations are solved within the restricted space of the central active region that implies the inclusion of the boundary or contact self-energies.

Let us start with the self-energies. The total self-energy is the sum of a contact contribution $\mathbf{\Sigma}_C(\epsilon)$ and the interaction one $\mathbf{\Sigma}_I(\epsilon)$:

$$\mathbf{\Sigma}(\epsilon) = \mathbf{\Sigma}_C(\epsilon) + \mathbf{\Sigma}_I(\epsilon). \qquad [6.127]$$

The contact part is the sum of two contributions $\alpha = L$ or R (left and right), or even more in many terminal devices:

$$\mathbf{\Sigma}_C(\epsilon) = \mathbf{\Sigma}_{CL}(\epsilon) + \mathbf{\Sigma}_{CR}(\epsilon), \qquad [6.128]$$

the components $\mathbf{\Sigma}^{\gtrless}_{C\alpha}$ being given by equation [6.83] while the retarded part is given by equation [6.68].

The next step is to define the interaction self-energy $\mathbf{\Sigma}_I(\epsilon)$. For most applications such as electron–phonon coupling or electron–photon coupling, it takes the form given by equation [6.100] for $\mathbf{\Sigma}^{\gtrless}_I(\epsilon)$. This can be further simplified to equation [6.107] and [6.108] in the case of the electron–phonon interaction treated in the EMA. However, in this case care must be taken, as shown later when writing these equations under matrix form in a discretized process. This leads to a matrix $\mathbf{\Sigma}^{\gtrless}_I(\mathbf{G^0}, \epsilon)$ corresponding to the LOA. In most cases, $\mathbf{G^0}$ is then replaced by the full $\mathbf{G}$ leading to $\mathbf{\Sigma}^{\gtrless}_I(\mathbf{G}, \epsilon)$, which is the basis of the SCBA.

Once $\mathbf{\Sigma}^{\gtrless}_I(\mathbf{G}, \epsilon)$ is known, we can determine $\mathbf{\Sigma}^r_I(\mathbf{G}, \epsilon)$ via the application of equation [6.28] to the self-energies. However, the principal part in the second term of equation [6.28] is usually dropped for the reason that it only leads to a shift in energies and also that it was shown to have a negligible effect. From this, we take:

$$\mathbf{\Sigma}^r_I(\epsilon) = \frac{\mathbf{\Sigma}^>(\epsilon) - \mathbf{\Sigma}^<(\epsilon)}{2}. \qquad [6.129]$$

With the knowledge of $\mathbf{\Sigma}_I$, it is then possible to determine the Green's function via equation [6.56] written in the energy domain under a slightly different form corresponding to Dyson's equation:

$$\begin{aligned} \mathbf{G}^r(\epsilon) &= \mathbf{G}^{0r}(\epsilon) + \mathbf{G}^{0r}(\epsilon)[\mathbf{\Sigma}^r_C(\epsilon) + \mathbf{\Sigma}^r_I(\epsilon)]\mathbf{G}^r(\epsilon), \\ \mathbf{G}^{\gtrless}(\epsilon) &= \quad \mathbf{G}^r(\epsilon)[\mathbf{\Sigma}^{\gtrless}_C(\epsilon) + \mathbf{\Sigma}^{\gtrless}_I(\epsilon)]\mathbf{G}^{r+}(\epsilon). \end{aligned} \qquad [6.130]$$

These expressions make explicit the fact that we can reduce the problem to the central active part with the condition of adding explicitly the contact self-energies.

We now proceed to the determination of the physical quantities of interest. For practical calculations, the equations are discretized, e.g. via the use of a local atomic like basis. In that case, the operator or matrix formulation used before adapts directly.

For instance, in quantum wires or multilayered systems, the current flowing between planes p and $p+1$ can be directly written from equation [6.124] under the form:

$$J_{p,p+1} = \frac{e}{h} \int d\epsilon Tr\Big\{\beta_{p,p+1}\mathbf{G}^{<}_{p+1,p} - \beta_{p+1,p}\mathbf{G}^{<}_{p,p+1}\Big\}, \qquad [6.131]$$

where $\beta_{p,p+1}$ represents the matrix $\mathbf{H}_0$ connecting planes p and $p+1$. A second basic quantity is the particle density needed to perform "Hartree" or "Schrödinger–Poisson" self-consistent calculations. From the definition of $G^{\gtrless}$ in equation [6.11] and in analogy with equation [6.13], we write within a discretized procedure for the electron (n_i) or hole (p_i) density at site j:

$$\begin{aligned} n_j &= -\tfrac{i}{V} \int \tfrac{d\epsilon}{2\pi} Tr_j \mathbf{G}^{<}(\epsilon), \\ p_j &= \tfrac{i}{V} \int \tfrac{d\epsilon}{2\pi} Tr_j \mathbf{G}^{>}(\epsilon), \end{aligned} \qquad [6.132]$$

where Tr_j is the trace over the states centered on site j and where V is the volume of the discretized cell. From this, the local density of states can be obtained as $n_i + p_i$, thereby involving $\mathbf{G}^{>} - \mathbf{G}^{<}$ equal to $\mathbf{G}^{r} - \mathbf{G}^{a}$.

The calculation of the discretized or tight-binding Green's function matrices usually proceed via the recursion technique [SVI 03]. This can be achieved for "n" or "p" type conduction in nanowire transistors. Finally, it is important to note that care should be taken when adapting the continuous electron–phonon self-energies of equations [6.107]–[6.108] to the discretized case. Indeed, both of them are diagonal and proportional to $G^{<}(\mathbf{r}, \mathbf{r})$. In a discretizing procedure, $G^{<}(\mathbf{r}, \mathbf{r})$ around site i is proportional to $(1/V)G^{<}_{ii}$ but G_{ii} itself is proportional to V so that the result is insensitive to V. Finally, we must recall (see section 6.7.4.1) that the size of the discretizing mesh must be optimized to allow the use of a diagonal approximation for $\mathbf{\Sigma}$ electron–phonon.

6.8. Bibliography

[AEB 11] AEBERHARDT U., "Quantum-kinetic theory of photocurrent generation via direct and phonon-mediated optical transitions", *Physical Review B*, vol. 84, p. 035454, 2011.

[AEB 12] AEBERHARDT U., "Nonequilibrium Green's function theory of coherent excitonic effects in the photocurrent response of semiconductor nanostructures", *Physical Review B*, vol. 86, p. 115317, 2012.

[BÜT 85] BÜTTIKER M., IMRY V., LANDAUER R. *et al.*, "Generalized many-channel conductance formula with application to small rings", *Physical Review*, vol. 31, p. 6207, 1985.

[BAY 61] BAYM G., KADANOFF L.P., "Conservation laws and correlation functions", *Physical Review*, vol. 124, p. 287, 1961.

[BAY 62] BAYM G., "Self-consistent approximations in many-body systems", *Physical Review*, vol. 127, p. 1391, 1962.

[CRA 68] CRAIG R., "Perturbation expansion for real-time Green's functions", *Journal of Mathematics and Physics*, vol. 9, p. 605, 1968.

[DAT 95] DATTA S., *Electronic Transport in Mesoscopic Systems*, Cambridge University Press, Cambridge, 1995.

[DEL 03] DELERUE C., LANNOO M., ALLAN G., "Concept of dielectric constant for nanosized systems", *Physical Review B*, vol. 68, p. 115411, 2003.

[DEL 04] DELERUE C., LANNOO M., *Nanostructures: Theory and Modelling*, Springer-Verlag, Berlin/Heidelberg, 2004.

[DIV 08] DI VENTRA M., *Electrical Transport in Nanoscale Systems*, Cambridge University Press, Cambridge, UK, 2008.

[GEL 51] GELL-MANN M., LOW F., "Bound States in Quantum Field Theory", *Review of Modern Physics*, vol. 84, p. 350, 1951.

[HAU 96] HAUG H., JAUHO A.-P., *Quantum Kinetics in Transport and Optics of Semi-Conductors*, Springer, New York, 1996.

[HYL 94] HYLDGAARD P., HERSHFIELD S., DAVIES J.H. *et al* "Resonant tunneling with an electron-phonon interaction", *Annals of Physics*, vol. 236, pp. 1-42, 1994.

[JIN 06] JIN S., PARK Y.J., MIN H.S., "A Three-dimensional simulation of quantum transport in silicon nanowire transistor in the presence of electron-phonon interactions", *Journal of Applied Physics*, vol. 99, p. 123719, 2006.

[KEL 65] KELDYSH L., "Diagram technique for nonequilibrium processes", *JETP*, vol. 20, p. 1018, 1965.

[LAN 57] LANDAUER R., "Spatial variation of currents and fields due to localized scatterers in metallic conduction", *IBM Journal of Research and Development*, vol. 1, p. 223, 1957.

[LAN 76] LANGRETH D.C., *Linear and Non Linear Electron Transport in Solids*, Plenum, New York, 1976.

[LAN 89] LANDAUER R., "Conductance determined by transmission: probes and quantised constriction resistance", *Journal of Physics: Condensed Matter*, vol. 1, p. 8099, 1989.

[LEV 93] LEVY YEYATI A., MARTIN-RODERO A., FLORES F., "Electron correlation resonances in the transport through a single quantum level", *Physical Review Letters*, vol. 71, p. 2991, 1993.

[LUT 60a] LUTTINGER J.M., "Fermi surface and some simple equilibrium properties of a system of interacting fermions", *Physical Review*, vol. 119, p. 1153, 1960.

[LUT 60b] LUTTINGER J.M., WARD J.C., "Ground-state energy of a many-Fermion system. II", *Physicl Review*, vol. 118, p. 1417, 1960.

[MAH 87] MAHAN G.D., "Quantum transport equation for electric and magnetic fields", *Physics Reports*, vol. 145, p. 145, 1987.

[MAH 00] MAHAN G.D., *Many Particles Physics*, Springer, New York, 2000.

[MAT 67] MATTUCK R.D., *A Guide to Feynman Diagrams in the Many-Body Problem*, McGraw-Hill, New-York, 1967.

[MEI 92] MEIR Y., WINGREEN N.S., "Landauer formula for the current through an interacting electron region", *Physical Review Letters*, vol. 68, p. 2512, 1992.

[MER 12] MERA H., LANNOO M., LI C. *et al.*, "Inelastic Scattering in nanoscale devices one-shot current-conserving lowest order approximation", *Physical Review B*, vol. 86, p. 161404, 2012.

[MER 13] MERA H., LANNOO M., CAVASSILAS N. *et al.*, "Nanoscale device modeling using a conserving analytic continuation technique", *Physical Review B*, vol. 88, p. 075147, 2013.

[RAM 86] RAMMER J., SMITH H., "Quantum field-theoretical methods in transport theory of metals", *Physical Review*, vol. 58, p. 323, 1986.

[SLA 63] SLATER J.C., *Quantum theory of molecules and solids*, McGraw-Hill, New-York, vol.1, 1963.

[SVI 03] SVIZHENKO A., ANANTRAM M.P., "Role of scattering in nanotransistors", *IEEE Transactions on Electron Devices*, vol. 50, p. 1459, 2003.

[THY 08] THYGESEN K.S., RUBIO A., "Conserving GW scheme for nonequilibrium quantum transport in molecular contacts", *Physical Review B*, vol. 77, p. 115333, 2008.

[WHI 92] WHITE J.A., "Self-consistent Green functions for the Anderson impurity model", *Physical Review B*, vol. 45, p. 1100, 1992.

7

Electron Devices Simulation with Bohmian Trajectories

7.1. Introduction: why Bohmian mechanics?

There are many different formulations of classical mechanics (Newtonian, Lagrangian, Hamiltonian, Poisson brackets, Hamilton–Jacobi, etc.). The various formulations differ mathematically and conceptually, yet each one makes identical predictions for all experimental results. The situation in quantum mechanics is certainly very similar. For example, elementary textbooks teach us that the harmonic oscillator problem is cleanly and easily solved through the creation and annihilation operators of the matrix (Heisenberg) formulation, while many other simple problems are better formulated directly with the wave function (or Schrödinger equation) [COH 78]. Another relevant example is the (Feynman) path integral formulation that is rarely the easiest way to approach a non-relativistic quantum problem, but it has innumerable and very successful applications in quantum field theory [FEY 65]. Some problems look difficult in one formulation (interpretation) of quantum mechanics and easy in another. The aim of this chapter is to discuss to what extent one of the formulations (interpretations) of the quantum theory, i.e. Bohmian mechanics, provides a singular tool to predict and explain the behavior of novel electron devices (an extended revision of the use of Bohmian mechanics to solve practical problems can be found in [BEN 14]).

Bohmian mechanics was originally proposed by Louis de Broglie in 1924 [DE 24] and fully developed into a consistent explanation of all quantum phenomena in terms of wave and particles by David Bohm in 1952 [BOH 52]. There are many good references where its basic ingredients can be easily understood [BEL 04,

Chapter written by Guillermo ALBAREDA, Damiano MARIAN, Abdelilah BENALI, Alfonso ALARCÓN, Simeon MOISES and Xavier ORIOLS.

BOH 93, DÜR 12, DÜR 09, ORI 11]. Nevertheless, since its mathematical structure for the non-relativistic quantum mechanics used in this chapter is quite simple, the brief introduction in section 7.2 is enough to understand the whole chapter (a connection with other simulation tools mentioned in this book can be found in Appendix D where a formal relation between the Wigner distribution function and the Bohmian trajectories is established).

Bohmian mechanics agrees with all quantum experiments done up to now. All researchers who analyze the ideas of de Broglie and Bohm with the pertinent scientific rigor conclude that there is no objective argument against them[1] Then, *why is it generally ignored by the scientific community?* There are several historical and sociological reasons that justify its marginal status. We mention two of them: first, there is a *vicious circle* with negative feedback. Since few people knows Bohmian mechanics, few people uses it for practical applications. Then, there are no much Bohmian contributions and, consequently, this formulation is not taught at universities or explained in research courses. As a result, few people know it and the *circle* starts again. Let us mention that the quantum chemistry community is an encouraging exception that has been able to escape from this *vicious circle* 15 years ago, mainly due to the works of Wyatt and co-workers [WYA 05, WYA 00a, WYA 00b]. Second, there is a widely spread belief that Bohmian mechanics, by construction, has a limited usefulness. It is argued that, apart from computing the wave function, Bohmian mechanics requires tracking a set of trajectories that, at the end of the day, will exactly reproduce the time evolution of the wave function, which was already known. Then, *what is the utility of the extra effort for computing Bohmian trajectories?* This criticism is valid for a single-particle problem where the wave function can be explicitly computed, but it is not pertinent at all for realistic quantum problems, where the wave function itself cannot be computed (because it lives in a R^{3N} configuration space), the so-called many-body problem[2]. In fact, any practical formulation of quantum transport is built up independently of the many-particle wave function in the configuration space and it evolves around some

1 John S. Bell used the following words to explain how simply the iconic double-slit experiment, in particular, and any other quantum phenomenon, in general, can be understood with Bohmian mechanics: "While the founding fathers agonized over the question *particle* or *wave*, de Broglie in 1925 proposed the obvious answer *particle* and *wave*. Is it not clear from the smallness of the scintillation on the screen that we have to do with a particle? And is it not clear, from the diffraction and interference patterns, that the motion of the particle is directed by a wave? De Broglie showed in detail how the motion of a particle, passing through just one of two holes in screen, could be influenced by waves propagating through both holes. And so influenced that the particle does not go where the waves cancel out, but is attracted to where they cooperate. This idea seems to me so natural and simple to resolve the wave–particle dilemma in such a clear and ordinary way, that it is a great mystery to me that it was so generally ignored" [BEL 04].

2 Dirac wrote in 1929: "The general theory of quantum mechanics is now almost complete. The underlying physical laws necessary for the mathematical theory of a large part of physics and the whole of chemistry are thus completely known, and the difficulty is only that the exact application of these laws leads to equations much too complicated to be soluble" [DIR 29].

mathematical entity living in the real space. For Bohmian mechanics, the conditional (Bohmian) wave function, described in section 7.2.3, is such an entity. Apart from the ability to approximate many-body problems with conditional wave functions defined in the real space, Bohmian mechanics allows a simple description of the quantum measurement process without invoking any special postulate for the collapse.

Because of the previously mentioned advantages, we argue that Bohmian mechanics is a well-suited computational tool for studying quantum transport, in general, and its high-frequency behavior, in particular. This effort on accurately predicting high-frequency quantum transport is somehow urgent nowadays because the International Technology Roadmap for Semiconductors[3] is expecting that quantum devices with THz operating frequencies will play an important role in the future electronic industry within a few years. For those readers familiar with electron transport with the Monte Carlo solution of the Boltzmann equation [JAC 89], Bohmian trajectories play the same role of the semiclassical Monte Carlo trajectories, but in a rigorous quantum regime. They provide a microscopic description (in terms of well-defined trajectories guided by waves) of the ensemble results obtained from other (wave alone) formulations. The rest of this introduction is devoted to revisit what, in our opinion, constitute two underlying difficulties that we have to face when modeling electronic devices beyond the DC regime. Such difficulties are the quantum measurement and the many-body problems. In this chapter, we will describe how Bohmian trajectories can help to solve them. In section 7.2, we summarize postulates, equations and the most important features of Bohmian mechanics regarding the description of electron transport including the measurement process and many-body correlations. Section 7.3 describes how Bohmian mechanics can be applied to simulate electron transport. We will discuss the main pieces required to build up a trajectory-based electronic device simulator. Although this simulator is able to provide any kind of dynamic property related with a quantum device, we will focus on the electrical current. Section 7.4 is devoted to provide a practical method to evaluate the electrical current. Thus, after a brief dissertation on the computation of DC currents in quantum systems and on the role of the multitime measurement to predict its fluctuations, we provide a detailed description of the equations required to evaluate DC, AC, transients and any moment of the current. We conclude in section 7.5.

7.1.1. *Quantum continuous measurement of the electrical current*

Before going into the Bohmian formulation of quantum transport, let us start by explaining why the quantum measurement at successive times plays a crucial role in the high-frequency behavior of quantum devices. This discussion will help us in

3 http://www.itrs.net.

understanding the practical and conceptual difficulties of high-frequency quantum transport modeling.

In 1926, Schrödinger published "*An Undulatory Theory of the Mechanics of Atoms and Molecules*" [SCH 26], where he described electrons in terms of a wave function $\psi(\vec{r},t) = \langle \vec{r}|\psi(t)\rangle$ solution of his famous equation. He interpreted the wave function $\psi(\vec{r},t)$ as a description of the electron charge density defined as $q|\psi(\vec{r},t)|^2$ being q the electron charge. Later, Born refined the interpretation of Schrödinger and defined $|\psi(\vec{r},t)|^2$ as the probability density of finding the electron in a particular position $\vec{r}$ and time t [WAE 68].

In order to realize why the interpretation of the wave function as a charge distribution is misleading[4], it will be relevant to discuss the DC and the partition noise generated by a flux of electrons impinging upon a tunneling barrier. The transport process is schematically depicted in Figure 7.1. For simplicity, we assume only injection of electrons from left to right. After the interaction with the barrier, the solution of the (time-dependent) Schrödinger equation [COH 78] provides a wave function that is spatially separated into transmitted, $\psi_T(\vec{r},t)$, and reflected, $\psi_R(\vec{r},t)$, parts with $\psi(\vec{r},t) = \psi_T(\vec{r},t) + \psi_R(\vec{r},t)$.

Naively, we could think that the charge q of each injected electron is divided into two smaller parts, $q_T = q\int|\psi_T(\vec{r},t)|^2 dr$ for the transmitted part and also $q_R = q\int|\psi_R(\vec{r},t)|^2 dr$ for the reflected one as seen in Figures 7.1(a)–(c). This wrong interpretation of the wave function would produce a DC value equal to $I_{DC} = \nu \cdot q_T$, ν being the number of electrons injected per unit of time. However, a measured electron is either fully transmitted or fully reflected as seen in Figure 7.1(d)–(f), not both[5]. From the probabilistic interpretation of the wave function, the transmission coefficient $T = \int|\psi_T(\vec{r},t)|^2 dr$ has to be defined as the ratio of transmitted electrons over the injected ones. An identical interpretation is required for the reflection coefficient $R = \int|\psi_R(\vec{r},t)|^2 dr$. This correct interpretation of the wave function provides a (ensemble-average) value of the DC current equal to $I_{DC} = \nu \cdot q \cdot T$, which becomes equal to the previous wrong result $I_{DC} = \nu \cdot q_T$ because $q_T = q \cdot T$.

4 The error of interpreting $|\psi(\vec{r},t)|^2$ as a charge density is clearly seen when dealing, for example, with three electrons whose (many-particle) wave function "lives" in a configuration space of nine variables, while the charge density is still defined in the real space with three variables. In scenarios with a very large number of particles, $n(\vec{r},t) = N\int d\vec{r}_2, d\vec{r}_3 \cdot d\vec{r}_N|\psi(\vec{r},\vec{r}_2,...,\vec{r}_N,t)|^2$ can be quite similar to the charge density, but not identical.

5 The *ontological* meaning of the dotted *empty* waves drawn in Figure 7.1 depends on which interpretation of quantum mechanics is selected (we mentioned some of them in [ZUR 05, COH 78, SCH 04, GRI 02, BOH 52]). In any case, such selection is not at all relevant in our argument. We only want to emphasize the experimental fact that, in the ammeter, we only detect a transmitted or reflected electron, but not both.

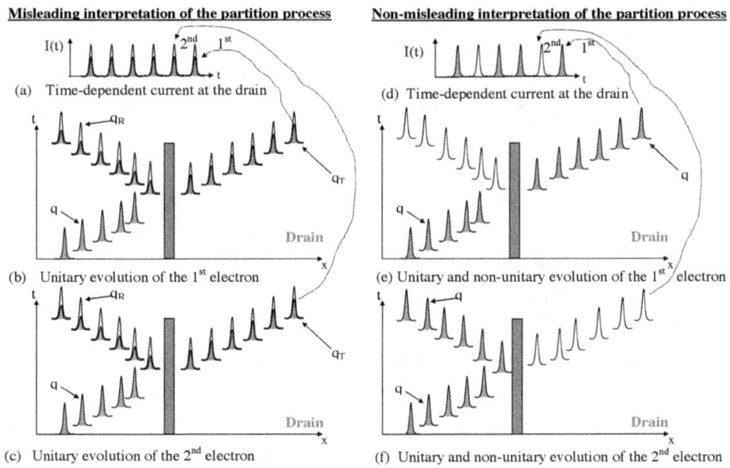

Figure 7.1. *Schematic representation of the partition noise present on the measured current in a) generated by a flux of electrons impinging on a tunneling barrier. Only the unitary (Schrödinger-like) evolution of the wave function is considered in b) and c). This unitary evolution alone provides a misleading explanation of quantum transport. The correct explanation with unitary and non-unitary (collapse-like) evolution is shown in d)–f). For a color version of this figure, see www.iste.co.uk/dollfus/nanodevices.zip*

A more formal explanation of why Figures 7.1(a)–(c) are intrinsically misleading to understand quantum transport is because, in principle, one cannot use the Schrödinger equation alone to discuss the time evolution of a quantum system. Orthodox textbooks [COH 78] point out that there are two dynamical laws for quantum systems. Between two measurements, the quantum system follows a unitary time-evolution determined by the Schrödinger equation. On the contrary, during the measurement process, the quantum system suffers a non-unitary evolution. For example, by being projected onto the measurement eigenstates. The unitary evolution in Figures 7.1(a)–(c) preserves the norm of the wave function, while the non-unitary evolution in Figures 7.1(d)–(f) "eliminates" those parts of the wave function that do not correspond to the measured data. Whenever the measurement of the current is relevant in the evolution of the quantum device, both laws are mandatory. However, for studying DC, a single measurement is enough and, then, the formal discussion on the role of measurement can be somehow relaxed (see section 7.4.1 for a detailed discussion on this point). On the contrary, the transport process described in Figures 7.1(d)–(f), which schematically includes both dynamical laws, becomes fully pertinent to understand the fluctuations of the current around its DC value. The first wrong interpretation of quantum transport in Figures 7.1(a)–(c) provides no noise (at zero frequency). At each time interval $1/\nu$, the transmitted charge is always equal to

$q_T = qT$. There are no fluctuations (see Figure 7.1(a)). The second interpretation of Figures 7.1(d)–(f) provides the experimental partition noise measured in the laboratory. The transmitted charge during the time interval $1/\nu$ fluctuates between the value q, when the electron is fully transmitted with probability T, and the value 0 when it is fully reflected with probability $R = 1 - T$. See Figure 7.1(d).

The situation can be even worse if we are interested in the power spectral density. Then, as it will be discussed in section 7.4.2, we have to consider multitime measurements (at least, time correlations between two values of the current measured at two different times). *How do we have to model the evolution of the system during the measurement process?* The answer is certainly not simple because of the inherent difficulties of the quantum measurement process (even at an ontological level [SCH 04]). In principle, the wave function has to suffer a non-unitary evolution in order to eliminate (at least modify) parts of the wave function during the measurement process[6]. As it will be shown in section 7.2.2 for generic measurements and in section 7.3.1 to compute the electrical current in particular, Bohmian mechanics provides a straightforward solution to the measurement problem.

7.1.2. *Coulomb correlations and displacement current*

The second issue that we want to discuss to emphasize the difficulties to correctly model high frequency quantum transport is the relevance of the displacement current in such scenarios. For DC transport, we generally deal only with the conduction current (or particle current) $I_p(t)$ related to the number of electrons crossing a particular surface S_i. Nevertheless, the electric field inside a quantum device is both inhomogeneous and time varying because of its time dependence on the external bias and the movement of electrons. Under such time-dependent scenarios, a displacement current $I_d(t)$, proportional to the time-derivative of the electric field, is always present.

The displacement current has no role when modeling DC because, by definition, the time-average value of $I_d(t)$ is zero. However, $I_d(t)$ has a fundamental role when modeling high-frequency transport. The total current is $I(t) = I_p(t) + I_d(t)$. The total current has to satisfy the current conservation law, meaning that $I(t)$ evaluated on a closed surface S must be zero at any time. This is just a consequence of the Maxwell equations [JAC 62]. This property allows us to argue that the current $I(t)$

6 It is notable that a non-unitary evolution of the wave function is also useful to model the process of *decoherence* in quantum systems [SCH 04]. For macroscopic systems, we cannot hope to keep track of all microscopic degrees of freedom. We trace out many variables (by integrating them) in order to keep only the most relevant ones (those belonging to electrons at the active region). Then, the equation of motion of this *open* system is not determined by the Schrödinger equation, but by different equations that allow simultaneous unitary and non-unitary evolutions of the open system.

measured by an ammeter (far from the simulation box) is equal to the current that we compute on a particular surface, S_i, of the simulation box. Therefore, special care must be taken with the computation of the electric field to obtain the displacement component, because the total time-dependent current cannot be known from merely counting transmitted electrons.

For electron devices, an additional clarification about the displacement current is relevant. Using the Gauss' equation, the current conservation law can be rewritten as a continuity equation $\partial\rho/\partial t + \vec{\nabla}\vec{J}_p = 0$, $\vec{J}_p$ being the conduction current density and ρ the charge density. Then, if we integrate this continuity equation over a very large volume V with boundaries deep inside the reservoirs, we can impose an additional charge neutrality requirement, $\int_V \rho dr^3 = 0$. This extra requirement is just a consequence of the fact that positive or negative deviations from charge neutrality inside the device tend to zero after time intervals proportional to the dielectric relaxation time [ALB 10b, ALB 13a].

Therefore, in most scenarios, the proper modeling of electron transport has to take into account the time dependency of the external bias together with the dynamics of electrons in a self-consistent way. This requires an (approximate) solution to the quantum many-body problem. The origin of this many-body problem is that, in the quantum world, the wave function that determines electron dynamics is defined in the N-particle configuration space, $\Psi(\vec{r}_1, \vec{r}_2, ..., \vec{r}_N, t)$. Therefore, the exact treatment of the Coulomb interaction among electrons in principle has to be defined in the N-particle configuration space as well, with a potential energy of the type $U(\vec{r}_1, \vec{r}_2, ..., \vec{r}_N, t)$. Solving the problem in the N-particle configuration space represents a very difficult (impossible) computational task. As a result, only reasonable approximations for the many-body problem are computationally accessible[7]. We will see in section 7.2.3 that the Bohmian formulation in terms of conditional wave functions provides an approximate solution to the many-body problem.

7.2. Theoretical framework: Bohmian mechanics

Bohmian mechanics is, most of the time, introduced through an alternative set of (nonlinear) equations that, together, play the role of the (linear) Schrödinger one. First, the continuity equation for the probability distribution leads to the definition of the Bohmian velocity field. Second, a modified Hamilton–Jacobi equation is found after introducing the so-called *quantum potential* [BEL 04, DÜR 12, ORI 11]. In this

7 The computational burden becomes even more dramatic if we include not only (free) electrons in the transport model but also the atoms that define the electronic band structure. Indeed, as devices are approaching the few-atoms limit, the quantum transport modeling becomes more and more interwoven with material modeling.

work, however, we will directly deal with the Schrödinger equation and its conditional form [ORI 07a, ALB 14, ALB 15].

In order to make the reader aware of the Bohmian formalism, in section 7.2.1 we first introduce in a very compact way the basic postulates and equations useful for this work. In section 7.2.2, we explain why and in which way Bohmian mechanics constitutes a quantum theory without observers, i.e. the quantum measurement in the Bohmian theory is described without invoking the wave function collapse postulate. Finally, in section 7.2.3 we explain in which way the concept of *conditional wave function* can be useful to tackle the many-body problem.

7.2.1. *Postulates and basic equations*

Consider a system of M_T (spinless) electrons described in the spatial coordinates $\vec{r} = \{\vec{r}_1, ..., \vec{r}_{M_T}\}$ by the many-particle wave function $\Psi(\vec{r}, t)$, which obeys the many-particle Schrödinger equation, i.e.:

$$i\hbar \frac{\partial \Psi(\vec{r}, t)}{\partial t} = \Big\{ - \sum_{a=1}^{M_T} \frac{\hbar^2}{2m^*} \nabla_a^2 + U(\vec{r}, t) \Big\} \Psi(\vec{r}, t), \qquad [7.1]$$

where for simplicity we have considered a solid-state system where the lattice–electron interaction is approximately included into the electron effective mass, m^*. The term $U(\vec{r}, t)$ is the potential energy that, here, defines the Coulomb interaction among electrons. From equation [7.1], it can be demonstrated that the probability distribution, $|\Psi(\vec{r}, t)|^2$, obeys the following continuity equation:

$$\frac{\partial |\Psi(\vec{r}, t)|^2}{\partial t} + \sum_{a=1}^{M_T} \nabla_a \vec{j}_a(\vec{r}, t) = 0, \qquad [7.2]$$

where $\vec{j}_a(\vec{r}, t)$ is the a-th component of the usual probability current density [COH 78]. From equation [7.2], the vector field defined as:

$$\vec{v}_a(\vec{r}, t) = \frac{\vec{j}_a(\vec{r}, t)}{|\Psi(\vec{r}, t)|^2}, \qquad [7.3]$$

can be interpreted as a *velocity field* for particle a. This velocity can be used to define trajectories in the configuration space:

$$\vec{r}_a^{\alpha}(t) = \vec{r}_a^{\alpha}(t_o) + \int_{t_o}^{t} \vec{v}_a(\vec{r}^{\alpha}(t'), t')dt', \quad [7.4]$$

with $\vec{r}^{\alpha}(t) = \{\vec{r}_1^{\alpha}(t), ..., \vec{r}_{M_T}^{\alpha}(t)\}$. The superindex α takes into account the uncertainity associated with the initial quantum state $\Psi(\vec{r}, t_o)$, and it is defined through the so-called *quantum equilibrium condition* [DÜR 04, ORI 11] at the initial time t_o:

$$|\Psi(\vec{r}, t_o)|^2 = \lim_{M_\alpha \to \infty} \frac{1}{M_\alpha} \sum_{\alpha=1}^{M_\alpha} \prod_{a=1}^{M_T} \delta(\vec{r}_a - \vec{r}_a^{\alpha}(t_o)). \quad [7.5]$$

It can be easily demonstrated that the evolution of the above infinite set of quantum trajectories, $\alpha = 1, 2, ..., M_\alpha \to \infty$, reproduces the probability distribution, $|\Psi(\vec{r}, t)|^2$, at any time. Equations [7.1] and [7.3]–[7.5] constitute a basic set of equations describing Bohmian mechanics [BEL 04, DÜR 12, DÜR 09, ORI 11]. As will be discussed in section 7.2.2, in Bohmian mechanics, any observable, for example the current I, is a function of the particle trajectories $\vec{r}^{\alpha}(t)$ in [7.4], i.e. $I(\vec{r}^{\alpha}(t))$. It is in this regard that Bohmian mechanics is a trajectory-based formulation of quantum mechanics.

7.2.2. *Bohmian explanation of the measurement process*

The Bohmian explanation of the quantum measurement is, perhaps, the most attractive (and also ignored) feature of the Bohmian explanation of the quantum nature [BEL 90, BEL 04, DÜR 12, ORI 11]. Although the Bohmian and the orthodox explanations of a measurement produce the same probabilistic predictions, the mathematical implementation of the equations of motion in each case is quite different.

In the standard interpretation of quantum theory, the projective measurement process is defined in a particular quantum region, the system (see Figure 7.2(b)). The state of the quantum system in this particular region is determined through the wave function $\psi_S(\vec{r}, t)$. The process of measuring a particular magnitude is mathematically defined through an operator, for example $\hat{G}$, acting on the wave function. The possible outcomes of the measurement process correspond to one of the possible eigenvalues g of this operator that satisfy the equation $\hat{G}\psi_g(\vec{r}) = g\psi_g(\vec{r})$, with $\psi_g(\vec{r})$ being an eigenvector of this operator. The set $\psi_g(\vec{r})$

forms an orthonormal basis of the Hilbert space of the quantum system so that the wave function at the initial time t can be written as:

$$\psi_S(\vec{r},t) = \sum_g c_g(t)\psi_g(\vec{r}), \qquad [7.6]$$

with $c_g(t)$ as a complex value with the only restriction that $\sum_g |c_g(t)|^2 = 1$, which ensures that $\psi_S(\vec{r},t)$ is well normalized. When measuring the eigenvalue g_a the total wave function $\psi_S(\vec{r},t)$ collapses into $\psi_{g_a}(\vec{r})$. Then, the probability of getting the value g_a in the measuring apparatus is just $P_{g_a} = |c_{g_a}(t)|^2$. In order to avoid unnecessary complications, we have assumed that this basis has no degeneracy.

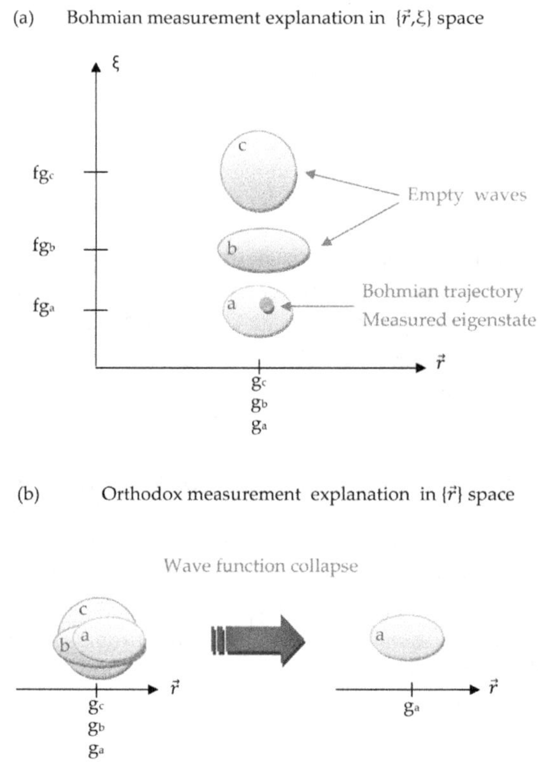

Figure 7.2. *a) Bohmian measurement in the $\{\vec{r},\xi\}$ configuration space: from the non-overlapping many-particle (system + apparatus) wave function, only the g_a part of the wave function where the Bohmian trajectory is present is needed to compute the evolution of the Bohmian system. b) Orthodox measurement in the $\{\vec{r}\}$ space: the (system) wave function collapses into the g_a part when the measurement takes place. For a color version of this figure, see www.iste.co.uk/dollfus/nanodevices.zip*

In order to mathematically define the measurement process in the Bohmian formalism, apart from the degrees of freedom $\vec{r}$ of the system, the degrees of freedom of the positions of the pointer $\vec{\xi}$ belonging to the measuring apparatus are required [HOL 93, BOH 93, BEL 04, DÜR 12, ORI 11]. Thus, we define a total wave function $\Phi(\vec{r},\vec{\xi},t)$ in a larger configuration space that includes the system plus the measuring apparatus, $\{\vec{r},\vec{\xi}\}$. According to the Bohmian postulates, we select a particular trajectory $\{\vec{r}^{\alpha}(t),\vec{\xi}^{\alpha}(t)\}$ of this larger configuration space. Then, in order to say that a measuring apparatus is able to correctly determine the eigenvalues g, there are some necessary conditions that the entire system has to satisfy.

First, the pointer positions $\vec{\xi}^{\alpha}(t)$ of such an apparatus have to be restricted to a particular region, $\vec{\xi}^{\alpha}(t) \in S_{g_1}$, every time that the quantum system is in the eigenstate $\psi_{g_1}(\vec{r})$. We define S_g as a restricted set of positions in the space of the ammeter position ξ. Let us define $\Phi_{g_1}(\vec{r},\vec{\xi},t)$ as the total wave function that fits with the property that any experiment whose quantum system is described by $\psi_{g_1}(\vec{r})$ implies that the pointer points in the particular region, $\vec{\xi}^{\alpha}(t) \in S_{g_1}$.

Second, the subspaces S_{g1} and S_{g2} of the whole configuration space must be non-overlaping during the measurement, i.e. $S_{g1} \cap S_{g2} = 0$. We have defined the restricted region allowed by the pointer positions associated with a second eigenstate, $\psi_{g_2}(\vec{r})$, as $\vec{\xi}^{\alpha}(t) \in S_{g_2}$. This implies that the states $\Phi_{g_1}(\vec{r},\vec{\xi},t)$ and $\Phi_{g_2}(\vec{r},\vec{\xi},t)$ do not overlap in the configuration space during the measurement.

Thus, given a "good" (projective) measurement apparatus, and given that the eigenstates $\psi_g(\vec{r})$ form a complete basis, during the measurement, the only good decomposition for $\Phi_g(\vec{r},\vec{\xi},t)$ is:

$$\Phi_g(\vec{r},\vec{\xi},t) = f_g(\vec{\xi},t)\;\psi_g(\vec{r}). \qquad [7.7]$$

We emphasize that $f_g(\vec{\xi},t)$ is a normalized function because $\Phi_g(\vec{r},\vec{\xi},t)$ and $\psi_g(\vec{r})$ are also normalized functions in their respective configuration spaces. By construction, $f_{g_1}(\vec{\xi},t) \cap f_{g_2}(\vec{\xi},t) = 0$ during the measuring time. Thus, even if $\psi_{g_1}(\vec{r})$ and $\psi_{g_2}(\vec{r})$ overlap, the states $\Phi_{g_1}(\vec{r},\vec{\xi},t)$ and $\Phi_{g_2}(\vec{r},\vec{\xi},t)$ do not overlap in the larger configuration space (see Figure 7.2(a)). We can then ensure that an arbitrary wave function of the quantum system [7.6] can be rewritten in the whole configuration space associated with a good measuring apparatus as:

$$\Phi(\vec{r},\vec{\xi},t) = \sum_g c_g(t)\;\; f_g(\vec{\xi},t)\;\psi_g(\vec{r}). \qquad [7.8]$$

In summary, during the measurement, the only total wave functions that can *live* in the entire quantum system that includes a good measuring apparatus of the eigenvalues

g are the ones written in [7.8]. An example of such wave functions is depicted in Figure 7.2(a). It is important to note that [7.3] implies no restriction on the wave function $\psi_S(\vec{r}, t)$ but only on the total wave function $\Phi(\vec{r}, \vec{\xi}, t)$. If these restrictions are not respected, we can find other types of total wave functions in the configuration space $\{\vec{r}, \vec{\xi}\}$, but they would be incompatible with stating that we have an apparatus that is able to measure the eigenavalue g with certainty at time t.

We can now show quite trivially how the projective measurement is exactly reproduced within Bohmian mechanics. As we have mentioned, apart from the wave function [7.8], we have to select an initial trajectory $\{\vec{r}^{\alpha}(0), \vec{\xi}^{\alpha}(0)\}$. Such a trajectory will evolve driven by the total wave function, and during the measurement, the particle trajectory $\{\vec{r}^{\alpha}(t), \vec{\xi}^{\alpha}(t)\}$ will be situated in only one of the non-overlapping wave packets of [7.8], for example $f_{g_a}(\vec{\xi}, t)\psi_{g_a}(\vec{r})$ as depicted in Figure 7.2(a). Thus, the pointer positions will be situated in $\vec{\xi}(t) \in S_{g_a}$ and we will conclude with certainty that the eigenvalue of the quantum system is g_a. In addition, the subsequent evolution of this trajectory can be computed from $f_{g_a}(\vec{\xi}, t)\psi_{g_a}(\vec{r})$ alone. In other words, we do not need the entire wave function [7.8] because the particle velocity can be computed from $f_{g_a}(\vec{\xi}, t)\psi_{g_a}(\vec{r})$. The rest of circles in Figure 7.2(a) are empty waves that do not overlap with $f_{g_a}(\vec{\xi}, t)\psi_{g_a}(\vec{r})$ so that they have no effect on the velocity of the Bohmian particle. This is how the orthodox collapse is interpreted within Bohmian mechanics. Let us mention that we have only considered projective measurements. Other types of measurements are also possible, which do not collapse the wave function into an eigenstate. Such measurements can also be explained within Bohmian mechanics with an extension of the ideas discussed here [BOH 93, DÜR 04, DÜR 12]. A simple numerical example about the quantum measurement of the total (displacement plus particle) current using Bohmian mechanics will be discussed in section 7.3.1.

A final remark about the quantum measurement is necessary. The Bohmian measurement process explained above implies increasing the number of degrees of freedom that we have to simulate from $\{\vec{r}\}$ to $\{\vec{r}, \vec{\xi}\}$. Sometimes, then, the use of Hermitian operators acting only on the wave function of the quantum system with the ability of providing the outcomes of the measurement process without the explicit simulation of the measuring apparatus is very welcomed. The reader is referred to Appendix C (section 7.9) for a detailed discussion on how to include operators in Bohmian mechanics. Let us emphasize, however, that we are talking only at the computational level. In simple words, operators are not needed in Bohmian mechanics, but they are a very helpful mathematical tool in practical computations. These ideas are emphasized by Daumer *et al.* [DAU 97] and Dürr *et al.* [DÜR 04, DÜR 12] when they talk about the "naive realism about operators".

7.2.3. *Bohmian mechanics for many-particle systems*

The active region of the electron device shown in Figure 7.4 can contain hundreds of electrons. However, as we mentioned, the many-particle Schrödinger equation in [7.1] can be solved only for very few degrees of freedom. A standard way to proceed consists of reducing the complexity of the problem by *tracing out* certain degrees of freedom. This process ends up with what is called the *reduced density matrix*. When the reduced density matrix is used, its equation of motion is no longer described by the Schrödinger equation but in general by a non-unitary operator. The reduced density matrix is no longer a pure state, but a mixture of states and its evolution is in general irreversible [DIV 08]. In this section, we discuss how Bohmian mechanics allows us reduce the complexity in a very different way. As it will be shown below, the concept of *conditional wave function* [DÜR 04] provides an original tool to deal with many-body open quantum systems [ORI 07a, DÜR 05, ALB 14].

7.2.3.1. *The conditional wave function*

Consider a bipartite quantum system $A + B$ whose spatial coordinates can be split as $\vec{r} = \{\vec{r}_a, \vec{r}_b\}$. We define $\vec{r}_a$ as the position of the $a-$electron in $\mathbf{R}^3$, while $\vec{r}_b = \{\vec{r}_1, ..., \vec{r}_{a-1}, \vec{r}_{a+1}, ..., \vec{r}_{M_T}\}$ are the positions of the rest of electrons in an $\mathbf{R}^{3(M_T-1)}$ space. The actual particle trajectories are accordingly denoted by $\vec{r}(t) = \{\vec{r}_a^{\alpha}(t), \vec{r}_b^{\alpha}(t)\}$. *How can one assign a wave function to the system A?* In general, this is not possible if the two subsystems are entangled, i.e. the total wave function cannot be written as a product $\Psi(\vec{r}) = \psi_a(\vec{r}_a)\psi_b(\vec{r}_b)$. However, we can modify our question and ask what is the wave function of the subsystem A that provides the exact velocity $\vec{v}_a$ given a particular position $\vec{r}_b^{\alpha}(t)$ in B. The answer given by Bohmian mechanics is the so-called *conditional wave function* [DÜR 04, DÜR 05]:

$$\phi_a(\vec{r}_a, t) = \Psi(\vec{r}_a, \vec{r}_b^{\alpha}(t), t), \qquad [7.9]$$

which constitutes a multidimensional slice of the whole wave function. In [7.9], we omit (for simplicity) the dependence of each conditional wave function $\phi_a(\vec{r}_a, t)$ on α.

In order to use the conditional wave function to reduce the degrees of freedom of a system, we must know how it evolves in time. It can be demonstrated [ORI 07a] that $\phi_a(\vec{r}_a, t)$ obeys the following wave equation:

$$i\hbar\frac{\partial\phi_a(\vec{r}_a, t)}{\partial t} = \Big\{ -\frac{\hbar^2}{2m}\nabla_a^2 + U_a(\vec{r}_a, \vec{r}_b^{\alpha}(t), t)$$
$$+G_a(\vec{r}_a, \vec{r}_b^{\alpha}(t), t) + iJ_a(\vec{r}_a, \vec{r}_b^{\alpha}(t), t)\Big\}\phi_a(\vec{r}_a, t). \qquad [7.10]$$

The explicit expressions of the potentials $G_a(\vec{r}_a, \vec{r}_b^{\alpha}(t), t)$ and $J_a(\vec{r}_a, \vec{r}_b^{\alpha}(t), t)$ that appear in [7.10] can be found in [ORI 07a]; however, their numerical values are in principle unknown and need some educated conjectures. On the other hand, the term $U_a(\vec{r}_a, \vec{r}_b^{\alpha}(t), t)$ can be any type of many-particle potential defined in the position representation; in particular, it can include short-range and long-range Coulomb interactions. For simplicity, we have divided the total electrostatic potential energy among the M_T electrons that appears in [7.1], into two parts $U(\vec{r}_a, \vec{r}_b^{\alpha}(t), t) = U_a(\vec{r}_a, \vec{r}_b^{\alpha}(t), t) + U_b(\vec{r}_b^{\alpha}(t), t)$. The remaining term $U_b(\vec{r}_b^{\alpha}(t), t)$ that do not involve the variable $\vec{r}_a$ is contained in the coupling potential G_a in [7.10].

From a practical point of view, all quantum trajectories $\vec{r}^{\alpha}(t) = \{\vec{r}_1^{\alpha}(t), \vec{r}_2^{\alpha}(t), .., \vec{r}_{M_T}^{\alpha}(t)\}$ have to be computed simultaneously. In order to gather all the above concepts, let us discuss an hypothetical computation with conditional wave functions by detailing a sequential procedure:

1) At the initial time t_o, we fix the initial position of all $a = 1, ..., M_T$ particles, $\vec{r}_a^{\alpha}(t_o)$, according to [7.5], and their associated single-particle wave function $\phi_a(\vec{r}_a, t_o)$. We define this set of positions with the superindex $\alpha = 1$;

2) From all particle positions, we compute the exact value of the potential $U_a(\vec{r}_a, \vec{r}_b^{\alpha}(t_o), t)$ for each particle. An approximation for the terms G_a and J_a is required at this point;

3) We then solve each single-particle Schrödinger equation [7.10] from t_o till $t_o + dt$;

4) From the knowledge of the single-particle wave function $\phi_a^{\alpha}(\vec{r}_a, t_o + dt)$, we can compute the velocities $\vec{v}_a^{\alpha}(t_o + dt)$ for each a-particle;

5) With the previous velocity, we compute the new position of each a-particle as $\vec{r}_a^{\alpha}(t_o + dt) = \vec{r}_a^{\alpha}(t_o) + \vec{v}_a^{\alpha}(t_o + dt)dt$;

6) Finally, with the set of new positions and wave functions, we repeat the whole procedure (steps 2 to 5) for another infinitesimal time dt till the total simulation time is finished.

Another experiment (or the same experiment at another time) will require selecting different initial positions in step 1 and repeating the whole loop. The advantage of the above algorithm using [7.10] instead of [7.1] is that, in order to find approximate trajectories, $\vec{r}_a^{\alpha}(t)$, we do not need to evaluate the wave function and potential energies in the whole configuration space in [7.1], but only over a smaller number of configuration points, $\{\vec{r}_a, \vec{r}_b^{\alpha}(t)\}$, associated with those trajectories defining the highest probabilities in [7.5].

For spinless electrons, the exchange interaction is naturally included in [7.10] through the terms G_a and J_a. Due to the Pauli exclusion principle, the modulus of the wave function tends to zero, $R(\vec{r}_a, \vec{r}_b^{\alpha}(t), t) \rightarrow 0$, in any neighborhood of $\vec{r}_{a_j}$

such that $|\vec{r}_{a_j} - \vec{r}^{\,\alpha}_{b_k}(t)| \rightarrow 0$ with j and k referring to the individual particles of systems A and B, respectively. Thus, both terms, $G_a(\vec{r}_a, \vec{r}^{\,\alpha}_b(t), t)$ and $J_a(\vec{r}_a, \vec{r}^{\,\alpha}_b(t), t)$, have asymptotes at $\vec{r}_{a_j} \rightarrow \vec{r}^{\,\alpha}_{b_k}(t)$ that *repel* the a-particle from other electrons. However, in order to exactly compute the terms G_a and J_a we must know the total wave function, which is in principle unknown. There are, however, a few ways to introduce the symmetry of the wave function without dealing directly with these two coupling terms [ORI 07a, ALA 13].

7.2.3.2. *An example: numerical results for a non-separable potential*

In order to numerically show the ability of the conditional wave functions discussed above to treat many-particle systems, we apply the above algorithm to a simple two-electrons system under a non-separable harmonic Hamiltonian. We consider two 1D particles so that the configuration space is $\mathbf{R}^2$. The object described by equation [7.9] can be easily understood in this simple case. Here, the conditional wave function $\phi_1(x_1, t)$ would represent a 1D slice of the whole 2D wave function centered on a particular configuration point of x_2, i.e. $\phi_1(x_1, t) = \Psi(x_1, x_2^\alpha(t), t)$. We use a non-separable potential energy:

$$U(x_1, x_2) = F \cdot (x_1 - x_2)^2, \qquad [7.11]$$

with $F = 10^{12} eV/m^2$ quantifying the strength of the many-body interaction. The many-body wave function $\Psi(x_1, x_2, t)$ can be solved exactly from equation [7.1] with $M_T = 2$. Once the exact 2D wave function $\Psi(x_1, x_2, t)$ is known, we can compute the exact 2D Bohmian trajectories straightforwardly from [7.3]–[7.5]. The initial wave function is a direct product, $\psi_1(x_1, 0) \cdot \psi_2(x_2, 0)$, of two Gaussian wave packets as the one defined in [7.45] of Appendix A. In particular, we consider $E_{o1} = 0.06$ eV, $x_{c1} = -50$ nm and $\sigma_{x1} = 25$ nm for the first wave packet, and $E_{o2} = 0.04$ eV, $x_{c2} = 50$ nm and $\sigma_{x2} = 25$ nm for the second. In this particular example, the exchange interaction among electrons has been disregarded. For a similar example with an exchange interaction, see [ALA 13].

In Figure 7.3, we have plotted the ensemble (Bohmian) kinetic energy for the two electrons $a = 1$ and $a = 2$ using a set of $M_\alpha = 160,000$ Bohmian trajectories. The exact expression of the Bohmian kinetic energy is discussed in Appendix C and defined in [7.72]. We first compute the results directly from the 2D exact wave function solution of [7.1]. We emphasize that there is an interchange of kinetic energies between the first and second particle (see their kinetic energy in the first and second oscillation) indicating the many-particle nature of the system. This effect clearly manifests that the Hamiltonian of that quantum system is non-separable. Alternatively, we can compute the Bohmian trajectories without knowing the many-particle wave function, i.e. using the conditional wave function $\phi_a(x_a, t)$ solution of [7.10] with a proper approximation of G_a and J_a. Here, we consider a zero-order Taylor expansion around $x_a^\alpha(t)$ for the unknown potentials G_a and

J_a. In other words, we consider them as purely time-dependent potentials $G_a(x_a, x_b^\alpha(t), t) \approx G_a''(x_a^\alpha(t), x_b^\alpha(t), t)$ and identically $J_a(x_a, x_b^\alpha(t), t) \approx J_a''(x_a^\alpha(t), x_b^\alpha(t), t)$. This constitutes the simplest approximation. Then, we know that these purely time-dependent terms only introduce a (complex) purely time-dependent phase in the solution of [7.10], so we can write $\phi_a(x_a, t)$ as:

$$\phi_a(x_a, t) = \tilde{\psi}_a(x_a, t) \exp(z_a^\alpha(t)), \qquad [7.12]$$

where the term $z_a^\alpha(t)$ is a (complex) purely time-dependent phase that has no effect on the trajectory $x_a^\alpha(t)$. Under the previous approximation, [7.10] can be simplified into the following equation for the computation of $\tilde{\psi}_a(x_a, t)$:

$$i\hbar \frac{\partial \tilde{\psi}_a(x_a, t)}{\partial t} = \Big(-\frac{\hbar^2}{2m} \frac{\partial^2}{\partial x_a^2} + U_a(x_a, x_b^\alpha(t)) \Big) \tilde{\psi}_a(x_a, t), \qquad [7.13]$$

where the potential energies are $U_1(x_1, x_2^\alpha(t)) = F(x_1 - x_2^\alpha(t))$ for $a = 1$ and $U_2(x_2, x_1^\alpha(t)) = F(x_1^\alpha(t) - x_2)$ for $a = 2$. For this particular scenario, our simplest approximation for the unknown terms G_a and J_a works perfectly and the agreement between 2D exact results and our 1D approximation is excellent (see Figure 7.3). We have computed the ensemble energies in order to justify that the algorithm is accurate not only for an arbitrarily selected set of Bohmian trajectories but also to ensemble them. We use steps 1 to 6 explained above for each trajectory $\{x_1^\alpha(t), x_2^\alpha(t)\}$. Each trajectory is also computed from a 2D version of the algorithm explained in Appendix A.

An improvement over the simple approximation used here for G_a and J_a, when constructing the conditional (Bohmian) wave functions, is necessary in other types of interacting potentials to get the same degree of accuracy as shown in Figure 7.3. A possibility that will be explored in future works is following the ideas presented in [NOR 10, NOR 15] where a full (infinite) set of equations for an exact description of the conditional wave functions is presented.

7.3. The BITLLES simulator: time-resolved electron transport

The previous section was devoted to discuss the ability of Bohmian mechanics to provide, on the one hand, a simple explanation of the measurement process, and, on the other hand, an algorithm to approximate the many-body problem. Now, we will focus on the application of this Bohmian machinery to build up a quantum electron devices simulator called BITLLES[8]. We have divided this section into three parts.

8 BITLLES is the acronym of Bohmian Interacting Transport for non-equiLibrium eLEctronic Structures. See http://europe.uab.es/bitlles.

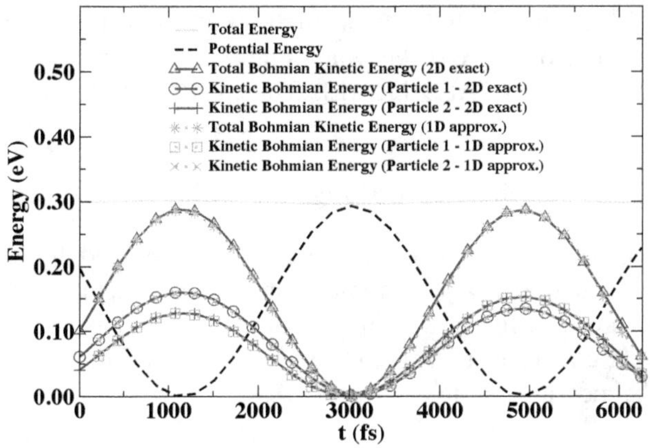

Figure 7.3. *Time evolution of individual (ensemble averaged) Bohmian kinetic energies of identical two-electron system without exchange interaction under a non-separable potential computed from* 2D *exact and* 1D *approximate solutions. For a color version of this figure, see www.iste.co.uk/dollfus/nanodevices.zip*

In section 7.3.1, we argue in detail how the effects of the measuring apparatus can be taken into account in the evolution of the active region of electronic devices. As a particular case of the measurements that can be made over an electron device, we focus here on the electrical current. By means of a simple model, we find a relationship between the current measured on the ammeter and the Bohmian trajectories of the system. By solving the system-apparatus Schrödinger equation for this simplified scenario, we argue that whenever the back action of the apparatus on the system trajectories is not much relevant, we can avoid the explicit computation of the pointer degrees of freedom and focus only on the system to compute the electrical current (without invoking any non-unitary evolution). Section 7.3.2 is devoted to explaining how to go beyond mean-field approaches and include full Coulomb correlations. We demonstrate that Coulomb interactions can be effectively included by defining a Poisson equation for each carrier and a complete set of time-dependent boundary conditions. We argue on the crucial role played by the leads in assuring overall-charge neutrality and current conservation. Finally, in section 7.3.3 we summarize the main pieces that define an electron injection model valid for systems with and without electron confinement. From a practical point of view, this model introduces, apart from the uncertainty in the initial position of the quantum trajectories, an additional randomness on the properties of the injected electrons related to their energies, velocities, etc. It is in this regard that the BITLLES simulator can be somehow understood as a quantum Monte Carlo algorithm. All the results presented in the following sections have been computed using BITLLES. It is worth noticing here that, because of the many analogies between the semiclassical and the Bohmian descriptions (both in terms of trajectories) of electron transport, the

BITLLES simulator also includes a semiclassical limit that corresponds to a many-particle version of the well-known Monte Carlo solution of the Boltzmann equation [ALB 08, ALB 10c, ALB 11, ALB 09a, BEN 13, ALB 10a].

7.3.1. *On the role of the measuring apparatus*

The functionality of any electronic device is determined by the relationship between the current measured by an ammeter and the voltage imposed at the external battery (see Figure 7.4 for a description of a typical electronic circuit). Any measurement of a classical or quantum device implies an interaction between the apparatus and the measured system.

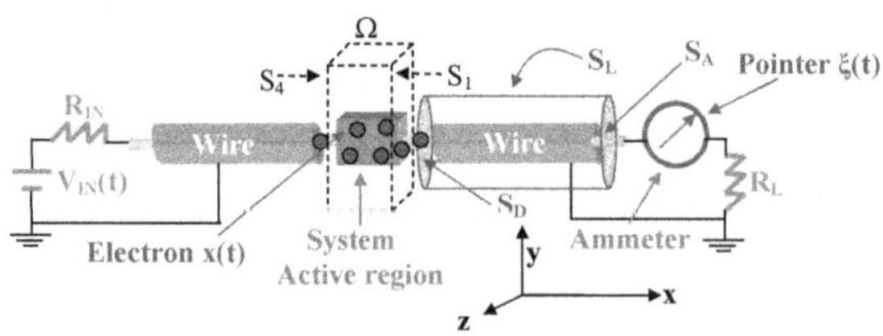

Figure 7.4. *Schematic representation of a typical electrical circuit used in this chapter for studying the current measurement in electrical device. Device simulators compute the current on the surface,* S_D*, of the active region, while the ammeter measures it on the surface,* S_A*. For a color version of this figure, see www.iste.co.uk/dollfus/nanodevices.zip*

As discussed in section 7.1.1, the *orthodox* time evolution of the wave function $\Psi(\vec{r},t)$ is governed by two different dynamical laws. First, there is a dynamical (deterministic) evolution according to [7.1] when the system is not measured. Second, there is an (stochastic) evolution known as *collapse* of the system wave function when it interacts with a measurement apparatus. On the contrary, Bohmian mechanics does not differentiate between measuring and non-measuring evolutions [DÜR 12, BEL 04, ORI 11]. In particular, we define $\vec{\xi} = \{\vec{\xi}_1, ..., \vec{\xi}_K\}$ as the K degrees of freedom that conform the pointer of a measuring apparatus (for example the ammeter). Because of the pointer, we have to deal with a wave function $\Phi(\vec{r},\vec{\xi},t)$ whose equation is equivalent to [7.1] but in the extended configuration space. The pointer position $\vec{\xi}^{\alpha}(t)$ and system position $\vec{r}^{\alpha}(t)$ move according to their equations of motion equivalent to [7.4]. The (Bohmian) position of a *good* pointer $\vec{\xi}^{\alpha}(t)$ are supposed to be correlated with the system position $\vec{r}^{\alpha}(t)$. The stochastic nature of the quantum measurement is recovered here because in the experimental setup we do not know the initial Bohmian positions $\vec{\xi}^{\alpha}(t_o)$ and $\vec{r}^{\alpha}(t_o)$, which have to be selected in the simulation according to quantum equilibrium condition [7.5].

The Bohmian explanation of the measurement process has, however, the following two technical difficulties:

– Firstly, we have to specify which Hamiltonian determines the evolution of the system plus apparatus. This difficulty is similar to specifying which operator provides good information about the measuring process in the *orthodox* quantum mechanics [DUR 04].

– The second difficulty is related to the computational limitations while solving the many-particle Schrödinger equation. The Schrödinger equation with the addition of the pointer is most of the times unsolvable. This technical difficulty is non existent when using operators because they act only on the system's wave function.

7.3.1.1. *The system plus apparatus Schrödinger equation*

The idea of including the pointer as an additional degree of freedom in the Schrödinger equation was already proposed by von Neumann in 1932 within *orthodox* quantum mechanics, when trying to provide a macroscopic explanation of the *collapse* of the wave function [VON 32]. Let us consider, for the moment, only two degrees of freedom. The variable $\vec{r}_1$ for the system and ξ for the center of mass of the apparatus (in a 1D system). The interaction among them is determined by the following Hamiltonian:

$$H_{int} = -i\hbar\lambda A(\vec{r}_1)\frac{\partial}{\partial\xi}, \qquad [7.14]$$

where $A(\vec{r}_1)$ is the magnitude of the system that we want to measure and λ is a coupling constant. From equation [7.14], the ξ component of the *local* momentum (velocity) of the wave function $\Phi(\vec{r}_1, \xi, t)$ will depend on the magnitude of $A(\vec{r}_1)$. Therefore, an initially localized wave packet $\Phi(\vec{r}_1, \xi, 0)$ will spread in the ξ direction because of the different ξ velocities. Then, the only new ingredient that we have to include when dealing with Bohmian mechanics is the presence of the pointer and system (Bohmian) trajectories, $\{\vec{r}_1^{\alpha}(t), \xi^{\alpha}(t)\}$. Next, we specify what expression we have to adopt for $A(\vec{r}_1)$ in order to effectively compute the electrical current [ALB 13b].

It is common to compute the electrical current on the (simulated) surface S_D in Figure 7.4, while a real measurement is performed on the (non-simulated) surface S_A. It is then crucial to understand in which extension the current on S_A is equal to that on S_D. In fact, these currents will be only equal if we consider the *total current* $I(t) = I_p(t) + I_d(t)$, where $I_p(t)$ and $I_d(t)$ are, respectively, the particle and displacement components discussed in section 7.1.2. Since the Maxwell equations ensure that the total current density $\vec{J}(\vec{r}_1,t)$ is a vector with a null divergence, then we can write $\int_S \vec{J}(\vec{r}_1,t)d\vec{s}=0$ for a closed surface $S=\{S_D,S_A,S_L\}$ where S_L is a the surface parallel to the transport direction in the cable as shown in Figure 7.4. In particular, for a cable we can assume $\int_{S_L} \vec{J}(\vec{r}_1,t)d\vec{s}=0$, so we finally get $\int_{S_D} \vec{J}(\vec{r}_1,t)d\vec{s}=-\int_{S_A} \vec{J}(\vec{r}_1,t)d\vec{s}$.

The function $A(\vec{r}_1)$ in equation [7.14] has then to be related to the total current $I(t)$ of the system, meaning that $I(t)$ has to be somehow linked to the positions $\vec{r}_1^{\alpha}(t)$ of the particles of the system. The total current measured by an ammeter, $I^{\alpha}(t)$, for a particular trajectory $\vec{r}_1^{\alpha}(t)=\{x^{\alpha}(t),y^{\alpha}(t),z^{\alpha}(t)\}$, can be defined as the time derivative of the following *particle* plus *displacement* charges:

$$I^{\alpha}(t)=\frac{d\left(Q_p(\vec{r}_1^{\alpha}(t))+Q_d(\vec{r}_1^{\alpha}(t))\right)}{dt}\equiv\frac{dQ\left(\vec{r}_1^{\alpha}(t)\right)}{dt}, \qquad [7.15]$$

where we define the *conduction* charge as:

$$Q_p(\vec{r}_1)=-q\int_{S_D} dy'\,dz'\int_{x'=x_D}^{\infty} dx'\delta(\vec{r}_1'-\vec{r}_1), \qquad [7.16]$$

q being the (unsigned) electron charge and x_D the x position of the lateral surface S_D. The particle charge is $Q_p(\vec{r}_1)=-q$ only if the electron is located at the right of x_D and inside the lateral surface S_D. Identically, we can interpret the *displacement* charge in [7.15] as:

$$Q_d(\vec{r}_1)=\int_{S_D}\varepsilon(\vec{r}_1')\vec{E}(x,y,z,x_D,y',z')d\vec{s}, \qquad [7.17]$$

being $\vec{E}(x,y,z,x_D,y',z')$ the electric field generated at $\vec{r}_1'=\{x_D,y',z'\}$ of the surface S_D by one electron at $\vec{r}_1=\{x,y,z\}$. It is important to emphasize that $Q_d(\vec{r}_1)$ is different from zero independently of the distance between the electron and the surface. The generalization of $Q_p(\vec{r}_1)$ and $Q_d(\vec{r}_1)$ to an arbitrary number of electrons $Q_p(\vec{r}_1,..,\vec{r}_{M_T})$ and $Q_d(\vec{r}_1,..,\vec{r}_{M_T})$ is quite simple. Finally, in our ammeter model, the von Neumann term A in equation [7.14] is the *conduction* plus *displacement* charges defined in equations [7.16] and [7.17], divided by the elementary charge q, i.e.:

$$A(\vec{r}_1,..,\vec{r}_{M_T})=-\frac{Q(\vec{r}_1,..,\vec{r}_{M_T})}{q}, \qquad [7.18]$$

where the (irrelevant) minus sign appears just to provide a positive pointer movement when an electron moves from left to right in Figure 7.4, which corresponds to a negative net current. Certainly, other models of the ammeter are possible; however, we will see next that the model proposed here implies that the total current $I^{\alpha}(t)$ is directly related to the acceleration of the pointer.

Using equations [7.14] and [7.18], we can describe the many-particle Schrödinger equation of the M_T electrons interacting with the K pointer particles, each one of mass m. In order to simplify the notation, we focus on the 1D center of mass of the pointer ξ whose mass is $M = K\,m$. Then, we can write:

$$i\hbar\frac{\partial\Phi(\vec{r},\xi,t)}{\partial t} = \Big(-\sum_{k=1}^{M_T}\frac{\hbar^2}{2m^*}\frac{\partial^2}{\partial\vec{r}_k^2} + U(\vec{r},t) - \frac{\hbar^2}{2M}\frac{\partial^2}{\partial\xi^2} + i\hbar\lambda\frac{Q(\vec{r})}{q}\frac{\partial}{\partial\xi}\Big)\Phi(\vec{r},\xi,t). \qquad [7.19]$$

At this point, it is relevant to compute the Bohmian velocity of the center of masses. By inserting the polar form of the wave function $\Phi(\vec{r},\xi,t) = R(\vec{r},\xi,t)e^{\frac{i}{\hbar}S(\vec{r},\xi,t)}$ into equation [7.19], we derive the corresponding Hamilton–Jacobi and the continuity equations from which we can define the pointer velocity as:

$$v_\xi(\vec{r},\xi,t) = \frac{1}{M}\frac{\partial S(\vec{r},\xi,t)}{\partial\xi} - \lambda\frac{Q(\vec{r})}{q}. \qquad [7.20]$$

Since the mass M is very large, the first term in the right-hand side of equation [7.20] can be neglected. Therefore, for a particular trajectory $\{\vec{r}^{\alpha}(t),\xi^{\alpha}(t)\}$, it turns out that the acceleration of the pointer, i.e. the time derivative of equation [7.20], is proportional to the total current of the system defined in equation [7.15]:

$$\frac{dv_\xi(\vec{r}^{\alpha}(t),\xi^{\alpha}(t),t)}{dt} \approx -\lambda\frac{dQ(\vec{r}^{\alpha}(t))/q}{dt} = -\frac{\lambda}{q}I^{\alpha}(t). \qquad [7.21]$$

7.3.1.2. *An example: numerical solution of system plus apparatus Schrödinger equation*

We solve here the 2D version of equation [7.19], where the system and the pointer are described by just one particle x and ξ. Consider that the initial wave function is a product of two Gaussian wave packets. The central kinetic energies, central positions and spatial dispersions being, respectively, $E_x = 0.1$ eV, $x_c = -100$ nm and $\sigma_x = 8$ nm for the particle, and $E_\xi = 0$ eV, $\xi_c = 0$ nm and $\sigma_\xi = 0.5$ nm for the pointer. The system consists of an electron (with m^* equal to 0.068 the electron free mass) impinging upon an Eckart barrier $U(x,t) = V_0/\cosh^2[(x - x_{bar})/W]$ with $V_0 = 0.3\ eV$, $x_{bar} = -50$ nm and $W = 1$ nm (see the line at $x = -50$ nm in

Figure 7.5). The pointer of the apparatus ξ (with $M \approx 75,000\ \text{m}^*$) interacts with the system through the term $i\hbar\lambda(Q(x)/q)\partial/\partial\xi$ with $\lambda = 50\ \text{nm/ps}$. We consider a lateral surface of $S_D = 900\ \text{nm}^2$ located at $x_D = 75\ \text{nm}$, defined in equation [7.15]–[7.17], is only different from zero on the right-hand side of the plots in Figure 7.5. We indicate this region by a (ammeter) rectangle in the configuration space.

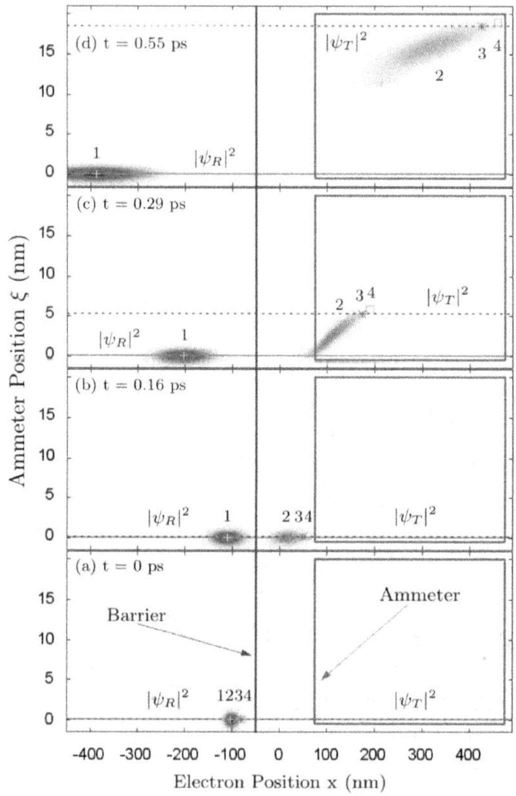

Figure 7.5. *Time evolution of the squared modulus of $\Phi(x, \xi, t)$ at four different times. The system barrier is indicated by a solid line and the region in the configuration space where the system–apparatus interaction is non-zero by a rectangle. The solid horizontal line indicates the modulus of the (reflected) conditional wave function $|\psi_R|^2 = |\Phi(x, \xi^{\alpha=1}(t), t)|^2$, while the dashed horizontal line corresponds to $|\psi_T|^2 = |\Phi(x, \xi^{\alpha=3}(t), t)|^2$. The time evolution of four trajectories $\{x^\alpha(t), \xi^\alpha(t)\}$ with different initial positions are presented with $\square$, $*$, $\times$ and $+$. The transmitted trajectories ($\square$, $*$ and $\times$) at (c) and (d) have different pointer positions associated with $\xi^\alpha(t)$ with $\alpha = 2, 3, 4$ because their evolution does only depend on the transmitted wave packet. The pointer position associated with the reflected trajectory ($+$) with $\alpha = 1$ does not move because there is no interaction between this trajectory and the apparatus. For a color version of this figure, see www.iste.co.uk/dollfus/nanodevices.zip*

The numerical solution of the modulus of $\Phi(x, \xi, t)$ is plotted at four different times. At the initial time, the entire wave function is at the left of the barrier. Later the wave function is splitted up into reflected and transmitted parts due to the barrier. In Figures 7.5(a) and (b), the velocity in the ξ direction remains zero because the wave function has not yet arrived to the apparatus. In Figures 7.5(c) and (d), the interaction of the apparatus with the transmitted part of the wave function shows up. The *local* velocity of the wave packet in ξ rises according to equation [7.20]. In Figure 7.5, we also plot the positions of the system and pointer $\{x(t), \xi(t)\}$ for four different trajectories. Each trajectory corresponds to a different experiment. There are three transmitted particles (in the x-direction). While the pointer (i.e. $\xi(t)$) does not move for the reflected particles, its evolution for the transmitted ones clearly shows the correlation with the electrical current of the system.

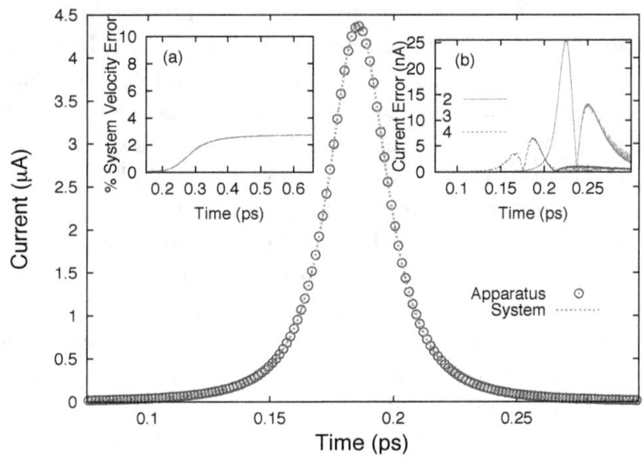

Figure 7.6. *Total current associated with the $*$ trajectory $\{x^3(t), \xi^3(t)\}$ in Figure 7.5 as a function of time computed from the acceleration of the pointer trajectory computed from equation [7.19] in (red) circles and from the system trajectory computed from equation [7.1] in dotted (green) line. For commodity, we reverse the sign of the current. a) Relative (ensemble) error for the system Bohmian velocity when computed from equation [7.1]. b) Absolute error of the system current of the $\alpha = 2$, $\alpha = 3$ and $\alpha = 4$ trajectories in Figure 7.5 when comparing the solution from equation [7.19] (with ammeter) and equation [7.1] (without ammeter). For a color version of this figure, see www.iste.co.uk/dollfus/nanodevices.zip*

At this point, it would be enlightening to compare the schematic results in Figure 7.1(e) and (f) with the numerical results in Figure 7.5. The transmitted trajectory in Figure 7.1(e) corresponds to any of the particles $\alpha = 2$, $\alpha = 3$ and $\alpha = 4$ in Figure 7.5. Let us take, for example, $\alpha = 3$, which is drawn in a horizontal dashed line in Figure 7.5. The wave packet ψ_T drawn in Figure 7.1(e) corresponds then to the conditional wave function $\Phi(x, \xi^{\alpha=3}(t), t)$. Following the conditional wave function $\psi_T = \Phi(x, \xi^{\alpha=3}(t), t)$ in Figure 7.5, we clearly see how the reflected part of the wave packet disappears in the (conditional) configuration space of the system $\{x, \xi^{\alpha=3}(t)\}$. The disappearance of part of the wave packet (collapse) due to the measurement of the electrical current is totally demystified in this (system–ammeter) picture. For the reflected particle in Figure 7.1(d), we follow identical arguments using the $\alpha = 1$ particle in Figure 7.5. See $\psi_R = \Phi(x, \xi^{\alpha=1}(t), t)$ in the horizontal solid line in Figure 7.5. Finally, in Figure 7.6, we plot the "measured" current in circles computed from the acceleration of the pointer (as indicated in equation [7.20]) for the third trajectory marked with an asterisk($*$) in Figure 7.5.

As seen in equation [7.21] and Figure 7.6, the evolution of the pointer $\xi^{\alpha=3}(t)$ describes the evolution of the total current $I^{\alpha=3}(t)$; however, the movement of the pointer (see Figure 7.6) is still not macroscopic (it only moves a few nanometers). It can be easily demonstrated that for other parameters λ and S_D, $\xi(t)$ would have a macroscopic movement. It is important to underline that for a realistic ammeter we could expect other features not included in the simple model just presented. Let us discuss this point in detail. As a matter of fact we can expect that the wave function of the system–apparatus will channel into a set of non-overlapping parts, each corresponding to a particular well-defined position of the pointer of the ammeter. As shown in Figure 7.5, our model splits the wave function, in the ξ direction, only into two non-overlapping channels. Although not complete, this feature is enough to calculate DC currents and low-frequency noise. If one is interested in going beyond this simple model and studying high-frequency regimes, other routes can be followed. For example, one could consider the Coulomb interaction between the electrons in the active region of the device and those in the cables by introducing a weak measurement protocol of the total current [MAR 14, TRA 13, ALB 16]. In any case, the model presented in this section gives enough information for our present purposes: it provides two channels (one for the reflected and one for the transmitted part of the system wave function) and a strict relation between the system and the ammeter (see equation [7.20]), from this we can infer the perturbation induced by the ammeter on the measured system.

Once the first technical difficulty (i.e. specifying how the ammeter is included in the Hamiltonian) is solved, we must discuss about the difficulties of solving the Schrödinger equation including the system and apparatus degrees of freedom. *Can we avoid the inclusion of the pointer degrees of freedom in the Schrödinger equation without losing much accuracy?* The answer to this question is affirmative whenever the apparatus induces a small distortion on the system. The distortion on the system's trajectories can be quantified by defining the relative (ensemble) error of the system velocity:

$$Error(t) = \lim_{M_\alpha \to \infty} \frac{\sum_{\alpha=1}^{M_\alpha} |v_\lambda^\alpha(t) - v_0^\alpha(t)|}{\sum_{\alpha=1}^{M_\alpha} |v_\lambda^\alpha(t)|}. \quad [7.22]$$

We define $v_\lambda^\alpha(t)$ as a Bohmian velocity of the system when we use equation [7.19] with $\lambda = 50\ \mathrm{nm/ps}$, while $v_0^\alpha(t)$ when we do not consider the apparatus ($\lambda = 0$) in [7.19]. We see in Figure 7.6(a) that the relative error on the velocity defined in [7.22] is less than 3%. Then, if we avoid the inclusion of the pointer, we can compute the total current directly from the system trajectory $x_0^\alpha(t)$ without apparatus, with a small error. This result is confirmed by the dashed line in Figure 7.6 that corresponds to the total current computed from [7.15] using the system trajectories in [7.19] when $\lambda = 0$. See the absolute error in Figure 7.6(b) defined as the difference (in absolute value) between the exact value in circles and the approximate value in a dotted line. The reason why the agreement between the current computed from the pointer trajectory and the system trajectory is so good because the main distortion of the system trajectory comes from the barrier, not by the apparatus. The former splits the initial wave packet into two separated parts (transmitted and reflected components), while the latter provides a small *adiabatic* perturbation on the system trajectories as seen in Figure 7.5. While the error is very small for the simple model pointer used in Figure 7.5, a larger error can be expected in real ammeters. In any case, it perfectly clarifies that we can use the Bohmian trajectories of the system alone (without the pointer) to compute the current when the back action of the apparatus on the system trajectories is not much relevant. It is very important to emphasize, however, that the change from equation [7.19] to [7.1] is only technical, without any fundamental implication. In summary, when the measurement apparatus has a small effect on the system (e.g. for the computation of DC and low-frequency noise) and the pointer position does perfectly specify the value of the total current of the system (as for our model system in Figure 7.6), then we can avoid the explicit simulation of the pointer in order to surpass computational burdens.

7.3.2. *Coulomb correlations beyond mean field*

In general, the Coulomb interaction introduces screening among electrons implying that the total charge in the whole circuit is zero, this is what we call *overall charge neutrality*. Moreover, the proper modeling of the total current has to take into account the dynamics of electrons in a self-consistent way to preserve the *conservation of the total current*. Both conditions require an (approximate) solution to the many-body Coulomb interaction. Below, we describe how using a small simulation box (including only the active region of the electronic device) both conditions can be preserved [ALB 09b, ALB 10b, LOP 08, ALB 13a].

7.3.2.1. *The many-particle Poisson equation in the active region*

The evaluation of the each term of $U_a(x_a, \vec{r}_b^{\alpha}[t], t)$ in equation [7.10] can be written as:

$$U_a(\vec{r}_a, \vec{r}_b^{\alpha}(t), t) = \sum_{j \neq a}^{N(t)} \frac{q^2}{4\pi\varepsilon|\vec{r}_a - \vec{r}_j^{\alpha}(t)|} + \sum_{j=N(t)+1}^{M_T} \frac{q^2}{4\pi\varepsilon|\vec{r}_a - \vec{r}_j^{\alpha}(t)|}, \qquad [7.23]$$

where the first ensemble $\{1, \ldots, N(t)\}$ corresponds to the electrons in the device active region and the second ensemble $\{N(t)+1, \ldots, M_T\}$ corresponds to those that are in the leads and reservoirs. We do not want to deal with the dynamics of the second ensemble. Thus, due to the uniqueness (electrostatic) theorem [JAC 62], the scalar potential inside the active region can be introduced through the solution of a Poisson equation with a proper definition of its boundary condition. Then, instead of using [7.23], we compute the following 3D Poisson equation inside the active region:

$$\nabla_{\vec{r}_a}^2 \left(\varepsilon(\vec{r_a}) U_a(\vec{r_a}, \vec{r}_b^{\alpha}(t), t)\right) = \rho_a(\vec{r}_a, \vec{r}_b^{\alpha}(t), t), \qquad [7.24]$$

where the charge density $\rho_a(\vec{r}_a, \vec{r}_b^{\alpha}(t), t)$ can be written as:

$$\rho_a(\vec{r}_a, \vec{r}_b^{\alpha}(t), t) = \sum_{\substack{j=1 \\ j \neq a}}^{N(t)} -q\delta(\vec{r}_a - \vec{r}_j^{\alpha}(t)). \qquad [7.25]$$

Each a-electron is associated with a different Poisson equation [7.24] with a different charge density [7.25]. See [ALB 09b] for a detailed discussion on this point. Now, the role of the second term of the right-hand side of [7.23], i.e. the interaction of the first ensemble $\{1, \ldots, N(t)\}$ of electrons with the second one $\{N(t)+1, \ldots, M_T\}$, can be, in principle, represented by a proper set of conditions at the borders of the active region [ALB 10b]. The achievement of overall-charge neutrality will ultimately depend on our ability to properly define these boundary conditions.

7.3.2.2. *Time-dependent boundary conditions of the Poisson equation*

Because the far from equilibrium conditions governing the dynamics of an electronic device, it is very difficult to anticipate an educated guess for the scalar potential (or the electric field) and the charge density on the boundaries of the active region when the leads and the reservoirs are not simulated. Our strategy consists of deriving analytical expressions for all these quantities along the leads and reservoirs to transfer the specification of the boundary conditions at the borders of the active region into a much simpler ones deep inside the reservoirs [ALB 10b, ALB 13a].

In order to derive a set of analytical equations, we only need to take into account two general considerations: first, the total charge in a large volume including the device active region, the leads and the reservoirs, tends to zero within the dielectric relaxation time, $\tau_c = \varepsilon/\sigma$. Then, it can be easily demonstrated that the electric field deep inside the reservoirs, $E^C_{S/D}(t)$, tends to its drift value $E^C_{S/D}(t) \rightarrow E^{\text{drift}}_{S/D}(t)$ within the same time $\tau_c = \varepsilon/\sigma$. Second, the scalar potentials deep inside the reservoir are fixed by the external bias $V^C_S(t) = 0$ and $V^C_D(t) = V_{\text{external}}(t)$. With these considerations in mind, the analytical (temporal and spatial) relations for the charge density, the electric field and the scalar potential must be then coupled to an injection model controlling the amount of charge on the borders of the active region. Such a coupled procedure results in a boundary conditions' algorithm for the Poisson equation in [7.24] that enforces overall-charge neutrality and current conservation [ALB 10b, ALB 13a].

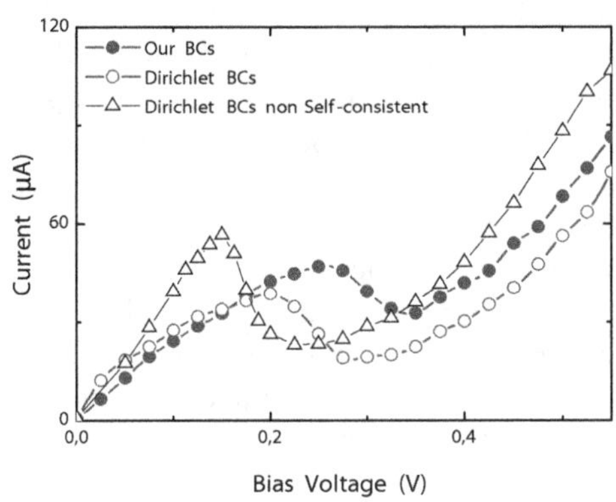

Figure 7.7. *RTD current-voltage characteristic. Results taking into account the Coulomb correlations between the leads and the device active region are presented in solid circles. Open circles refer to the same results neglecting the lead-active region interaction. Open triangles refer to a single-particle scenario. Transport takes place from source to drain in the x direction with lateral dimensions $L_y = L_z =$ $48.6\,\text{nm}$. Room temperature is assumed. For a color version of this figure, see www.iste.co.uk/dollfus/nanodevices.zip*

In order to highlight the importance of the Coulomb interaction in the prediction of the current–voltage characteristic of nanoelectronic devices, let us apply the above concepts to simulate a simple resonant tunneling device (RTD) [ALB 10b]. We consider two highly doped drain-source $GaAs$ regions (the leads), two $AlGaAs$ barriers and a quantum well (the device active region). All Schrödinger equations [7.10] are coupled to its own Poisson equations [7.24] with the boundary conditions given above. How the electrical current is computed will be explained in detail in section 7.4, here we will directly focus on the results. We distinguish between (1) the Coulomb interaction among electrons inside the active region and (2) the Coulomb correlations among these electrons and those outside the active region. In Figure 7.7, we compare the characteristic DC obtained by including all Coulomb correlations with those obtained, first, by neglecting the dynamical Coulomb correlations between electrons inside and outside the active region (i.e. assuming standard Dirichlet boundary conditions) and, second, switching off all Coulomb correlations (i.e. single-particle treatment of electron transport). The differences between the three approaches appear not only in the magnitude of the current but also in the position of the resonant region. More details can be found in [ALB 10b].

7.3.3. *Stochastic injection of electrons*

Since a large part of the degrees of freedom of a circuit shown in Figure 7.4 are neglected (we want to explicitly simulate only the active region), we cannot completely specify the initial $N(t)$-particle wave function inside the active region (we do not know with certainty the number of electrons $N(t)$, their energies, etc.). We can assume, however, that the mean energy of injected electrons follows Fermi–Dirac statistics. We thus include the probabilities of these states by introducing an additional probability distribution h. From a practical point of view, this means that apart from the uncertainty in the initial position of the quantum trajectories (the α-distribution mentioned in section 7.1), we do also have an additional uncertainty on the properties of the injected electrons, $h = \{1, ..., M_h\}$, that arises because the active region is an open system. Strictly speaking, because of the h distribution, we are no longer dealing with a single pure $N(t)$-particle state, but with a mixed quantum system prepared by statistically combining different pure states of $N(t)$ particles. The microscopic description of the system, however, implies that we have a perfect knowledge of each of the $N(t)$ particle states and that the statistical ensemble of the density matrix is explicitly taken into account.

Our Bohmian protocol requires each electron to be described by a (conditional) wave function (h distribution) plus a Bohmian trajectory (α distribution). Every time an electron with a particular initial wave function is selected to enter the device active region according to the h distribution, an initial position $x^{\alpha,h}(0)$ associated with this (h selected) wave packet has to be randomly selected according to [7.5]. See Appendix A for numerical details of the wave packet and trajectories computation. Next, we explain how the h distribution of the wave packets is defined. For each contact, we select a flat potential region in an (non-physical) extension (for $x < 0$ in the source and $x > L$ in the drain) of the simulation box (see Figure 7.8). The initial Gaussian wave packet[9] is defined in this flat potential region (deep inside the contact). In particular, the central position of all wave packets is selected $x_c = 100$ nm far from the border of the active region and the spatial dispersion of the wave packet is $\sigma_x = 25$ nm (i.e. the wave packet is somehow similar to a scattering state). The only two additional h-parameters that we still have to fix to fully define the wave packet are the central kinetic energy E_o of the wave packet and the injecting time t_o when the electron *effectively* enters the simulation box. The selection of the energy E_o has to satisfy the Fermi–Dirac occupation function $f(E_o)$ that depends on the (quasi) Fermi energy and temperature. The selection of the time when the electron is injected is a bit more complex [ORI 07b].

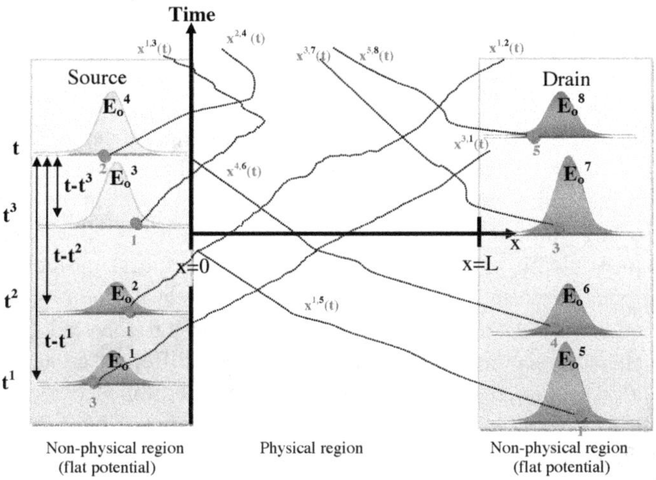

Figure 7.8. *Schematic representation of the (physical $0 < x < L$ and unphysical) simulation box for a particular selection of the parameters E_o^h and entering times t^h of the initial Gaussian wave packets (h-distribution) and the initial positions $x^{\alpha,h}(0)$ of the Bohmian trajectories (α-distribution). The wave packets $h = 1$ and $h = 2$ have identical energies $E_o^1 = E_o^2$ but different injecting times $t^1 \neq t^2$. The injection model does also assume that the exact h-distribution (except for an irrelevant time offset in t^h) plotted here will be repeated in a later time, but with different $x^{\alpha,h}(0)$, accomplishing the quantum equilibrium of the α- distribution. For a color version of this figure, see www.iste.co.uk/dollfus/nanodevices.zip*

9 In a flat potential region without interaction, the single-particle wave packets are exactly the normalized conditional (Bohmian) wave functions discussed in this work.

In order to simplify our explanation, let us assume a 1D system with parabolic bands where the (central) kinetic energy E_o of the wave packet is related to the (central) wave vector k_o by $E_o = (\hbar k_o)^2/(2m^*)$. Let us define t_o as the minimum temporal separation between the injection of two wave packets whose central wave vectors and central positions fit into the following particular phase-space cell $k_o \in [k_b,\ k_b + \Delta k)$ and $x_c \in [x_b,\ x_b + \Delta x)$, x_b being the border of the simulation region. For a 1D system, the value of t_o can be easily estimated. The number of electrons n_{1D} in the particular phase space cell $\Delta k \cdot \Delta x$ is $n_{1D} = 2 \cdot \Delta k \cdot \Delta x/(2\pi)$ where the factor 2 takes into account the spin degeneracy. These electrons have been injected into Δx during the time interval Δt defined as the time needed for electrons with velocity $v_x = \Delta x/\Delta t = \hbar k_o/m^*$ to travel a distance Δx. Therefore, the minimum temporal separation, t_o, between the injection of two electrons into the previous cell is Δt divided by the maximum number n_{1D} of electrons:

$$t_o = \frac{\Delta t}{n_{1\text{D}}} = \left(\frac{1}{\pi} \frac{\hbar\, k_o}{m^*}\ \Delta k \right)^{-1}. \qquad [7.26]$$

It is very instructive to understand the minimum temporal separation t_o in [7.26] as a consequence of the wave packet version of the Pauli principle. The simultaneous injection of two electrons with similar central positions and central momentums would require such a huge amount of energy that its probability is almost zero. In other words, subsequent electrons with central positions and central momentums equal to the preceding ones can only be injected after a time interval given by t_o. This is the time interval necessary to ensure that the first electron has traveled a distance $t_o \times \hbar\ k_o/m^*$ so that the second electron is located in a different central position (no Pauli principle). The injection of electrons (from the mentioned phase-space cell) at multiple times of t_o depends finally on the statistics imposed by the Fermi–Dirac function mentioned above. During each attempt of injection at multiples of t_o, we select a random number r, and the electron is effectively injected only if $f(E_o) > r$. The mathematical definition of the rate and randomness of the injection process are given by the following binomial probability $P(E_o, N, \tau)$ [ORI 07b]:

$$P(E_o, N, \tau) = \frac{M_\tau!}{N! \cdot (M_\tau - N)!} f(E_o)^N \left(1 - f(E_o)\right)^{M_\tau - N}. \qquad [7.27]$$

This expression defines the probability that N electrons (from the mentioned phase-space cell) are *effectively* injected into the active region during the time interval τ. The parameter M_τ is the number of attempts of injecting electrons during this time interval τ, defined as a natural number that rounds the quotient τ/t_o to the nearest natural number toward zero. The number of injected electrons can be $N = 1, 2, ... \leq M_\tau$. In order to satisfy the requirements of section 7.3.2.2, the final injection model can be a little more elaborated. More details can be found in [ALB 10b].

According to Figure 7.8, we can understand the contacts as large reservoirs of electrons that are waiting to enter the active region. The parameters x_c, σ_x, E_o (or central wave vector) and a time t^h, defined as the entering times, are assigned to each initial wave packet. Additionally, we give an initial position $x^{\alpha,h}(0)$ to each electron. The measured value of the current related with the Bohmian trajectories is thus ultimately associated with two sources of uncertainty, α and h, i.e. $I^{\alpha,h}(t)$. In order to give a precise method to compute the electrical current and its moments, from now on we explicitly write the whole indexation.

7.4. Computation of the electrical current and its moments with BITLLES

This section is devoted to provide a practical method to evaluate the electrical current with the BITLLES simulator. Let us recall here that any kind of information related with the electron transport taking place in the electronic device (such as power consumption and kinetic and potential distributions) can be easily computed with BITLLES. After a brief dissertation on the limitations of the information contained in the values of the DC currents in quantum systems and the importance of multitime measurements to predict its fluctuations, in section 7.4.2 we will provide a detailed description of the equations required to evaluate DC, AC, transients and current fluctuations. Let us recall here that the mathematical formalism introduced in this section does not differ much (in form) from the semiclassical one. The reason is simple, both descriptions use ensembles of trajectories to compute observables. It is in this regard that the BITLLES simulator includes also a semiclassical Monte Carlo package to simulate electron devices by solving the Boltzmann transport equation [ALB 10c, ALB 11, ALB 09a, BEN 13].

7.4.1. *Preliminary considerations*

The prediction of the DC current measured in a laboratory can be computed using two different protocols [ORI 13]. First, we can compute I_{DC} by time averaging the measured value of the total current $I(t)$ from a *unique device* during a large (ideally infinite) period of time τ :

$$I_{DC} = lim_{\tau\to\infty}\frac{1}{\tau}\int_0^{\tau} I(t)dt. \qquad [7.28]$$

We can use the conduction current $I_p(t)$ instead of the total current $I(t)$ in equation [7.28] because $\frac{1}{\tau}\int_0^{\tau} I_d(t)dt \to 0$ for large τ (the displacement current is proportional to a time derivative of the electric field).

Second[10], we can compute I_{DC} from an ensemble-average of all possible eigenvalues of the current I_i that satisfy the equation $\hat{I}|\psi_i\rangle = I_i|\psi_i\rangle$ for a particular operator $\hat{I}$, at one particular time t:

$$I_{DC} = \langle\psi(t)|\hat{I}|\psi(t)\rangle = \sum_{j}^{M} c_j^*(t)\langle\psi_j|\sum_{i=1}^{M} I_i\hat{A}\cdot c_i(t)|\psi_i\rangle = \sum_{i=1}^{M} I_i P(I_i), \qquad [7.29]$$

where we have used the orthonormal property of the eigenstates $\langle\psi_j|\psi_i\rangle = \delta_{i,j}$ and the definition of the (Born) probability $P(I_i) = |c_i(t)|^2$. When the operator $\hat{I}$ is the conduction current density operator, $\hat{I}_p = |r\rangle\langle r|\hat{p} + \hat{p}|r\rangle\langle r|$ with $|r\rangle\langle r|$ and $\hat{p}$ the position and momentum operators, then the ensemble-average value $\langle\psi(t)|\hat{I}|\psi(t)\rangle$ gives the well-known result [COH 78]:

$$\begin{aligned} I_{DC} = \langle I_p(t)\rangle &= \int_{S_i} \vec{J}(\vec{r},t)d\vec{s} = \int_{S_i} \langle\psi(t)|\hat{I}_p|\psi(t)\rangle d\vec{s} \\ &= \int_{S_i} \frac{\hbar}{m} Im\left(\psi^*(\vec{r},t)\vec{\nabla}\psi(\vec{r},t)\right) d\vec{s}. \end{aligned} \qquad [7.30]$$

The quantum expression of the current probability is defined in expression [7.48] of Appendix C. Expression [7.30] is extraordinarily useful and simple because it allows us to predict the DC of quantum devices without knowing the eigenstates $|\psi_i\rangle$ or eigenvalues I_i of the current operator $\hat{I}_p$. However, the amount of information contained in [7.30] is limited.

In order to illustrate how reliable is the information provided by expression [7.30] to the electronic industry, let us discuss the transient from "OFF" to "ON" values in the drain-source current of a digital quantum FET when the gate contact voltage suddenly changes from V_{OFF} to V_{ON}. In Figure 7.9, we plot the measured current $I(t_1)$ and the ensemble current $\langle I(t_1)\rangle$. The first problem appears because $\langle I(t_1)\rangle$ is not the measured current in a laboratory in a single transistor. The difference between $I(t_1)$ and $\langle I(t_1)\rangle$ can be irrelevant for a large signal-to-noise ratio (for example in FETs with many electrons) as seen in Figure 7.9(a), but it will be certainly meaningful for low signal-to-noise ratios (for example in FETs with very few electrons) as seen in Figure 7.9(b). The second problem when using [7.30] for AC, transients and noise at high frequencies is that the total time-dependent current cannot be computed only from $\int_{S_i} \vec{J}(\vec{r},t)\, d\vec{s}$ alone, but we have to include the displacement current, with all its computational difficulties. The third problem appears because in scenarios where

10 Strictly speaking, no ergodic theorem exists for an out of equilibrium system [PRI 65]. Therefore, the ergodic connection between [7.28] and [7.29] has to be considered as only a very reasonable approximation for DC transport, but not as an exact result.

the system is measured many times, we have to take into account the unitary time evolution determined by the Schrödinger equation plus the non-unitary evolution due to the measurement. We briefly discuss this issue below.

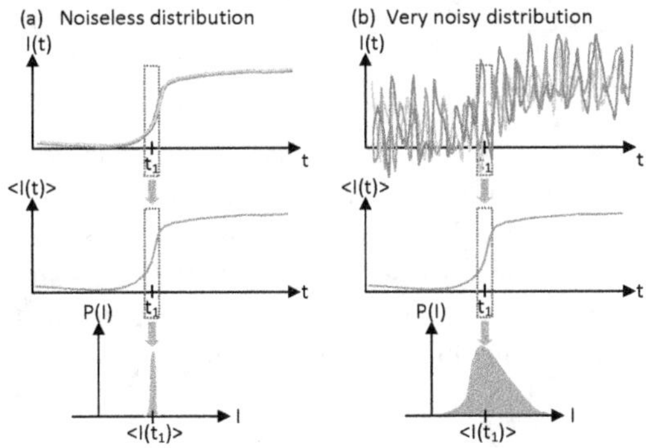

Figure 7.9. *Schematic representation of the measurement of transient currents $I(t)$ on a sample. The ensemble average current $\langle I(t_1)\rangle$ at time t_1 can provide misleading information because different current probability distributions provide identical average values. The current probability distribution P(I) at time t_1 provides the complete information of the transport process. However, its computation requires a proper treatment of unitary and non-unitary evolution of the quantum system. For a color version of this figure, see www.iste.co.uk/dollfus/nanodevices.zip*

In general, we could argue that the ensemble-average value of such fluctuations can be quantified as:

$$\Delta I = \sum_{i=1}^{M} (I_i - I_{DC})^2 \cdot P(I_i). \qquad [7.31]$$

This expression can be related to an ensemble of systems with a unique time measurement of the squared current. Thus, in principle, no multitime measurement is needed for evaluating [7.31]. However, all electronic devices can work properly only below a particular cutoff frequency, behaving as a low-pass filter. Therefore, electronic systems are not able to measure "all" the noise appearing in [7.31], but only the noise whose components have a frequency lower than the measuring device (or the system under test) bandwidth. Therefore, what we can measure in a laboratory is not expression [7.31], but the power spectral density of the noise below the mentioned cutoff frequency [ORI 13]. Such a power spectral density is related, through a Fourier transform, to the correlation function. The correlation function is the ensemble value of an event defined as measuring the current $I_i(t_1)$ at time t_1 and the current $I_j(t_2)$ at time t_2:

$$\langle I(t_2)I(t_1)\rangle = \sum_i^M \sum_j^M I_j(t_2)I_i(t_1)P\left(I_j(t_2), I_i(t_1)\right). \qquad [7.32]$$

The computation of the probability $P\left(I_j(t_2), I_i(t_1)\right)$ of this two-time measurement process has to be done carefully. The mandatory perturbation of the state because of the measurement is the reason why modeling the measurement process plays a fundamental role in determining the noise, even the low (zero) frequency noise. In what follows, we show how the BITLLES simulator computes, in practice, any moment of the total, particle plus displacement, current without the necessity of introducing any kind of non-unitary evolution or using second quantization tools.

7.4.2. *Practical method to compute DC, AC, transients and higher moments*

As mentioned in section 7.1.2, at GHz or THz frequencies, the computation of the current and its fluctuations requires to deal with both conduction and displacement components. The total current $I(t)$ can be computed from the following two terms:

$$I(t) = \int_S \vec{J}(\vec{r}, t) \cdot d\vec{s} + \int_S \epsilon(\vec{r}) \frac{\partial \vec{E}(\vec{r}, t)}{\partial t} \cdot d\vec{s}, \qquad [7.33]$$

with $\epsilon(\vec{r})$ the (inhomogeneous) electric permittivity. Alternatively, the Ramo–Shockley–Pellegrini theorem [SHO 38, RAM 39, PEL 86, PEL 93a, PEL 93b], briefly described in Appendix B, provides an excellent tool to rewrite expression [7.33] with an explicit dependence on the electron (Bohmian) velocity $v_x^k(t)$. For a two-terminal device with (ideal) metallic contacts, the total current generated by $N(t)$ electrons inside the active region of length L at time t can be written as:

$$I(t) = q \sum_{k=1}^{N(t)} \frac{v_x^k(t)}{L}. \qquad [7.34]$$

Outside the active region, in the metallic contacts, the k-electron is immediately thermalized and it does no longer contribute to the current. Expression [7.34] will be used in the rest of the chapter for the computation of the current (its derivation can be found in Appendix A).

7.4.2.1. *Computing DC, AC and transients*

Taking into account the quantum equilibrium [7.5] and the additional statistical distribution h, we can determine the expectation value of the electrical current at time t_1 from the following ensemble average:

$$\langle I(t_1)\rangle = \lim_{M_\alpha, M_h \to \infty} \frac{1}{M_\alpha M_h} \sum_{\alpha=1}^{M_\alpha} \sum_{h=1}^{M_h} I^{\alpha,h}(t_1). \qquad [7.35]$$

Therefore, the procedure to compute the average current would be the following:

1) At $t = 0$, we select a particular realization of the h-distribution and a particular realization of the α-distribution (see Figure 7.8).

2) We solve the (conditional) Schrödinger equation from time $t = 0$ till $t = t_1$ (see section 7.2.3.1).

3) From [7.34], we compute the value $I^{\alpha,h}(t_1)$.

4) We repeat steps 1 to 3 for the whole ensemble $\alpha = \{1, ..., \infty\}$ and $h = \{1, ..., \infty\}$ to evaluate [7.35].

When the bias is fixed to a constant value, the whole circuit becomes stationary. For a stationary process, the mean current in [7.35] is independent of time. Then, if the process is ergodic, we can compute the mean current from the following (first-order) time average expression:

$$\overline{I^{\alpha,h}(t)} = \lim_{T\to\infty} \frac{1}{T} \int_{-T/2}^{T/2} I^{\alpha,h}(t)dt. \qquad [7.36]$$

In this case, the practical procedure for the computation of the mean current is simpler. Before beginning the simulation, we select only one particular realization of the h-distribution for an *infinite*[11] number of electrons. Simultaneously, we fix the α-distribution of the initial positions for the previous (infinite) realization of wave packets. See Figure 7.8 where we have represented a scheme of a single h- and α-element of the distributions in the simulation box in position and time.

A single sample function often provides little information about the statistics of the process. However, if the process is assumed to be ergodic, i.e. time averages equal ensemble averages, then all statistical information can be derived from just one sample element of the h- and α-distributions. To compute transients or AC, the circuit is

11 The practical procedure for the *infinite* number is selecting a number large enough so that the mean current remains practically unchanged for successive times.

no longer stationary. Then, ergodicity cannot be assumed and the mean value of the current at each particular time, t_1, can be only computed from the ensemble average in [7.35].

7.4.2.2. *Computing current fluctuations and higher moments*

Let us consider now the problem of providing reliable information to the electronic industry about the switching time needed to differentiate the "OFF" and "ON" values of the current in a digital quantum FET (drawn schematically in Figure 7.9)[12]. As discussed at the end of section 7.4.1, we feel uncomfortable with providing ensemble-average information about the transient performance of the FET. The reason is because the ensemble-average value $\langle I(t_1)\rangle$ at a particular time t_1 can be misleading. We can get the value $\langle I(t_1)\rangle$ because the measured currents $I(t_1)$ in all samples are very close to $\langle I(t_1)\rangle$, as we plot in Figure 7.9(a), or because $I(t_1)$ fluctuates a lot among different values, as plotted in Figure 7.9(b). Certainly, it would be mandatory to provide the electronic industry with new complementary information. For example, the statistical variations around the value $\langle I(t_1)\rangle$. This difference is well understood in the classical logic world where one is concerned specifically about noise margins in logic gates. However, the importance of knowing the fluctuations is far less appreciated when dealing with quantum devices. For example, we can be interested in knowing whether the fluctuations greater than the mean value $\langle I(t_1)\rangle$ occur more or less often than the lower ones. The skewness is a measure of the asymmetry of the probability distribution of the (random variable) current. See the symmetry of the function $P(I)$ in Figure 7.9(a) and the asymmetry in Figure 7.9(b) with positive skew. The quantum prediction of the skewness requires computing correlations among the value of the current measured during three consecutive times. This means that we must have a good operator representation for these currents at different times and have confidence that we understand exactly how these operators evolve over the time of interest in the correlation functions.

The consideration of the transport dynamics affected by the measuring process becomes fully pertinent to understand the fluctuations of the current around its DC value. While the *orthodox* explanation of quantum phenomena would have to be dealt with using the second law of quantum mechanics (i.e. the collapse of the wave function), our trajectory-based formulation does not require any non-unitary additional evolution because the evolution of the system is considered within an enlarged configuration space that includes the measuring apparatus (see the discussion in section 7.2.2). Therefore, once we know $I^{\alpha,h}(t)$ at any time, the probability of each current value and any higher moment of the current distribution can be straightforwardly computed. This is another very relevant advantage of using quantum trajectories to study quantum electron transport over *orthodox* techniques. The algorithm to compute the current fluctuations is quite simple. The fluctuating

12 It should be noted that this switching time is quite different from the delay time discussed in VLSI, and this switching is considerably longer than the delay time.

signal of the current can be defined from $\Delta I^{\alpha,h}(t) = I^{\alpha,h}(t) - \langle I^{\alpha,h}(t)\rangle$. We can obtain information of the noise from the variance (or the mean square or the second moment) defined as $\langle \Delta I(t)^2\rangle = \langle I(t)^2\rangle - \langle I(t)\rangle^2$. However, experimentalists are interested in having information on how the noise is distributed along the different frequencies[13]. The fluctuations of the current are computed from the covariance:

$$\langle \Delta I(t_1)\Delta I(t_2)\rangle = \lim_{M_\alpha,M_h\to\infty} \frac{1}{M_\alpha M_h} \sum_{\alpha=1}^{M_\alpha}\sum_{h=1}^{M_h} \Delta I^{\alpha,h}(t_1)\Delta I^{\alpha,h}(t_2). \quad [7.37]$$

If the process is ergodic, i.e. $\langle \Delta I^{\alpha,h}(t)\Delta I^{\alpha,h}(t+\tau)\rangle = \overline{\Delta I_i(t)\Delta I_i(t+\tau)}$, we can compute the noise equivalently from the autocorrelation function:

$$\overline{\Delta I(t)\Delta I(t+\tau)} = \lim_{T\to\infty}\frac{1}{T}\int_{-T/2}^{T/2} \Delta I^{\alpha,h}(t)\Delta I^{\alpha,h}(t+\tau)dt. \quad [7.38]$$

In addition, a process is called wide-sense (or weakly) stationary if its mean value is constant and its autocorrelation function depends only on $\tau = t_2 - t_1$. Then, we define the autocorrelation function $R(\tau)$ as:

$$R(\tau) = \overline{\Delta I(t)\Delta I(t+\tau)}, \quad [7.39]$$

because it depends only on $\tau = t_2 - t_1$. Wide-sense stationary processes are important because the autocorrelation function in [7.39] and the power spectral density function $S(f)$ (measured by experimentalists) form a Fourier transform pair:

$$S(f) = \int_{-\infty}^{\infty} R(\tau)e^{-j2\pi f\tau}\, d\tau. \quad [7.40]$$

This is known as the Wiener–Khinchine theorem. In many systems, we obtain the well-known Schottky's result [SCH 18] or Poissonian shot noise, $S_{shot}(0) = 2q\,\langle I\rangle$.

As a practical example of the computation of the fluctuations, we show here the current response to a the step input voltage in the negative differential conductance region of an RTD. The input signal is the step voltage $V(t) = V_1 u(t) + V_2\,[1 - u(t)]$

13 Most of electronic apparatuses, and the ammeter itself, have to be interpreted as low-pass filters. Therefore, they are not able to measure *all* noise of the spectrum, but only up to a maximum frequency

where $u(t)$ is the Heaviside (step) function. The voltages V_1 and V_2 are constant. Then, the current response can be expressed as $I(t) = I_{tran}(t)+I_1u(t)+I_2[1-u(t)]$ where I_1 and I_2 are the stationary currents corresponding to V_1 and V_2, respectively, and I_{tran} is the *intrinsic* transient current. The results are reported in Figure 7.10 where $I_{tran}(t)$ manifests a delay with respect to the step input voltage due to the dynamical adjustment of the electric field in the leads. After this delay, the current response becomes a RLC-like response (dot-dashed line), i.e. purely exponential. Performing the Fourier transform of $I_{tran}(t)$ in Figure 7.11 and comparing with the single pole spectra (Fourier transform of the RLC-like responses, dashed and dashed dotted lines), we are able to estimate the cutoff frequency and the frequency offset due to the delay [TRA 11].

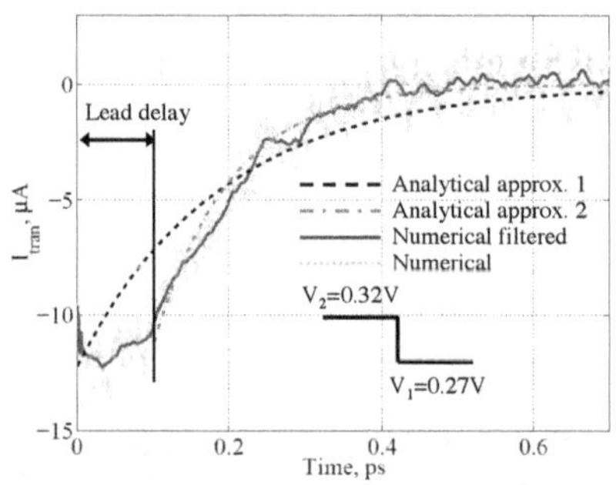

Figure 7.10. *Transient current $I_{tran}(t)$ computed analytically and numerically from [7.35]. For a color version of this figure, see www.iste.co.uk/dollfus/nanodevices.zip*

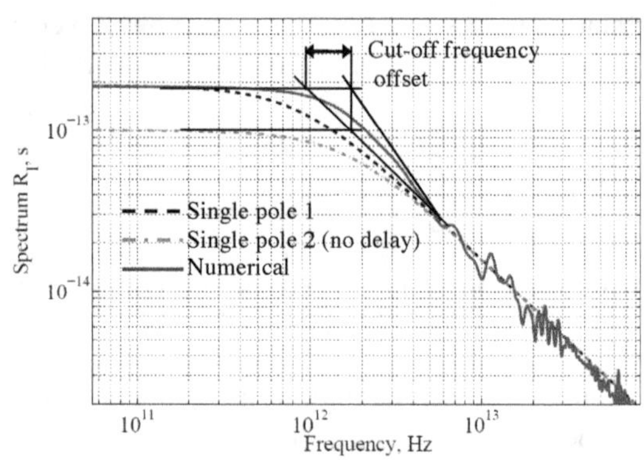

Figure 7.11. *Fourier transform of $I_{tran}(t)$ in Figure 7.10. A logarithmic scale is used to resolve the cutoff frequency offset. For a color version of this figure, see www.iste.co.uk/dollfus/nanodevices.zip*

In order to understand how the many-body Coulomb interaction affects the noise in RTDs, we also investigate the correlation between an electron trapped in the resonant state during a dwell time τ_d and those remaining in the left reservoir. This correlation exists essentially because the trapped electron perturbs the potential energy felt by the electrons in the reservoir. In the limit of noninteracting electrons, the Fano factor will be essentially proportional to the partition noise; however, if the dynamical Coulomb correlations are included in the simulations (see Figure 7.12), this result is no longer reached, and hence the Fano factor is super-poissonian. Finally, we are also interested in the high-frequency spectrum $S(f)$ given by [7.40] revealing information about the internal energy scales of the RTD that is not available from DC transport (see Figure 7.13).

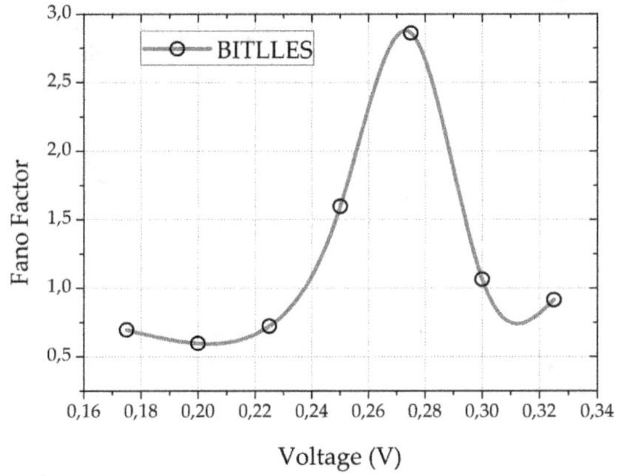

Figure 7.12. *Fano factor F defined as $F = S(0)/(2q\langle I \rangle)$, evaluated using the current fluctuations directly available from the BITLLES simulator. For a color version of this figure, see www.iste.co.uk/dollfus/nanodevices.zip*

Since expressions [7.55] and [7.56] provide information on the *continuously measured* (total) electrical current at any time, any higher moment can be trivially computed from them by simply rewriting expression [7.37] in terms of a series of times. It is in this regard that the trajectory-based approach described in this work provides fully time-resolved access to electron transport dynamics and electrical current statistics.

7.5. Conclusion

The scientific community has an acceptable capability to predict the I-V characteristics (DC) of nanoelectronic devices. However, several difficulties are found when trying to complement these studies with AC, transient or noise behavior. The latter predictions are equally (or even more) important than DC predictions when assessing the final role that new devices will play in the near future. The drawbacks found when trying to go beyond the DC regime are mainly related with the measurement and the many-body problems.

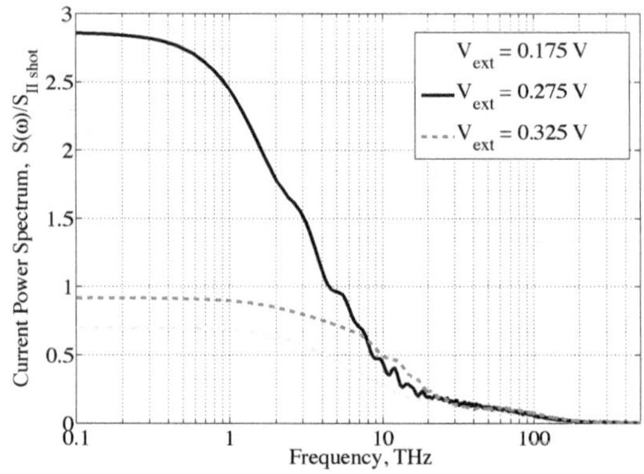

Figure 7.13. *Current noise power spectrum referred to Poissonian shot noise at different biases. For a color version of this figure, see www.iste.co.uk/dollfus/nanodevices.zip*

Firstly, the necessity to compute current–current correlations to provide information about the higher moments of the electrical current necessitates the need to deal with the effect of measurement process on the system, and it is not obvious at all how to determine the effects of the measuring process on the behavior of the electronic system in terms of the collapse of the wave function. We have shown in sections 7.2.2 and 7.3.1 that Bohmian mechanics provides a reasonable solution to this problem by explicitly merging waves and particles and its equations of motion. Moreover, we argued that the back action of the ammeter on the system can be sometimes disregarded without losing much accuracy.

Secondly, the sensitivity of the electrical current to many-body correlations increases dramatically when reducing the dimensions of the electronic devices below a few nanometers and increasing their operation frequencies above the GHz regime. Physical constrictions such as overall charge neutrality or total current conservation, closely related to carrier–carrier correlations, become crucial to make reasonable predictions of the conduction and displacement currents. The use of the so-called conditional (Bohmian) wave function (see section 7.2.3) and its equation of motion has been demonstrated to be a powerful strategy to deal with the many-body Coulomb correlations when combined with a proper definition of the (many-body) Poisson equation (see section 7.3.2).

Finally, the use of the above trajectory-based concepts within a quantum open system allows us to build a quantum Monte Carlo algorithm where the stochastic nature of the electrical current (section 7.3.3) is finally attributed to the number of electrons, their energies and the initial positions of the Bohmian trajectories. Based on this machinery, we have developed an electron transport simulator (BITLLES) that gives access to *time-resolved* electron dynamics and, in particular, to any moment of the electrical current (section 7.4). Certainly, this quantum (Bohmian) trajectory approach is still in its infancy and much work is still needed. In particular, a proper modeling of decoherent phenomena and a description of band structure beyond effective mass are still missing. Some preliminary works in these directions can be found in [ORI 16a, ORI 16b]. In any case, we have shown that Bohmian mechanics offers a powerful formalism to study quantum devices with the capabilities (DC, AC, noise, transients) that the semiclassical Monte Carlo solution of the Boltzmann equation has provided for traditional semiclassical devices.

7.6. Acknowledgments

The authors acknowledge discussion with D.K. Ferry, D. Pandey, F.L. Traversa, X. Cartoixà and D. Jiménez. This work has been partially supported by the Fondo Europeo de Desarrollo Regional (FEDER) and the Ministerio de Economia y Competitividad through the Spanish Projects TEC2012-31330 and TEC2015-67462-C2-1-R, the Beatriu de Pinós program through the Project 2010BP-A00069, the Generalitat de Catalunya (2014 SGR-384) and by the European Union Seventh Framework Program under the Grant Agreement no. 604391 of the Flagship initiative "Graphene-Based Revolutions in ICT and Beyond". D.M. is supported in part by INFN and acknowledges the support of COST action (MP1006) through STSM.

7.7. Appendix A: Practical algorithm to compute Bohmian trajectories

In this appendix, we present the simplest numerical algorithm to compute the wave function and the trajectories needed for the Bohmian formalism. Let us start by the wave function solution of the following time-dependent 1D Schrödinger equation:

$$i\hbar\frac{\partial\psi(x,t)}{\partial t} = -\frac{\hbar^2}{2m^*}\frac{\partial^2\psi(x,t)}{\partial x^2} + U(x,t)\psi(x,t), \qquad [7.41]$$

where $\psi(x,t)$ can be understood as a single-particle wave function or the conditional wave function discussed in [7.9]. Identically, this equation [7.41] can be understood as a single-particle version of [7.1] or alternatively as equation [7.13].

The first step to numerically solve the wave function is to define a mesh in time $t_j = j\Delta t$ and space $x_k = k\Delta x$ variables. Then, we define $\psi_j(x_k) = \psi(x,t)|_{x=x_k;t=t_j}$. The second step is to provide a finite-difference approximation for the temporal and spatial derivatives present in [7.41]. In particular, we use:

$$\left.\frac{\partial\psi(x,t)}{\partial t}\right|_{x=x_k;t=t_j} \approx \frac{\psi_{j+1}(x_k)-\psi_{j-1}(x_k)}{2\Delta t}, \qquad [7.42]$$

$$\left.\frac{\partial^2\psi(x,t)}{\partial x^2}\right|_{x=x_k;t=t_j} \approx \frac{\psi_j(x_{k+1})-2\psi_j(x_k)+\psi_j(x_{k-1})}{\Delta x^2}. \qquad [7.43]$$

Inserting equations [7.42] and [7.43] into [7.41], we obtain the following simple recursive expression:

$$\psi_{j+1}(x_k) = \psi_{j-1}(x_k) + i\frac{\hbar\Delta t}{\Delta x^2 m^*}\left(\psi_j(x_{k+1}) - 2\psi_j(x_k) + \psi_j(x_{k-1})\right) - i\frac{2\Delta t}{\hbar}V_j(x_k)\psi_j(x_k). \qquad [7.44]$$

Once we know the wave function at the particular times t_j and t_{j-1} for all spatial positions in the mesh, we can compute the wave function for all positions at next time t_{j+1} using [7.44]. The recursive application of [7.44] provides the entire time evolution of the wave packet. However, we have to clarify that [7.44] is not valid for the first, x_1, and last, x_N, points. To avoid discussions about the boundary conditions, we can use a very large spatial simulation box so that the entire wave packet is contained in it at any time. Then, the wave function at the borders is negligible.

This explicit solution can be unstable, and its error grows in each recursive application of [7.44] [ORI 11]. To provide a (conditional) stable solution, we have to deal with small values of Δt and Δx. For example, to study electron transport in nanoscale structures as depicted in Figures 7.14 and 7.15, this recursive procedure provides accurate results (the norm of the wave packets is conserved with high precision) when Δx is on the order of 1–2 $\dot{A}$ and the temporal step, Δt, is around 10^{-16}s.

In order to define the initial value of the wave function, we can consider that the wave packet at the two initials times $t = \{t_1, t_1 + \Delta t\}$ evolves in a flat potential region contained in a much larger simulation box. Then, for example, we can define the initial wave function as a time-dependent Gaussian wave packet [COH 78]:

$$\psi(x,t) = \left(\frac{2a^2}{\pi}\right)^{1/4} \frac{e^{i\phi} e^{i(k_c(x-x_0))}}{\left(a^4 + \frac{4\hbar^2(t-t_1)^2}{m^{*2}}\right)^{1/4}} \times \exp\left(-\frac{\left[x - x_0 - \frac{\hbar k_c}{m^*}(t-t_1)\right]^2}{a^2 + \frac{2i\hbar(t-t_1)}{m^*}}\right), \qquad [7.45]$$

where a is the spatial dispersion of the wave packet, m^* is the particle effective mass, x_0 is the central position of the wave packet at the initial time t_1, $k_c = \sqrt{\frac{2m^*E}{\hbar^2}}$ is the central wave vector in the x direction related with the central energy E and $\phi = -\theta - \hbar k_C{}^2 t/(2m^*)$ with $\tan(2\theta) = 2\hbar t/(m^* a^2)$ (see [COH 78]). In particular, at the initial time $t = t_1$, we obtain the simplified expression:

$$\psi(x, t = t_1) = \left(\frac{2}{\pi a^2}\right)^{1/4} e^{i(k_c(x-x_0))} \exp\left(-\frac{(x-x_0)^2}{a^2}\right). \qquad [7.46]$$

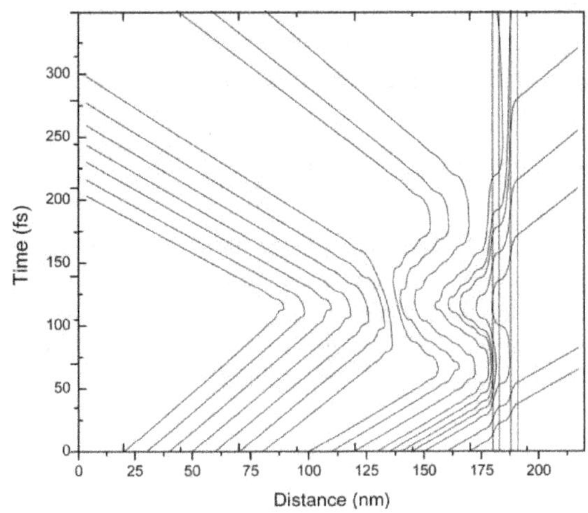

Figure 7.14. *Representative Bohmian trajectories associated with double-packet scattering in a double-barrier potential. The position of the barriers is indicated by vertical lines*

Once we know the wave function at any position and time, we can compute the Bohmian trajectories very easily. We rewrite expression [7.3] to compute the velocity in a simpler form:

$$v(x,t) = \frac{J(x,t)}{|\psi(x,t)|^2} = \frac{\hbar}{m^*} Im \left(\frac{\frac{\partial \psi(x,t)}{\partial x}}{\psi(x,t)} \right), \tag{7.47}$$

where $J(x,t)$ is defined as:

$$\begin{aligned} J(x,t) &= i\frac{\hbar}{2m}\left(\psi(x,t)\frac{\partial \psi^*(x,t)}{\partial x} - \psi^*(x,t)\frac{\partial \psi(x,t)}{\partial x}\right) \\ &= \frac{\hbar}{m^*} Im\left(\psi^*(x,t)\frac{\partial \psi(x,t)}{\partial x}\right), \end{aligned} \tag{7.48}$$

the square modulus $|\psi(x,t)|^2 = \psi^*(x,t)\psi(x,t)$ and $Im(..)$ takes the imaginary part of a complex number.

From the previous velocity [7.47], we compute each Bohmian trajectory $x(t)$ from [7.4]. The initial position of the Bohmian trajectory, defined here as $x(t_o)$, is selected according to the initial distribution $|\psi(x_o, t_o)|^2$. From a practical point of view, the Bohmian trajectory is not defined in a mesh neither in position nor in time. The trajectory is computed during each simulation step Δt as:

$$x(t_{j,k}) = x(t_{j-1,k}) + v(x_k, t_{j-1})\Delta t_k, \tag{7.49}$$

where Δt_k is the minimum of Δt and the time spent by the trajectory $\{t_{j,k}, t_{j-1,k}\}$ in the cell $\{x_k, x_{k+1}\}$. If needed, we repeat the computation with a new $\Delta t'_k$ in another cell $\{x'_k, x_{k'+1}\}$ and velocity $v(x'_k, t_{j-1.k'})$ until the total time step Δt is finished. The Bohmian velocity in each time and position is numerically computed from the discrete wave function $\psi_j(x_k)$ using [7.47] as:

$$v(x_k, t_j) = \frac{\hbar}{2\, m^* \Delta x} Im\left(\frac{\psi_j(x_{k+1}) - \psi_j(x_{k-1})}{\psi_j(x_k)}\right). \tag{7.50}$$

More complex algorithms can be found in the appendix of Chapter 1 in [ORI 11]. Finally, we emphasize again that the use of these trajectories extracted from [7.41] can seem not very useful when $\psi(x,t)$ is single-particle wave function. However, when $\psi(x,t)$ is the conditional wave function discussed in [7.9], then such trajectories are extremely useful because the many-particle wave function cannot be known. See the discussion in section 7.2.3.

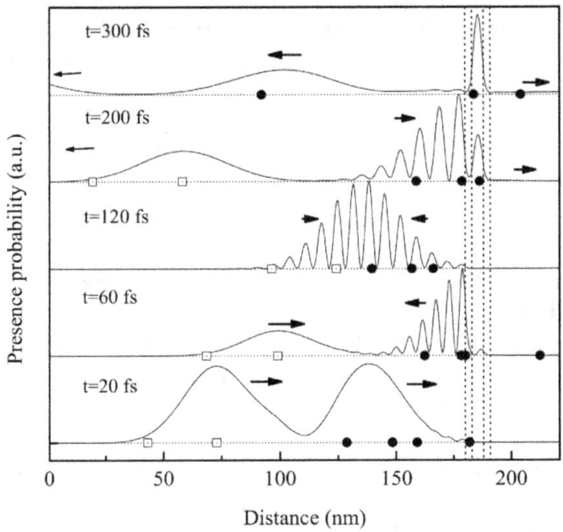

Figure 7.15. *Picture motion of the same double-packet wave function considered in Figure 7.14, calculated by numerical integration of the time-dependent Schrödinger equation. Five representative "snapshots" obtained at different times are shown with the vertical scale arbitrarily changed in each case for clarity. Although the norm of the wave function is always unity, it does not seem so because of the changes in the scale. The marks are visual aids that indicate the position of some related Bohmian trajectories shown in Figure 7.14. The double-barrier position is indicated by the vertical dashed lines, and the arrows indicate the sense of motion of the two packets*

To finish this practical explanation on how to obtain Bohmian trajectories from the wave function, we want to discuss some features of Bohmian trajectories related to the fact that they do not cross in the configuration space. We consider a quite "exotic" initial wave packet. We use a sum of two Gaussian wave packets defined by the expression [7.45] with different central positions and central (momenta) wave vectors [ORI 96, ORI 05a]. The wave packet is certainly quite exotic because it describes just one particle. In Figures 7.14 and 7.15, we see that Bohmian trajectories can be reflected for two different reasons: first, because of their interaction with the classical potential (the particles collide with the barrier) and, second, because of the collision with other trajectories traveling in the opposite direction (Bohmian trajectories do not cross). The second process is responsible for the reflection of those particles of the first packet, which never reach the barrier, and for the reflection of the entire second packet. These collisions between Bohmian particles are related to the quantum potential (see [BOH 52]) in regions where the classical potential is zero, but for them to occur, there should be particles coming from right to left. In this regard, if the initial wave packet is prepared as a superposition of eigenstates incident from left to right (as it is always assumed in scattering thought experiments), and the classical potential is zero on the left-hand side of the barrier, then finding particles coming from the left-hand side of this region will be at the very least very uncommon.

7.8. Appendix B: Ramo–Shockley–Pellegrini theorems

We describe here an algorithm to compute the total current based on the Ramo–Shockley–Pellegrini theorems [SHO 38, RAM 39, PEL 86, PEL 93a, PEL 93b]. From a practical point of view, an algorithm based on the Ramo–Shockley–Pellegrini theorems is preferred in front of [7.33] because it avoids some spurious numerical effects [ALA 09] and provides an intuitive picture of the connection between the total electrical current and the geometry of the active region [BEN 12, ALB 12].

Consider a rectangular volume $\Omega = L_x \cdot L_y \cdot L_z$ containing the whole active region drawn in Figure 7.4 and defined by six orthogonal surfaces $\{S_1, ..., S_6\}$. A vector function $\vec{F}_i(\vec{r})$ inside the volume Ω (i running from 1 to 6) is defined through the expression $\vec{F}_i(\vec{r}) = -\vec{\nabla} B_i(\vec{r})$, where the scalar function $B_i(r')$ is the solution of the Laplace equation for the particular boundary condition on the surface $B_i(\vec{r}) = 1$; $\vec{r} \,\epsilon\, S_i$ and zero elsewhere, i.e. $B_i(\vec{r}) = 0$; $\vec{r} \,\epsilon\, S_{h \neq i}$:

$$\vec{\nabla}\left(\varepsilon(\vec{r})\vec{F}_i(\vec{r})\right) = -\vec{\nabla}\left(\varepsilon(\vec{r})\vec{\nabla}B_i(\vec{r})\right) = 0. \qquad [7.51]$$

The total time-dependent current through the surface S_i can be then divided into two terms [ORI 05b]:

$$I_i(t) = \Gamma_i^q(t) + \Gamma_i^e(t), \qquad [7.52]$$

where:

$$\Gamma_i^q(t) = -\int_\Omega \vec{F}_i(\vec{r})\vec{j}_c(\vec{r}_i, t)d^3\vec{r}, \qquad [7.53]$$

$$\Gamma_i^e(t) = \int_S \vec{F}_i(\vec{r})\varepsilon(\vec{r})\frac{\partial}{\partial t}V(\vec{r}, t)d\vec{s}. \qquad [7.54]$$

We use the subindex i in equations [7.52]–[7.54] because the current through a surface different from S_i leads to a different definition of $\vec{F}_i(\vec{r})$. Let us remark that these expressions compute the total current in a surface S_i through a volume integral in [7.53] and an integral over all surfaces of the volume Ω in [7.54]. The computation of $\Gamma_i^q(t)$ and $\Gamma_i^e(t)$ with trajectories can be obtained numerically from [ALA 09]:

$$\Gamma_i^q(t) = \sum_{j=1}^{N(t)} \vec{F}_i(\vec{r}_j^{\alpha}(t)) q\vec{v}_j(\vec{r}_j^{\alpha}(t), t), \qquad [7.55]$$

and

$$\Gamma_i^e(t) = \int_S \vec{F}_i(\vec{r})\varepsilon(\vec{r})\frac{\partial V(\vec{r},t)}{\partial t} d\vec{s}. \qquad [7.56]$$

Let us mention that [7.55] not only contains the conduction current but also contains part of the displacement current. Therefore, the numerical evaluation of the total current through a particular surface S_i due to a set of $N(t)$ quantum trajectories (as discussed in section 7.3.1) can be computed from equations [7.55] and [7.56] [ALA 09, ALB 12]. In fact, for a two terminal device we have to consider $L_x \ll L_y, L_z$ in order to ensure that all electric fields end finally in one of the two cables and are captured by the two surfaces of Ω close to the cables where the current is computed. Then, the shape of $B_1(\vec{r})$ in the x direction is linear (we have consider $S_1 = S_D$ in Figure 7.4) and a good approximation for the function $\vec{F}_1(\vec{r})$ is:

$$\vec{F}_1(\vec{r}) \cdot \vec{x} \approx \frac{1}{L_x}. \qquad [7.57]$$

This is exactly the geometry discussed by Ramo and Shockley in the vacuum tube [RAM 39, SHO 38]. Finally, if we assume that the surfaces S_1 and S_4 are in a metallic contact where there are no variations of potential and the rest of surfaces of S are far from the active region, we can neglect [7.56] and taking [7.57] in [7.55] we obtain the final result of [7.34].

7.9. Appendix C: Bohmian mechanics with operators

According to the final remark in section 7.2.2, sometimes, the use of a Hermitian operator acting only on a quantum system with the ability of providing the outcomes of the measurement process without the explicit simulation of the measuring apparatus is highly appreciated. Operators are not needed in Bohmian mechanics, but they can be very helpful mathematical tricks in practical computations. In this Appendix, we develop the expressions for commutating ensemble results from operators as a (infinite) sum of Bohmian trajectories. We consider a Hermitian operator $\hat{A}$ and its mean value $\left\langle \hat{A} \right\rangle_\psi$ in the position representation. Then, the mean value of this operator over the wave function $\psi(\vec{r},t)$ is given by:

$$\left\langle \hat{A} \right\rangle_\psi = \int_{-\infty}^{\infty} \psi^*(\vec{r},t)\hat{A}\left(\vec{r}, -i\hbar\frac{\partial}{\partial\vec{r}}\right)\psi(\vec{r},t)d\vec{r}. \qquad [7.58]$$

Alternatively, the same mean value can be computed from Bohmian mechanics by defining a spatial average of a "local" magnitude $A_B(\vec{r})$ weighted by $R^2(\vec{r},t)$:

$$\left\langle \hat{A} \right\rangle_\psi = \int_{-\infty}^{\infty} R^2(\vec{r},t)A_B(\vec{r})d\vec{r}. \qquad [7.59]$$

In order to obtain the same value using equations [7.58] and [7.59], we can easily identify the local mean value $A_B(\vec{r})$ as:

$$A_B(\vec{r}) = Real\left(\left[\frac{\psi^*(x\vec{r},t)\hat{A}\left(\vec{r},-i\hbar\frac{\partial}{\partial\vec{r}}\right)\psi(\vec{r},t)}{\psi^*(\vec{r},t)\psi(\vec{r},t)}\right]_{\psi(\vec{r},t)=R(\vec{r},t)e^{i\frac{S(\vec{r},t)}{\hbar}}}\right). \qquad [7.60]$$

It is important to emphasize that the local Bohmian operators $A_B(\vec{r})$ are not the eigenvalues of the operator $\hat{A}$. In general, the eigenvalues are not position dependent, while $A_B(\vec{r})$ are. The expression $A_B(\vec{r})$ is what is needed to compute the mean values of $\hat{A}$ with [7.59].

For practical purposes, we will compute the mean value using [7.59] with a large $\alpha = 1, \ldots, M_\alpha$ number of Bohmian trajectories with different initial positions. We will select the initial position $\vec{r}^\alpha(t_o)$ of the Bohmian trajectories according to the quantum equilibrium condition. Therefore, we can use [7.5] to write $R^2(\vec{r},t)$ in [7.59]. Finally, we obtain:

$$\left\langle \hat{A} \right\rangle_\psi = \lim_{M_\alpha\to\infty}\frac{1}{M_\alpha}\sum_{\alpha=1}^{M_\alpha} A_B(\vec{r}^\alpha(t)). \qquad [7.61]$$

By construction, in the limit $M_\alpha \to \infty$, the value of [7.61] is identical to the value of [7.59]. Now, we provide a few examples of how some common mean values are computed from the orthodox quantum formalism and Bohmian trajectories. First, we compute the mean value of the position:

$$\langle x \rangle_\psi = \int_{-\infty}^{\infty} \psi^*(x,t)x\psi(x,t)dx, \qquad [7.62]$$

with $x_B(x) = x$ so that:

$$\langle x \rangle_\psi = \int_{-\infty}^{\infty} R^2(x,t) x dx. \qquad [7.63]$$

Identically, the mean value of the momentum:

$$\langle p \rangle_\psi = \int_{-\infty}^{\infty} \psi^*(x,t) \left(-i\hbar \frac{\partial}{\partial x} \right) \psi(x,t) dx, \qquad [7.64]$$

with $p_B(x) = \partial S(x,t)/\partial x$:

$$\langle p \rangle_\psi = \int_{-\infty}^{\infty} R^2(x,t) \frac{\partial S(x,t)}{\partial x} dx. \qquad [7.65]$$

For the classical potential, we have:

$$\langle V \rangle_\psi = \int_{-\infty}^{\infty} \psi^*(x,t) V(x,t) \psi(x,t) dx, \qquad [7.66]$$

with $V_B(x) = V(x,t)$ so that:

$$\langle V \rangle_\psi = \int_{-\infty}^{\infty} R^2(x,t) V(x,t) dx. \qquad [7.67]$$

Now, we compute the mean value of the kinetic energy:

$$\langle K \rangle_\psi = \int_{-\infty}^{\infty} \psi^*(x,t) \left(-\frac{\hbar^2}{2m} \frac{\partial^2}{\partial x^2} \right) \psi(x,t) dx. \qquad [7.68]$$

It is important to note that the local mean value of the kinetic energy takes into account the Bohmian kinetic energy plus the quantum potential. In particular, $K_B(x)$ can be obtained from the expression:

$$K_B(x) = Real \left(- \frac{R(x,t) e^{-iS(x,t)/\hbar} \frac{\hbar^2}{2m} \left(\frac{\partial}{\partial x} \right)^2 R(x,t) e^{iS(x,t)/\hbar}}{R^2(x,t)} \right). \qquad [7.69]$$

The real part[14] of K_B is:

$$K_B = \frac{1}{2m}\left(\frac{\partial S(x,t)}{\partial x}\right)^2 + Q(x,t), \qquad [7.70]$$

where $Q(x,t) = -\frac{\hbar}{2m}\frac{\nabla^2|R(x,t)|}{|R(x,t)|}$. Finally, we obtain the Bohmian expression of the mean kinetic energy of the ensemble of trajectories:

$$\langle K\rangle_\psi = \int_{-\infty}^{\infty} R^2(x,t)\left(\frac{1}{2m}\left(\frac{\partial S(x,t)}{\partial x}\right)^2 + Q(x,t)\right)dx. \qquad [7.71]$$

In particular, if we want to compute the ensemble (Bohmian) kinetic energy (without the quantum potential $\langle Q\rangle$), using [7.61], we obtain:

$$\langle K_B\rangle - \langle Q\rangle = \lim_{M_\alpha\to\infty}\frac{1}{M_\alpha}\sum_{\alpha=1}^{M_\alpha}\frac{1}{2}m^* v^2(x^\alpha(t),t). \qquad [7.72]$$

Finally, we emphasize that this way of computing ensemble values can be very useful because we avoid the inclusion of the apparatus degree of freedom as we have done in section 7.2.2. However, it is not always possible to know that we have selected a *good* operator that perfectly describes the measuring apparatus as we have done in section 7.2.2. These ideas are emphasized by Daumer *et al.* [DAU 97] and Dürr *et al.* [DÜR 04, DÜR 12] when they refer to the "naive realism about operators".

7.10. Appendix D: Relation between the Wigner distribution function and the Bohmian trajectories

In this appendix, we illustrate the formal relation between the Wigner distribution function and Bohmian mechanics. Obviously, since both formulations reproduce orthodox quantum mechanics, the observable results of the Wigner distribution function are identical to the ones obtained with the Bohmian trajectories. For computational purposes, the merit of the Wigner distribution function is its ability to deal with mixed states (defined as a statistical ensemble of pure states) through a Wigner–Weyl transformation of the density matrix. The density matrix of a mixed

14 It can be demonstrated quite easily that the imaginary part of [7.69] is equal to the spatial derivative of the current that becomes 0 when integrated over all space. We use $J(x = \pm\infty, t) = 0$, which is always valid for wave functions that are normalized to unity, but it is not true for other types of wave functions such as plane waves.

state can be written as $\rho(x,x') = \sum_j c_j\psi_j(x)\psi_j^*(x')$ where c_j specifies the fraction of the ensemble in the pure state $\psi_j(x)$. For the sake of simplicity, we avoid the explicit time dependence of the wave function and c_j. The need for dealing with mixed states comes from the fact that we are dealing with an open system (i.e. the active region) with a partial knowledge of the many-particle wave function in the whole (active region, cables, battery, etc.) quantum system [FRE 90]. In the Bohmian language used in this chapter, the pure state $\psi_j(x)$ is just the conditional wave function discussed in section 7.2.3.1 and c_j is just the h distribution defined in section 7.3.3.

The Wigner distribution function for a mixed state can be defined as a Wigner–Weyl transformation of the density matrix:

$$W(x,p) = \frac{1}{h}\sum_j c_j \int_{-\infty}^{\infty} e^{-ipy/\hbar}\psi_j(x+\frac{y}{2})\psi_j^*(x-\frac{y}{2})dy, \qquad [7.73]$$

where p is the classical momentum. If we write the wave function of each pure state in polar form, i.e. $\psi_j(x) = R_j(x)e^{iS_j(x)/\hbar}$, equation [7.73] becomes:

$$W(x,p) = \frac{1}{h}\sum_j c_j \int_{-\infty}^{\infty} e^{-ipy/\hbar}R_j(x+\frac{y}{2})e^{iS_j(x+\frac{y}{2})/\hbar}R_j(x-\frac{y}{2})e^{-iS_j(x-\frac{y}{2})/\hbar}dy. \qquad [7.74]$$

which provides a phase-space description of the quantum system. In addition to the orthodox position distribution $\rho(x) = R^2(x)$, the Wigner distribution function provides a position-momentum distribution of the probabilities of the electrons with many similitudes with the classical Boltzmann distribution function used for the simulation of semiclassical devices.

At this point, a relevant question appears: *what is the relation between the momentum information provided by the Wigner distribution function and the (local) velocity provided by the Bohmian theory?* In this appendix, we want to establish this connection. We show that the average (linear) momentum at a given position x computed from the Wigner distribution coincides with the Bohmian momentum at the same position, i.e. $\bar{p}(x)_W = p_B(x) = \partial S(x)/\partial x$.

The space conditional momentum $\bar{p}(x)_W$ can be written in terms of the Wigner function [HIL 09, MOY 49] as:

$$\bar{p}_W(x) = \frac{\int_{-\infty}^{\infty} pW(x,p)dp}{\int_{-\infty}^{\infty} W(x,p)dp} = \frac{\int_{-\infty}^{\infty} pW(x,p)dp}{\rho(x)}. \qquad [7.75]$$

where $\rho(x) = R^2(x)$. In order to avoid unnecessary complex notation, we just focus on one pure state. The generalization to mixed states can be done straightforwardly. Using the chain rule for a function $F(x,y)$ derivable and zero-valued at $y \to \pm\infty$, we can use the following relation:

$$\int_{-\infty}^{\infty} dy e^{-ipy/\hbar} \frac{\partial}{\partial y} F(x,y) = \frac{i}{\hbar} p \int_{-\infty}^{\infty} dy e^{-ipy/\hbar} F(x,y), \quad [7.76]$$

and then rewrite the numerator of the right-hand side of equation [7.75] as:

$$\int_{-\infty}^{\infty} pW(x,p)dp = \frac{-i}{2\pi} \int_{-\infty}^{\infty} dp \int_{-\infty}^{\infty} dy e^{-ipy/\hbar}$$
$$\cdot\Big[R(x+\frac{y}{2})R(x-\frac{y}{2})e^{\frac{i}{\hbar}[S(x+\frac{y}{2})-S(x-\frac{y}{2})]} \frac{i}{\hbar}\Big(\frac{\partial S(x+\frac{y}{2})}{\partial y} - \frac{\partial S(x-\frac{y}{2})}{\partial y}\Big)$$
$$+e^{\frac{i}{\hbar}[S(x+\frac{y}{2})-S(x-\frac{y}{2})]} \frac{\partial}{\partial y}\Big(R(x+\frac{y}{2})R(x-\frac{y}{2})\Big)\Big]. \quad [7.77]$$

To proceed, let us focus on the following term:

$$\frac{\partial S(x+\frac{y}{2})}{\partial y} - \frac{\partial S(x-\frac{y}{2})}{\partial y} =$$
$$= \lim_{t\to 0} \frac{S(x+\frac{y+t}{2}) - S(x+\frac{y}{2})}{t} - \lim_{t\to 0} \frac{S(x-\frac{y+t}{2}) - S(x-\frac{y}{2})}{t}$$
$$= \frac{1}{2}\lim_{t\to 0} \frac{S(x+\frac{t}{2}+\frac{y}{2}) - S(x+\frac{y}{2})}{t/2} + \frac{1}{2}\lim_{t\to 0} \frac{S(x-\frac{y}{2}) - S(x-\frac{t}{2}-\frac{y}{2})}{t/2}$$
$$= \frac{1}{2}\left[\frac{\partial S(x+\frac{y}{2})}{\partial x} + \frac{\partial S(x-\frac{y}{2})}{\partial x}\right], \quad [7.78]$$

which can be used to rewrite equation [7.77] as:

$$\int_{-\infty}^{\infty} pW(x,p)dp = \int_{-\infty}^{\infty} dy \delta(y)$$
$$\cdot\Big[R(x+\frac{y}{2})R(x-\frac{y}{2})e^{\frac{i}{\hbar}[S(x+\frac{y}{2})-S(x-\frac{y}{2})]} \frac{1}{2}\Big(\frac{\partial S(x+\frac{y}{2})}{\partial x} + \frac{\partial S(x-\frac{y}{2})}{\partial x}\Big)$$
$$-i\hbar e^{\frac{i}{\hbar}[S(x+\frac{y}{2})-S(x-\frac{y}{2})]} \frac{\partial}{\partial y}\Big(R(x+\frac{y}{2})R(x-\frac{y}{2})\Big)\Big]$$
$$= R^2(x)\frac{\partial S(x)}{\partial x}. \quad [7.79]$$

In equation [7.79], we have used the following property:

$$\int_{-\infty}^{\infty} dp e^{-ipy/\hbar} = 2\pi\hbar\delta(y), \qquad [7.80]$$

and the fact that the second term within the integral in equation [7.79] is 0, i.e.:

$$\frac{\partial}{\partial y}\left(R(x+\frac{y}{2})R(x-\frac{y}{2})\right)\Big|_{y=0} =$$

$$= \left[\frac{1}{2}\frac{\partial R(x+\frac{y}{2})}{\partial x}R(x-\frac{y}{2}) - \frac{1}{2}R(x+\frac{y}{2})\frac{\partial R(x-\frac{y}{2})}{\partial x}\right]\Big|_{y=0} = 0. \qquad [7.81]$$

Thus, plugging the result obtained in equation [7.79] into equation [7.75], we can write the position-dependent averaged momentum in the Wigner formalism as:

$$\bar{p}_W(x) = p_B(x) = \frac{\partial S(x)}{\partial x}. \qquad [7.82]$$

In other words, the (average) momentum at a given position x computed from the Wigner function coincides with the (local) momentum defined in Bohmian mechanics. Let us come back to the mixed states mentioned at the beginning of this appendix. It is easy to realize that a mixed state will satisfy:

$$\bar{p}_W(x) = \frac{\sum_j c_j R_j^2(x)\frac{\partial S_j(x)}{\partial x}}{\sum_j c_j R_j^2(x)}. \qquad [7.83]$$

The rigth-hand side of [7.83] is just the weighted sum of the Bohmian momentums at the position x.

Finally, let us discuss the kind of phase-space distribution that arises by gathering all Bohmian trajectories. When dealing simultaneously with several (pure) states, there will be several Bohmian velocities at each position x (one velocity for each pure state). Therefore, as for the Wigner distribution, we will also have a (Bohmian) phase space. Both, Wigner and Bohmian phase-space distributions are closely related. As discussed in [7.83], the averaged momentums are identical in each position x. Let us emphasize that, by construction, the phase-space distribution arising from the Bohmian velocities is always non-negative. The number of Bohmian trajectories with momentum $p_B(x)$ at the position x has to be positive (or zero if there are no particles). This desirable feature of a phase-space probability distribution is not always found for the Wigner distribution. It is in this sense that we claimed throughout the chapter that the Bohmian trajectories in the BITLLES simulator for quantum devices are closely related to the Monte Carlo trajectories obtained from the Boltzmann equation for semiclassical devices.

7.11. Bibliography

[ALA 09] ALARCÓN A., ORIOLS X., "Computation of quantum electron transport with local current conservation using quantum trajectories", *Journal of Statistical Mechanics: Theory and Experiment*, vol. 2009, p. P01051, 2009.

[ALA 13] ALARCÓN A., YARO S., CARTOIXÀ X. *et al.*, "Computation of many-particle quantum trajectories with exchange interaction: application to the simulation of nanoelectronic devices", *Journal of Physics: Condensed Matter*, vol. 25, p. 325601, 2013.

[ALB 08] ALBAREDA G., SUÑÉ J., ORIOLS X., "Monte Carlo simulations of nanometric devices beyond the mean-field approximation". *J. Comp. Elect.*, vol. 7, p. 213, 2008.

[ALB 09a] ALBAREDA G., JIMENEZ D., ORIOLS X., "Intrinsic noise in aggressively scaled field-effect transistors", *Journal of Statistical Mechanics: Theory and Experiment*, vol. 2009, p. P01044, 2009.

[ALB 09b] ALBAREDA G., SUÑÉ J., ORIOLS X., "Many-particle hamiltonian for open systems with full coulomb interaction: application to classical and quantum time-dependent simulations of nanoscale electron devices", *Physical Review B*, vol. 79, p. 075315, 2009.

[ALB 10a] ALBAREDA G., ALARCÓN A., ORIOLS X., "Electric power in nanoscale devices with full Coulomb interaction", *International Journal of Numerical Modelling: Electronic Networks, Devices and Fields*, vol. 23, pp. 354–363, 2010.

[ALB 10b] ALBAREDA G., LÓPEZ H., CARTOIXÀ X. *et al..*, "Time-dependent boundary conditions with lead-sample coulomb correlations: application to classical and quantum nanoscale electron device simulators", *Physical Review B*, vol. 82, p. 085301, 2010.

[ALB 10c] ALBAREDA G., SAURA X., ORIOLS X. *et al.*, "Many-particle transport in the channel of quantum wire double-gate field-effect transistors with charged atomistic impurities", *Journal of Applied Physics*, vol. 108, p. 043706, 2010.

[ALB 11] ALBAREDA G., TRAVERSA F.L., BENALI A. *et al.*, *Many-particle Monte Carlo Approach to Electron Transport, Applications of Monte Carlo Method in Science and Engineering*, InTech, 2011.

[ALB 12] ALBAREDA G., TRAVERSA F.L., BENALI A. *et al.*, "Computation of quantum electrical currents through the Ramo-Shockley-Pellegrini theorem with trajectories", *Fluctuation and Noise Letters*, vol. 11, p. 1242008, 2012.

[ALB 13a] ALBAREDA G., BENALI A., ORIOLS X., "Self-consistent time-dependent boundary conditions for static and dynamic simulations of small electron devices", *Journal of Computational Electronics*, vol. 12, pp. 730-742, 2013.

[ALB 13b] ALBAREDA G., MARIAN D., BENALI A. *et al.*, "Time-resolved electron transport with quantum trajectories", *Journal of Computational Electronics*, vol. 12, p. 405, 2013.

[ALB 14] ALBAREDA G., APPEL H., FRANCO I. *et al.*, "Correlated electron-nuclear dynamics with conditional wave functions", *Physical Review Letters*, vol. 113, p. 083003, 2014.

[ALB 15] ALBAREDA G. , BOFILL J.M. , TAVERNELLI I. *et al.*, "Conditional Born-Oppenheimer Dynamics: Quantum Dynamics Simulations for the Model Porphine". *J. Phys. Chem. Lett.*, vol. 6, p. 1529, 2015.

[ALB 16] ALBAREDA G., TRAVERSA F.L., BENALI A., "Sequential measurement of displacement and conduction currents in electronic devices". *Fluct. Noise Lett.*, vol. 15, p. 1640004, 2016.

[BEL 90] BELL J.S., "Against measurement", *Physics World*, vol. 3, p. 33, 1990.

[BEL 04] BELL J.S., *Speakable and Unspeakable in Quantum Mechanics*, Cambridge University Press, 2004.

[BEN 12] BENALI A., TRAVERSA F.L., ALBAREDA G. *et al.*, "Effect of gate-all-around transistor geometry on the high-frequency noise: analytical discussion", *Fluctuation and Noise Letters*, vol. 11, p. 1241002, 2012.

[BEN 13] BENALI A., TRAVERSA F.L., ALBAREDA G. *et al.*, "Improving the intrinsic cut-off frequency of gate-all-around quantum-wire transistors without channel length scaling", *Applied Physics Letters*, vol. 102, p. 173506, 2013.

[BEN 14] BENSENY A., ALBAREDA G., SANZ A. *et al.*, "Applied Bohmian mechanics", *European Physical Journal D*, vol. 68, p. 286, 2014.

[BOH 52] BOHM D., "A suggested interpretation of the quantum theory in terms of "hidden" variables. I", *Physical Review*, vol. 85, pp. 166–179, 1952.

[BOH 93] BOHM D., HILEY B.J. *et al.*, *The Undivided Universe: An Ontological Intepretation of Quantum Theory*, Routledge & Kegan Paul, London, 1993.

[COH 78] COHEN-TANNOUDJI C., DIU B., LALOË F., *et al.*,*Quantum Mechanics (Vol I and II)*, Wiley VCH, 1978.

[DAU 97] DAUMER M., DÜRR D., GOLDSTEIN S. *et al.*, "Naive realism about operators", *Probability, Dynamics and Causality*, vol. 45, nos. 2–3, p. 379, 1997.

[DE 24] DE BROGLIE L., Recherches sur la théorie des quantas, PhD thesis, University of Paris, 1924.

[DIR 29] DIRAC P.A.M., "Quantum mechanics of many-electron systems", *Proceedings of the Royal Society of London*, vol. A , no. 123, p. 714, 1929.

[DIV 08] DI VENTRA M., *Electrical Transport in Nanoscale Systems*, Cambridge University Press, 2008.

[DÜR 04] DÜRR D., GOLDSTEIN S., ZANGHI N. *et al.*, "Quantum equilibrium and the role of operators as observables in quantum theory", *Journal of Statistical Physics*, vol. 116, p. 9595, 2004.

[DÜR 05] DÜRR D., GOLDSTEIN S., TUMULKA R. *et al.*, "On the role of density matrices in Bohmian mechanics", *Foundations of Physics*, vol. 35, p. 449, 2005.

[DÜR 09] DÜRR D., TEUFEL S., *Bohmian Mechanics: The Physics and Mathematics of Quantum Theory*, Spinger, Germany, 2009.

[DÜR 12] DÜRR D., GOLDSTEIN S., ZANGHÌ N., *Quantum Physics Without Quantum Philosophy*, Spinger, Germany, 2012.

[FEY 65] FEYNMAN R.P., HIBBS A.R., *Quantum Mechanics and Path Integrals*, McGraw-Hill, New York, 1965.

[FRE 90] FRENSLEY W.R., "Boundary conditions for open quantum systems driven far from equilibrium", *Review of Modern Physics*, vol. 62, pp. 645–744, 1990.

[GRI 02] GRIFFITHS R.B., *Consistent Quantum Theory*, Cambridge University Press, 2002.

[HIL 09] HILEY B.J., "On the relationship between the Wigner-Moyal and Bohm approaches to quantum mechanics: a step to a more general theory?", *Foundations of Physics*, vol. 40, p. 356, 2009.

[HOL 93] HOLLAND P.R., *The Quantum Theory of Motion: An account of the de Broglie-Bohm Causal Interpretation of Quantum mechanics*, Cambridge University Press, 1993.

[JAC 62] JACKSON J., *Classical Electrodynamics*, John Wiley, New York, 1962.

[JAC 89] JACOBONI C.L.P., *The Monte Carlo Method for Semiconductor Device Simulation*, Springer, New York, 1989.

[LOP 08] LOPEZ H., ALBAREDA G., CARTOIXÀ X. *et al.*, "Boundary conditions with Pauli exclusion and charge neutrality: application to the Monte Carlo simulation of ballistic nanoscale devices", *Journal of Computational Electronics*, vol. 7, p. 213, 2008.

[MAR 14] MARIAN D., ZANGHÌ N., ORIOLS X., "On the back-action of THz measurement on the total current of quantum devices", *Proceedings of the IWCE*, p. 117, 2014.

[MOY 49] MOYAL J.E., "Quantum mechanics as a statistical theory", *Mathematical Proceedings of the Cambridge Philosophical Society*, vol. 45, p. 99, 1949.

[NOR 10] NORSEN T., "The theory of (exclusively) local beables", *Foundations of Physics*, vol. 40, p. 1858, 2010.

[NOR 15] NORSEN T., MARIAN D., ORIOLS X., "Can the wave function in configuration space be replaced by single-particle wave functions in physical space?", *Synthese*, vol. 192, p. 3125, 2015.

[ORI 96] ORIOLS X., MARTÍN F., SUÑÉ J., *et al.*, "Implications of the noncrossing property of Bohm trajectories in one-dimensional tunneling configurations", *Physical Review A*, vol. 54, p. 2594, 1996.

[ORI 05a] ORIOLS X., "Comment on "biphoton double-slit experiment"", *Physical Review A*, vol. 71, p. 017801, 2005.

[ORI 05b] ORIOLS X., ALARCÓN A., FERNÀNDEZ-DÍAZ E., *et al.*, "Time-dependent quantum current for independent electrons driven under nonperiodic conditions", *Physical Review B*, vol. 71, p. 245322, 2005.

[ORI 07a] ORIOLS X., "Quantum-trajectory approach to time-dependent transport in mesoscopic systems with electron-electron interactions", *Physical Review Letters*, vol. 98, p. 066803, 2007.

[ORI 07b] ORIOLS X. FERNANDEZ-DIAZ E. A.A., ALARCÓN A., "An Electron injection model for time-dependent simulators of nanoscale devices with electron confinement: Application to the comparison of the intrinsic noise of 3D-, 2D- and 1D- ballistic transistors", *Solid State Electron.*, vol. 51, p. 306, 2007.

[ORI 11] ORIOLS X., MOMPART J., *Applied Bohmian Mechanics: From Nanoscale Systems to Cosmology*, Pan Stanford Publishing, Singapore, 2011.

[ORI 13] ORIOLS X., FERRY D., "Quantum transport beyond DC", *Journal of Computational Electronics*, vol. 12, p. 317, 2013.

[ORI 16a] ORIOLS X., ZHAN Z., COLOMÉS E. *et al.*, "Dissipative quantum transport using one-particle time-dependent (conditional) wave functions", *Proceedings of the IWCE*, 2015.

[ORI 16b] ORIOLS X., "Can decoherence make quantum theories unfalsifiable? Understanding the quantum-to-classical transition without it", *J. Phys. Conf. Series*, vol. 701, p. 012004, 2016.

[PEL 86] PELLEGRINI B., "Electric charge motion, induced current, energy balance, and noise", *Physical Review B*, vol. 34, p. 5921, 1986.

[PEL 93a] PELLEGRINI B., "Elementary applications of quantum-electrokinematics theorem", *Il Nuovo Cimento*, vol. D 15, p. 881, 1993.

[PEL 93b] PELLEGRINI B., "Extension of the electrokinematics theorem to the electromagnetic field and quantum mechanics", *Il Nuovo Cimento*, vol. D 15, p. 855, 1993.

[PRI 65] PRICE P., *Fluctuation Phenomena in Solids*, Academic Press, New York, 1965.

[RAM 39] RAMO S., "Currents induced by electron motion", *Proceedings of IRE*, vol. 27, p. 584, 1939.

[SCH 18] SCHOTTKY Q., "Über spontane stromschwankungen in verschiedenen elektrizitätsleitern", *Annalen der Physik (Leipzig)*, vol. 57, p. 541, 1918.

[SCH 26] SCHRÖDINGER E., "An undulatory theory of the mechanics of atoms and molecules", *Physical Review*, vol. 28, pp. 1049–1070, 1926.

[SCH 04] SCHLOSSHAUER M., "Decoherence, the measurement problem and interpretations of quantum mechanics", *Reviews of Modern Physics*, vol. 76, pp. 1267–1305, 2004.

[SHO 38] SHOCKLEY W., "Currents to conductors induced by a moving point charge", *Journal of Applied Physics*, vol. 9, p. 635, 1938.

[TRA 11] TRAVERSA F.L., BUCCAFURRI E., ALARCÓN A. *et al.*, "Time-dependent many-particle simulation for resonant tunneling diodes: interpretation of an analytical small-signal equivalent circuit", *IEEE Transaction on Electron Devices*, vol. 58, pp. 2104–2112, 2011.

[TRA 13] TRAVERSA F.L., ALBAREDA G., DI VENTRA M. *et al.*, "Robust weak-measurement protocol for Bohmian velocities", *Physical Review A*, vol. 87, p. 052124, 2013.

[VON 32] VON NEUMANN J., *Mathematische Grundlagen der Quantenmechanik*, Springer Verlag, Berlin, 1932.

[WAE 68] VAN DER WAERDEN B.L., *Sources of Quantum Mechanics*, Dover Publications, 1968.

[WYA 00a] WYATT R.E., BITTNER E.R., "Quantum wave packet dynamics with trajectories: implementation with adaptive Lagrangian grids", *Journal of Chemical Physics*, vol. 113, p. 8898, 2000.

[WYA 00b] WYATT R.E., KOURI D.J., HOFFMAN D.K., "Quantum wave packet dynamics with trajectories: implementation with distributed approximating functionals", *Journal of Chemical Physics*, vol. 112, p. 10730, 2000.

[WYA 05] WYATT R.E., *Quantum Dynamics with Trajectories: Introduction to Quantum Hydrodynamics*, Springer, New York, 2005.

[ZUR 05] ZUREK W., "Probabilities from entanglement, Borns rule $p_k = |\Psi_k|^2$ from envariance", *Physical Review A*, vol. 71, p. 052105, 2005.

8

The Monte Carlo Method for Wigner and Boltzmann Transport Equations

For many years the semiclassical Boltzmann approach to transport in semiconductors has been very successful in interpreting the physics of electron devices. In this approach, based on a distribution function defined in the phase-space, electrons and holes are considered as localized particles that commonly interact with different kinds of scatterers. For accurate investigations with deep physical foundation, the stochastic solution of the Boltzmann transport equation (BTE) using the particle Monte Carlo (MC) method has been shown to be very powerful.

For quantum transport problems, it is possible to use the concept of the Wigner function that is defined in the phase-space as a Fourier transform of the density matrix (DM). In the classical limit, this function reduces to the classical distribution function. The dynamical equation of the Wigner function, that is the Wigner transport equation (WTE), is very similar to the Boltzmann counterpart, except in the influence of the potential whose rapid space variations generate quantum effects. Hence, a similar MC method can be developed to solve the WTE.

This chapter focuses on both the semiclassical Boltzmann and quantum Wigner approaches to transport in semiconductor systems, and on the MC method to solve the BTE and the WTE for device simulation.

Chapter written by Philippe DOLLFUS, Damien QUERLIOZ and Jérôme SAINT MARTIN.

8.1. The WTE

Consider an electron gas in a potential $V(\mathbf{r})$ and a band structure $E(\mathbf{k})$, described by its envelope functions $\psi(\mathbf{r})$. The first idea to perform a full quantum treatment of the envelope functions may be to directly solve their Schrödinger equation, that may be written:

$$\left[E(\hat{\mathbf{k}})+V(\hat{\mathbf{r}})\right]\psi=i\hbar\frac{\partial\psi}{\partial t} \qquad [8.1]$$

where $\hat{\mathbf{r}}$ and $\hbar\hat{\mathbf{k}}$ are the position and momentum operators, respectively. However, a device is an open system that cannot be fully described by the Schrödinger equation [8.1]. The contacts and the interactions with the environment (phonons, doping impurities, other electrons, etc.) influence the electron states in a way that cannot be really described by [8.1]. Alternatively, different formalisms have been developed to model an open device. The key idea consists of using a formalism that is able to include the statistical physics of the device and thus to make up for quantum transport an equivalent to the BTE for the semiclassical approach. We will focus here on two quantum formalisms able to describe the statistical physics of a system. They are based on the *DM* and the *Wigner function*, respectively. They are briefly presented in this chapter to capture the essential physics they contain and the relationship between them.

8.1.1. *Density matrix and Liouville equation*

The most natural approach to rigorously modeling the statistics of a quantum system is the DM formalism. The DM is defined as follows. We consider a basis $|\psi_i\rangle$ of the electron quantum states with an occupation probability p_i related to the statistical physics of the system. The DM operator [COH 06] is defined by:

$$\hat{\rho}=\sum_i p_i|\psi_i\rangle\langle\psi_i|, \qquad [8.2]$$

or in real-space representation by:

$$\rho(\mathbf{r},\mathbf{r}',t)=\langle\mathbf{r}|\hat{\rho}(t)|\mathbf{r}'\rangle=\sum_i p_i(t)\psi_i(\mathbf{r})\psi_i^*(\mathbf{r}') \qquad [8.3]$$

The occupation probability density $n(\mathbf{r})$ is obtained in a straightforward way from the diagonal elements of the DM:

$$n(\mathbf{r}) = \rho(\mathbf{r},\mathbf{r}) = \sum_i p_i(t)\psi_i(\mathbf{r})\psi_i^*(\mathbf{r}) \qquad [8.4]$$

The off-diagonal elements are called coherences and characterize the real-space delocalization of electrons, that is the quantum uncertainties only. This formalism is thus able to describe the quantum physics of a system submitted to statistics.

To study actual problems of transport a dynamic equation for the DM is required. First, consider the simple case of a closed system described by the Schrödinger equation [8.1]. By differentiating the density operator [8.2] with respect to time, we obtain:

$$\begin{aligned}\frac{\partial \hat{\rho}}{\partial t} &= \sum_i p_i \left[\frac{\partial |\psi_i\rangle}{\partial t}\langle\psi_i| + |\psi_i\rangle \frac{\partial \langle\psi_i|}{\partial t} \right] \\ &= \frac{1}{i\hbar}\sum_i p_i \left[\hat{\mathrm{H}}|\psi_i\rangle\langle\psi_i| - |\psi_i\rangle\langle\psi_i|\hat{\mathrm{H}} \right]\end{aligned} \qquad [8.5]$$

which leads to the *Liouville equation:*

$$i\hbar \frac{d\hat{\rho}}{dt} = \left[\hat{\mathrm{H}}, \hat{\rho}\right] \qquad [8.6]$$

For the Hamiltonian of equation [8.1] in the effective mass approximation m, the equation [8.6] becomes:

$$i\hbar \frac{\partial \rho(\mathbf{r}_1,\mathbf{r}_2)}{\partial t} = -\frac{\hbar^2}{2m}\left(\frac{\partial^2}{\partial \mathbf{r}_1^2} - \frac{\partial^2}{\partial \mathbf{r}_2^2}\right)\rho(\mathbf{r}_1,\mathbf{r}_2) + \left(V(\mathbf{r}_1) - V(\mathbf{r}_2)\right)\rho(\mathbf{r}_1,\mathbf{r}_2) \qquad [8.7]$$

This looks promising and effective. However, can we approach advanced topics with this formalism? When expressing the DM in the reciprocal space, this formalism may model the electron–phonon interaction accurately, including collisional broadening (CB) and collisional retardation (CR) and intracollisional field effect (ICFE) [BRU 89, JAC 92, ROS 92a]. However, it does not allow the accurate study of real space-dependent problems. DM-based device simulation is possible using the Pauli master equation that takes into account only the diagonal elements of the DM [FIS 98, FIS 99]. However, in spite of improvements [GEB 04a, GEB 04b], the modeling of terminal contacts in an open system is difficult within this formulation that is thought to be valid only for devices smaller than the electron dephasing length [FIS 99].

Nowadays, in most research works on nanodevices, in spite of recent efforts to simulate the electron transport in quantum cascade lasers with several approximations [DUP 10, DIN 12], the DM tends to be replaced by the Wigner quasi-distribution function or by the Green's functions whose formalisms make it possible to include the contacts, interactions and self-consistence in a more simple and rigorous way than in the DM approach.

8.1.2. *The Wigner function*

For a statistical ensemble of particles described by a density operator ρ the Wigner formalism of quantum transport is based on the Wigner function f_w originally introduced by Wigner in 1932 [WIG 32]. It is defined in the phase-space $(\mathbf{r},\mathbf{k})$ as a differential Fourier transform of the DM $\rho(\mathbf{r},\mathbf{r}')=\langle\mathbf{r}|\hat{\rho}(t)|\mathbf{r}'\rangle$, also called *Weyl-Wigner* transform, that is [JAC 01]:

$$f_w(\mathbf{r},\mathbf{k})=\frac{1}{(2\pi)^d}\int d\mathbf{r}'\, e^{-i\mathbf{k}.\mathbf{r}'}\, \rho(\mathbf{r}+\mathbf{r}'/2,\mathbf{r}-\mathbf{r}'/2) \qquad [8.8]$$

where d is the real-space dimension of the transport problem. The fact that this quantum function depends on the same phase-space coordinates as the Boltzmann distribution function $f_b(\mathbf{r},\mathbf{k})$ of semi-classical transport is remarkable. It is all the more surprising that the space probability density of electrons $n(\mathbf{r})$ can be obtained by the same way as in the semiclassical case, that is:

$$\begin{aligned}\int d\mathbf{k}\, f_w(\mathbf{r},\mathbf{k})&=\frac{1}{(2\pi)^3}\int d\mathbf{r}'\left\langle\mathbf{r}+\frac{\mathbf{r}'}{2}\middle|\hat{\rho}(t)\middle|\mathbf{r}-\frac{\mathbf{r}'}{2}\right\rangle\int d\mathbf{k}e^{-i\mathbf{k}.\mathbf{r}'}\\&=\langle\mathbf{r}|\hat{\rho}(t)|\mathbf{r}\rangle=n(\mathbf{r})\end{aligned} \qquad [8.9]$$

It may also be shown that the average value of an operator $\hat{\mathrm{A}}$ depending only on operator $\hat{\mathbf{r}}$, such as we may define $\hat{\mathrm{A}}=V(\hat{\mathbf{r}})$, is [QUE 10a]:

$$\langle\hat{\mathrm{A}}\rangle=\int d\mathbf{r}\int d\mathbf{k}\,\mathrm{V}(\mathbf{r})\, f_w(\mathbf{r},\mathbf{k}) \qquad [8.10a]$$

Similarly, the average value of an operator $\hat{A}$ depending only on operator $\hat{\mathbf{k}}$, so that we may define $\hat{A} = W(\hat{\mathbf{k}})$, is:

$$\langle \hat{A} \rangle = \int d\mathbf{r} \int d\mathbf{k}\, W(\mathbf{k}) f_w(\mathbf{r},\mathbf{k}) \qquad [8.10b]$$

An average value for these two kinds of observables – which include many observables of interest in electron devices – can thus be computed the same way as in the semiclassical case, which will be discussed later. In the more general case, it has been shown that the average value of $\hat{A}$ writes [JAC 01, QUE 10a]:

$$\langle \hat{A} \rangle = \int d\mathbf{r} \int d\mathbf{k}\, \mathcal{A}_w(\mathbf{r},\mathbf{k}) f_w(\mathbf{r},\mathbf{k}) \qquad [8.11]$$

where $\mathcal{A}_w(\mathbf{r},\mathbf{k}) = \int d\mathbf{r}' \exp(-i\mathbf{k}\cdot\mathbf{r}')\, A(\mathbf{r}-\mathbf{r}'/2, \mathbf{r}+\mathbf{r}'/2)$ is the Weyl–Wigner transform of the operator $\hat{A}$.

The Wigner function thus shares many properties with the Boltzmann distribution function and in many cases it may be used as a distribution function, though it is not. This important point will be discussed later.

8.1.3. *Dynamic equation: The Wigner equation*

To derive the dynamic equation of the Wigner function, we first consider a closed system without any scattering. The Liouville equation [8.7] is re-written in the so-called center-of-mass coordinates $\mathbf{r} = (\mathbf{r}_1 + \mathbf{r}_2)/2$ and $\mathbf{r}' = \mathbf{r}_1 - \mathbf{r}_2$, that is:

$$\begin{aligned} i\hbar \frac{\partial \rho\left(\mathbf{r}+\frac{\mathbf{r}'}{2}, \mathbf{r}-\frac{\mathbf{r}'}{2}\right)}{\partial t} &= -\frac{\hbar^2}{m}\left(\frac{\partial}{\partial \mathbf{r}}\frac{\partial}{\partial \mathbf{r}'}\right)\rho\left(\mathbf{r}+\frac{\mathbf{r}'}{2}, \mathbf{r}-\frac{\mathbf{r}'}{2}\right) \\ &+ \left(V\left(\mathbf{r}+\frac{\mathbf{r}'}{2}\right) - V\left(\mathbf{r}-\frac{\mathbf{r}'}{2}\right)\right)\rho\left(\mathbf{r}+\frac{\mathbf{r}'}{2}, \mathbf{r}-\frac{\mathbf{r}'}{2}\right) \end{aligned} \qquad [8.12]$$

The differential Fourier transform of [8.12] gives, for a 3D problem ($d = 3$):

$$\begin{aligned} \frac{\partial f_w(\mathbf{r},\mathbf{k})}{\partial t} &= -\frac{\hbar^2}{mi\hbar}\frac{1}{(2\pi)^3}\int d\mathbf{r}' e^{-i\mathbf{k}.\mathbf{r}'}\left(\frac{\partial}{\partial \mathbf{r}}\frac{\partial}{\partial \mathbf{r}'}\right)\rho\left(\mathbf{r}+\frac{\mathbf{r}'}{2}, \mathbf{r}-\frac{\mathbf{r}'}{2}\right) \\ &+ \frac{1}{(2\pi)^3 i\hbar}\int d\mathbf{r}' e^{-i\mathbf{k}.\mathbf{r}'}\left(V\left(\mathbf{r}+\frac{\mathbf{r}'}{2}\right) - V\left(\mathbf{r}-\frac{\mathbf{r}'}{2}\right)\right)\rho\left(\mathbf{r}+\frac{\mathbf{r}'}{2}, \mathbf{r}-\frac{\mathbf{r}'}{2}\right) \end{aligned} \qquad [8.13]$$

We consider the two right-hand side terms of [8.13] separately. The first one is integrated by parts, which leads to:

$$\begin{aligned}
&-\frac{\hbar^2}{mi\hbar}\frac{1}{(2\pi)^3}\int d\mathbf{r}'e^{-i\mathbf{k}.\mathbf{r}'}\left(\frac{\partial}{\partial \mathbf{r}}\frac{\partial}{\partial \mathbf{r}'}\right)\rho\left(\mathbf{r}+\frac{\mathbf{r}'}{2},\mathbf{r}-\frac{\mathbf{r}'}{2}\right)\\
&=-\frac{\hbar^2}{mi\hbar}\frac{1}{(2\pi)^3}\frac{\partial}{\partial \mathbf{r}}\int d\mathbf{r}'e^{-i\mathbf{k}.\mathbf{r}'}\left(\frac{\partial}{\partial \mathbf{r}'}\right)\rho\left(\mathbf{r}+\frac{\mathbf{r}'}{2},\mathbf{r}-\frac{\mathbf{r}'}{2}\right)\\
&=-\frac{\hbar\mathbf{k}}{m}\cdot\frac{\partial f_w(\mathbf{r},\mathbf{k},t)}{\partial \mathbf{r}}
\end{aligned} \qquad [8.14]$$

By introducing the inverse Fourier transform of the Wigner function [8.8] in the second term of [8.13], we obtain:

$$\begin{aligned}
&\frac{1}{(2\pi)^3 i\hbar}\int d\mathbf{r}'\,e^{-i\mathbf{k}.\mathbf{r}'}\left(V\left(\mathbf{r}+\frac{\mathbf{r}'}{2}\right)-V\left(\mathbf{r}-\frac{\mathbf{r}'}{2}\right)\right)\rho\left(\mathbf{r}+\frac{\mathbf{r}'}{2},\mathbf{r}-\frac{\mathbf{r}'}{2}\right)\\
&=\frac{1}{(2\pi)^3 i\hbar}\int d\mathbf{r}'\int d\mathbf{k}'\,e^{-i\,\mathbf{k}.\mathbf{r}'+i\mathbf{k}'.\mathbf{r}'}\left(V\left(\mathbf{r}+\frac{\mathbf{r}'}{2}\right)-V\left(\mathbf{r}-\frac{\mathbf{r}'}{2}\right)\right)f_w(\mathbf{r},\mathbf{k}')
\end{aligned} \qquad [8.15]$$

The WTE in the ballistic limit is finally

$$\frac{\partial f_w}{\partial t}+\frac{\hbar}{m}\mathbf{k}\cdot\nabla_{\mathbf{r}}f_w=Q\,f_w(\mathbf{r},\mathbf{k}) \qquad [8.16]$$

which includes the quantum evolution term for the potential V

$$Q\,f_w(\mathbf{r},\mathbf{k})=Q_V\,f_w(\mathbf{r},\mathbf{k})=\int d\mathbf{k}'\,V_w(\mathbf{r},\mathbf{k}-\mathbf{k}')\,f_w(\mathbf{r},\mathbf{k}') \qquad [8.17]$$

and the Wigner potential V_w defined as:

$$V_w(\mathbf{r},\mathbf{k})=\frac{1}{i\hbar(2\pi)^d}\int d\mathbf{r}'e^{-i\,\mathbf{k}.\mathbf{r}'}\left(V\left(\mathbf{r}+\frac{\mathbf{r}'}{2}\right)-V\left(\mathbf{r}-\frac{\mathbf{r}'}{2}\right)\right) \qquad [8.18]$$

At this point, it is important to observe and keep in mind that the effect of the potential V on the Wigner function is non-local, which makes the description of quantum effects possible, in contrast with the case of semiclassical transport that will be introduced below.

Up to now, we only considered the case of ballistic transport, that is free of collisions with phonons or impurities. The effect of scattering in the Wigner equation will be introduced later, after describing the semiclassical Boltzmann formalism and its connection with the quantum Wigner formalism.

8.2. The semiclassical limit: BTE

The difficulty of transport physics lies in the solution of the Schrödinger equation [8.1] for envelope functions including the effects of scattering that play a fundamental role. Here, we present the simplest approach: the semiclassical approach. It has been successfully used for a considerable number of device problems in the past.

This approach is based on the idea that electrons are in well-localized wave packets of space extension σ, formed from Bloch waves. If we consider the well-known Gaussian type wave packet, the envelope function can be written as:

$$\psi(\mathbf{r}) = N e^{-\frac{(\mathbf{r}-\mathbf{r}_0)^2}{2\sigma^2}} e^{i\mathbf{k}_0 \mathbf{r}} \text{ and } \psi(\mathbf{k}) = N' e^{-\frac{\sigma^2(\mathbf{k}-\mathbf{k}_0)^2}{2}} e^{-i\mathbf{r}_0 \mathbf{k}} \qquad [8.19]$$

in the real and reciprocal spaces, respectively. N and N' are here normalization constants. The dynamics of such wave functions are well known [KIT 96]. The real space velocity of the wave packet center is the group velocity, that is:

$$\frac{d\mathbf{r}_0}{dt} = \mathbf{v}_g = \frac{\partial E}{\partial \hbar \mathbf{k}} \qquad [8.20]$$

The potential energy of the wave packet is:

$$\langle E_P \rangle = \int d\mathbf{r} \left|\psi(\mathbf{r})\right|^2 V(\mathbf{r}) \approx V(\mathbf{r}_0) \qquad [8.21]$$

The approximation in [8.21] is correct if the potential $V(\mathbf{r}) = -eU(\mathbf{r})$ is linear, quadratic or varies slowly on the space extension σ of the wave function. It follows from the law of energy conservation that $dE = -dE_P$, that is $\frac{\partial E_C}{\partial \hbar \mathbf{k}} \cdot d\hbar \mathbf{k}_0 = \frac{-\partial V}{\partial \mathbf{r}} \cdot d\mathbf{r}_0$, thus $\mathbf{v} \cdot \frac{d\hbar \mathbf{k}_0}{dt} = -\nabla V \cdot \mathbf{v}$. In absence of a magnetic field, the center of the wave packet then propagates in **k**-space with a velocity:

$$\frac{d\hbar\mathbf{k}_0}{dt} = -\nabla V = e\nabla U \qquad [8.22]$$

The velocity equations [8.20] and [8.22] are similar to Newton's laws for a point particle, which leads to the terminology "semiclassical" transport. From this point of view, electrons are considered as point particles moving with the velocity of their wave packet center. It is however well known that the wave packets tend to spread during their propagation (i.e. their spatial width increases) [COH 06], so that [8.20] and [8.22] do not give a full picture of wave packet dynamics. We now consider that this warping is negligible and that wave packet extension is unchanged. These two assumptions are not obvious and have been discussed in the context of the theory of decoherence [QUE 08]. The fact that electron wave packets are spontaneously Gaussian will be discussed too [QUE 10a].

8.2.1. *Distribution function and Boltzmann equation: the standard point of view*

The transport phenomena result from an average over the behavior of a large number of electrons. The model of semiclassical transport is thus based on the definition of the Boltzmann distribution function $f_b(\mathbf{r},\mathbf{k},t)$, which corresponds to the probability density for finding a wave packet centered on **r** and **k** at time t (or equivalently to find an electron at positions **r** and **k** of the phase-space at time t). The dynamic equation of this distribution function without collision is readily obtained from the probability conservation:

$$0 = \frac{d f_B(\mathbf{r},\mathbf{k},t)}{dt} = \frac{\partial f_b(\mathbf{r},\mathbf{k},t)}{\partial t} + \nabla_{\mathbf{r}} f_B(\mathbf{r},\mathbf{k},t)\frac{d\mathbf{r}}{dt} + \nabla_{\mathbf{k}} f_b(\mathbf{r},\mathbf{k},t)\frac{d\mathbf{k}}{dt} \qquad [8.23]$$

which leads to the BTE:

$$\begin{aligned}\frac{\partial f_b(\mathbf{r},\mathbf{k},t)}{\partial t} &= -\mathbf{v}(\mathbf{k})\cdot\nabla_{\mathbf{r}} f_b(\mathbf{r},\mathbf{k},t) \\ &\quad + \frac{1}{\hbar}\nabla V\cdot\nabla_{\mathbf{k}} f_b(\mathbf{r},\mathbf{k},t) + C\,f_b(\mathbf{r},\mathbf{k},t)\end{aligned} \qquad [8.24]$$

where a generic term $C\,f_b(\mathbf{r},\mathbf{k},t)$ has been added to include the collision or scattering effects on the Boltzmann distribution function. C is the collision operator. Within the semiclassical approach this collision term is commonly evaluated using the Fermi golden rule, that is the first-order perturbation theory. Collisions are considered as instantaneous scattering events that modify the wave vector. Between two consecutive scattering events, electrons experience ballistic free flights whose duration is determined from the rate of scattering $s(\mathbf{k},\mathbf{k}')$ for an electron initially in state $\mathbf{k}$ to $\mathbf{k}'$. The collision term for a non-degenerate semiconductor is then written as:

$$C\,f_b(\mathbf{r},\mathbf{k},t)=\sum_i\int d\mathbf{k}'\left[s_i(\mathbf{k}',\mathbf{k})\,f_b(\mathbf{r},\mathbf{k}',t)-s_i(\mathbf{k},\mathbf{k}')\,f_b(\mathbf{r},\mathbf{k},t)\right]\qquad[8.25]$$

where the subscript i labels the type of scattering process. For an interaction Hamiltonian $H_{int,i}$, the rate of scattering according to the Fermi golden rule is:

$$s_i(\mathbf{k},\mathbf{k}')=\frac{2\pi}{\hbar}\left|\langle\psi_{\mathbf{k}'}|H_{int,i}|\psi_{\mathbf{k}}\rangle\right|^2 D_s(\mathbf{k}')\,\delta(E'-E)\qquad[8.26]$$

where $D_s(\mathbf{k}')$ is the density of states, with the energies $E=E(\mathbf{k})$ and $E'=E(\mathbf{k}')$. In the case of degenerate carrier gas, we should also properly include the exclusion principle and the actual availability of final states. The full derivation of most scattering rates may be found in many textbooks, for example in [JAC 89] and [LUN 00]. We will return to some of them in section 8.3.2. The two main approximations of the semiclassical approach (which will be further discussed later) should be reminded here:

– the potential is linear, quadratic or slowly variable on a distance σ;

– the electrons are in well-localized wave packets of Gaussian typical form.

It should be noted that in the Boltzmann equation [8.24] the effect of potential $\frac{1}{\hbar}\nabla V\cdot\nabla_{\mathbf{k}}f_b(\mathbf{r},\mathbf{k},t)$ is purely local, in contrast to the case of the quantum evolution term [8.17] of the WTE. That is why it cannot naturally include the quantum effects such as tunneling. It is often called the drift term of the BTE.

The semiclassical approximation (either using accurate or simplified treatments) has, until recently, been the foundation of most device simulations. It has shown tremendous success in explaining and modeling most phenomena seen in electron devices. The solution of the Boltzmann equation (using direct solution or, more often, a statistical MC approach) has been used for studies requiring detailed understanding of the transport. A modern application is the study of ballistic effects in ultrasmall metal–oxide–semiconductor field-effect transistors (MOSFETs) [SAI 04, PAL 05] or high electron mobility transistors (HEMTs) [LUS 05, MIL 08].

Approximations of the Boltzmann equation (hydrodynamic or drift–diffusion equations) are now industry standards for device design and are also used for many fundamental studies.

8.2.2. *The connection between WTE and BTE*

Now, let us return to the WTE [8.16] and particularly its right-hand side, that is the quantum evolution term [8.17] that includes the Wigner potential [8.18]. In the particular case of linear or quadratic potential V the Wigner potential becomes:

$$V_w(\mathbf{r},\mathbf{k}) = \frac{1}{i\hbar(2\pi)^d} \nabla V \int d\mathbf{r}' \, e^{-i\mathbf{k}.\mathbf{r}'} \, \mathbf{r}' = -\frac{1}{\hbar} \nabla V \, \delta'(\mathbf{k}) \qquad [8.27]$$

The quantum evolution term is then:

$$Q f_w(\mathbf{r},\mathbf{k}) = \frac{1}{\hbar} \nabla V \, \nabla_{\mathbf{k}} f_w(\mathbf{r},\mathbf{k}) \qquad [8.28]$$

which is nothing but the drift term of the BTE [8.24].

In the case of a potential varying slowly with respect to typical values of $1/k$, equations [8.27] and [8.28] are still a good approximation. Near thermal equilibrium, $1/k$ is typically close to $\hbar/\sqrt{2\, m\, k_b T}$ that is also close to the "natural" delocalization length σ as discussed [QUE 10a]. Hence, the Wigner equation tends to the Boltzmann equation in the case for which the Boltzmann equation was established, that is for linear, quadratic or slowly varying potential. The Boltzmann equation may thus be seen as the semiclassical limit, that is a particular case, of the Wigner equation.

We then immediately understand the potential advantage of the Wigner formalism for studying the transport in nanodevices operating between the ballistic and the diffusive limits, and in particular for investigating the transition between the well-known semiclassical regime and the less understood quantum transport regime, provided that we can introduce the effects of scattering in the WTE, which will be discussed in the next section.

Finally, it should be noted that if the potential can be separated in rapidly and slowly varying parts, both of them can be treated separately [NED 04], that is for:

$$V(\mathbf{r}) = V_{slow}(\mathbf{r}) + V_{rapid}(\mathbf{r}) \qquad [8.29]$$

The Wigner equation in the ballistic limit [8.16] may then be rewritten as:

$$\frac{\partial f_W}{\partial t}+\frac{\hbar}{m}\mathbf{k}\cdot\nabla_{\mathbf{r}}f_W-\frac{1}{\hbar}\nabla V_{slow}\cdot\nabla_{\mathbf{k}}f_W=Q_{V_{rapid}}f_W \qquad [8.30]$$

The WTE has been used in different domains of quantum physics, for example in atomic physics [LUT 97], in quantum optics [BER 02, DEL 08], in time-varying signal processing [HLA 92] or in particle physics and cosmology [CAL 88]. The history and state-of-the-art of its utilization in electronic nanodevices is also pretty rich and will be presented briefly later.

8.3. Scattering in Boltzmann and Wigner equations

Many sources of scattering may act on the behavior of charged particles (electrons and holes) in a semiconductor crystal. Among them the most important particles are phonons and ionized impurities. The scattering mechanisms of interest can be classified as intravalley when the scattering lets the particle in the initial valley of the band structure, or intervalley if the initial and final states lie in different valley. Particles, especially holes, may also change band, which helps to distinguish between intraband and interband transitions.

The interaction of charged particles with phonons is due to the phonon induced by the crystal through either a deformation potential mechanism or the electrostatic force resulting from the polarization wave induced by the coherent displacements of crystal atoms, that is the phonons themselves. The deformation potential mechanism is typical of covalent materials (Si, Ge, ...), while the electrostatic mechanisms are typical of polar materials (GaAs, InP, ...). We also make a distinction between acoustic and optical phonons. In the case of electrostatic scattering potential, the interaction with acoustic phonons is called piezoelectric interaction, while the interaction with optical phonons is called polar optical interaction. Emission and absorption of phonons by carriers play an important role in transport since it is the source of energy exchange with the crystal lattice and thus of carrier relaxation.

The scattering with ionized impurities is of course of Coulomb nature. It is always elastic and thus cannot be a source of energy dissipation. However, it tends to randomize the velocity direction of particles and thus plays an important role in the drift velocity relaxation and in the control of the carrier mobility. Its effect decreases as the temperature of the crystal increases and when electrons are heated by the electric field.

Other more specific scattering mechanism may have an important role in some materials and/or devices. In particular, the interface between two different materials might be rough, which may cause a perturbation of the potential and induce a source of additional scattering, called "surface roughness scattering". It is in particular of strong importance at the SiO_2/Si interface of MOSFETs [LIU 05, MIC 11]. In disordered alloys, as SiGe or GaAlAs, the random atomistic composition generates fluctuations of potential with respect to the homogeneous material, which is another source of scattering, usually called "alloy scattering" [HAR 76]. This scattering mechanism is elastic and isotropic. It has an important influence on the carrier mobility in alloyed materials [HAU 76, MUR 11].

In the following sections, we first briefly describe the main features of scattering rates with phonons and ionized impurities in the way they are usually considered in semiclassical transport through the collision operator of the Boltzmann equation. Then, in section 8.3.2, we discuss the assumptions and approximations that allow us to give the same form to the scattering rates in the Wigner equation as in the Boltzmann equation, as a kind of semiclassical limit of scattering. Finally, we summarize in section 8.3.3 the so-called Wigner–Boltzmann equation that consists of introducing the Boltzmann collision operator in the Wigner equation with the full quantum treatment of the effect of potential.

8.3.1. *The Boltzmann collision operator: phonon and impurity scattering*

As explained in 8.2.1, in the semiclassical Boltzmann approach, the collisions are considered within the first-order perturbation theory, which assumes that each scattering event is perfectly localized in space and time, with a rate given by the Fermi golden rule [8.26] and used to derive the collision term [8.25] of the Boltzmann equation [8.24]. In the scattering rates $s_i(\mathbf{k},\mathbf{k}')$ [8.26] for the interaction mechanism i characterized by the interaction Hamiltonian $H_{int,i}$, the most important term is obviously the square matrix element:

$$\left|M_i(\mathbf{k},\mathbf{k}')\right|^2 = \left|\langle\psi_{\mathbf{k}'}|H_{int,i}|\psi_{\mathbf{k}}\rangle\right|^2 \qquad [8.31]$$

Here, we briefly summarize the main information on the matrix elements and scattering rates of impurity scattering and some important phonon scattering mechanisms with acoustic and optical phonons in both cases of deformation potential and electrostatic interactions. In the very simple case of an isotropic and parabolic band of effective mass m in bulk material of volume V with density of states $D_s(\mathbf{k}') = V/8\pi^3$, we will also give the equation of the scattering rates $\Gamma_i(E)$ integrated over all final states, that is:

$$\Gamma_i(\mathbf{k}) = \int s_i(\mathbf{k},\mathbf{k}')d\mathbf{k}' = \frac{2\pi}{\hbar}\int |M_i(\mathbf{k},\mathbf{k}')|^2 D_s(\mathbf{k}')\,\delta(E-E')d\mathbf{k}' \qquad [8.32]$$

which may be conveniently calculated as a function of energy, that is as $\Gamma_i(E)$. Indeed, these quantities Γ_i make up the relevant information for treating the scattering processes of individual particles using a MC algorithm, as explained in section 8.4. More detailed information on other scattering mechanisms and in the case of more general description of energy bands can be found in [JAC 89] and [MOG 93].

8.3.1.1. *Phonon scattering*

To describe the electron–phonon interaction, the matrix element must involve the full system, the electrons and the phonons. The state of the combined system formed by an electron and a phonon takes the form $|\psi_\mathbf{k}\rangle = |\mathbf{k}\rangle|c\rangle$, where $|\mathbf{k}\rangle$ and $|c\rangle$ describe the unperturbed states of the electron and the lattice (phonon), respectively. After an expansion of $H_{int,i}$ in a Fourier series and considering standard Bloch states for electrons, the matrix element can be written as [JAC 89]:

$$|M_i(\mathbf{k},\mathbf{k}')| = \left(\frac{2\pi}{V}\right)^{3/2} \sum_{\mathbf{q}} \langle c'|\mathcal{H}_i^{\mathbf{q}}|c\rangle\,\mathcal{R}(\mathbf{k},\mathbf{k}') \qquad [8.33]$$

where $\mathcal{R}(\mathbf{k},\mathbf{k}') = \int d\mathbf{r}\, u_{k'}^*(\mathbf{r})\, u_k(\mathbf{r})\exp(i\mathbf{G}\cdot\mathbf{r})$ is the overlap integral of Bloch functions in the unit cell, with **G** a vector of the reciprocal lattice. For any electron–phonon scattering mechanism, when writing the Hamiltonian interaction in terms of creation and annihilation operators, that is in the second quantization framework, we arrive at the general equation of the scattering rate:

$$s_i(\mathbf{k},\mathbf{k}') = \frac{2\pi}{\hbar}\frac{V}{8\pi^3}F(\mathbf{q})\left[N_\mathbf{q}(\omega_\mathbf{q}) + \frac{1}{2} - \frac{i}{2}\right]\mathcal{G}(\mathbf{k},\mathbf{k}')\,\delta(E' - i\hbar\omega_\mathbf{q} - E) \qquad [8.34]$$

where $i = 1$ stands for an absorption and $i = -1$ for an emission of a phonon of wave vector $\mathbf{q} = \mathbf{k}' - \mathbf{k}$ and of angular frequency $\omega_\mathbf{q}$. The energies E and E' are the electron energy of the initial state $E = E(\mathbf{k})$ and final state $E' = E(\mathbf{k}')$, respectively. The function $F(\mathbf{q})$ depends on both the interaction process (deformation potential or electrostatic interaction) and the phonon mode (acoustic intravalley or intervalley). The function $N_{\mathbf{q}} = \left[\exp\left(\frac{\hbar\omega_q}{k_b T}\right) - 1\right]^{-1}$ is the average phonon number, assuming the phonon gas remains in equilibrium state at the temperature of the crystal. The function $\mathcal{G}(\mathbf{k},\mathbf{k}') = |\mathcal{R}(\mathbf{k},\mathbf{k}')|^2$ is called the overlap factor. This function depends mainly on the angle θ between $\mathbf{k}$ and $\mathbf{k}'$ measured from the center of the Brillouin zone. In valleys that have a minimum far from Γ point, it is nearly constant and is often considered equal to 1, its actual value being including in the coupling constant. In the case of holes and for electrons in Γ valley, the overlap factor cannot be considered as a constant. Accurate equations may be found in the literature, for both electrons [FAW 70] and holes [WIL 71].

In the case of deformation potential interaction, [8.34] takes the form:

$$\begin{aligned} s_i(\mathbf{k},\mathbf{k}') = {} & \frac{V}{8\pi^3}\frac{\pi}{\rho V \omega_{\mathbf{q}}} |\mathbf{D}\cdot\mathbf{q}|^2 \left[N_{\mathbf{q}}(\omega_{\mathbf{q}}) + \frac{1}{2} - \frac{i}{2}\right] \\ & \times \mathcal{G}(\mathbf{k},\mathbf{k}')\,\delta(E' - i\hbar\omega_{\mathbf{q}} - E) \end{aligned} \qquad [8.35]$$

Where $\mathbf{D}$ is a coupling tensor describing the shift in the electron band per unit deformation of the crystal and ρ is the density of the crystal.

8.3.1.1.1. Deformation potential interaction – Intravalley acoustic phonons

For intravalley acoustic phonons, in most cases the tensor $\mathbf{D}$ can be treated as a scalar quantity and the squared term in [8.35] reduces to $D_{ac}^2 q^2$ for longitudinal phonons, where D_{ac} is called acoustic deformation potential and has the unit of energy. In this process the phonon energy is usually small, that is much smaller than the electron kinetic energy at room temperature. The phonon energy can be well approximated as $\hbar\omega_{\mathbf{q}} \simeq \hbar u q$, where u is the longitudinal sound velocity, and the phonon number can be simplified to:

$$N_{\mathbf{q}} \simeq \frac{k_B T}{\hbar\omega_{\mathbf{q}}} - \frac{1}{2} \simeq \frac{k_B T}{\hbar q u} - \frac{1}{2} \qquad [8.36]$$

The phonon energy being small, the acoustic phonon scattering is often treated as an elastic process. Emission and absorption of phonons are thus merged into a single process, the rate of which is finally given by:

$$s_{ac}(\mathbf{k},\mathbf{k}') = \frac{k_B T}{4\pi^2 \hbar \rho u^2} D_{ac}^2 \, \delta(E' - E) \qquad [8.37]$$

The integration of [8.37] over all final states for a parabolic band of effective mass m leads to the energy-dependent scattering rate:

$$\Gamma_{ac}(E) = \frac{\sqrt{2}}{\pi} \frac{k_B T \, m^{3/2}}{\hbar^4 \rho u^2} D_{ac}^2 \sqrt{E} \qquad [8.38]$$

The extension of this expression to non-parabolic and ellipsoidal valleys is straightforward [JAC 89]. The non-elastic treatment of this process leads to more complicated expressions and numerical algorithms in an MC simulation [JAC 89]. Though non-elastic effects may be neglected at room temperature, they play an important role in the energy relaxation at low temperature and low electric field [HES 84, AUB 10].

8.3.1.1.2. Deformation potential interaction – Intervalley phonons

For intravalley scattering with optical phonons of high-energy and low-wave vector (e.g. in L valleys of III-V semiconductors), and, more important, for intervalley scattering with both optical and acoustic phonons of nearly constant energy $\hbar\omega_0$, the squared coupling term in [8.35] reduces to a coupling constant D_0^2, where D_0 has the unit of energy per length unit. It is thus a zero-order process of rate:

$$s_0(\mathbf{k},\mathbf{k}') = \frac{V}{8\pi^3} \frac{\pi}{\rho V \omega_0} D_0^2 \left[N_{\mathbf{q}}(\omega_0) + \frac{1}{2} - \frac{i}{2} \right] \delta(E' - i\hbar\omega_0 - \Delta E_{iv} - E) \qquad [8.39]$$

where ΔE_{iv} is the energy separation between initial and final valleys. The integration of [8.39] over all final states for spherical parabolic bands of effective mass m gives the scattering probability per unit of time:

$$\Gamma_0(E) = Z_{iv} \frac{1}{\sqrt{2}\,\pi} \frac{m^{3/2}}{\hbar^3 \rho \omega_0} D_0^2 \left[N_{\mathbf{q}}(\omega_0) + \frac{1}{2} - \frac{i}{2} \right] \sqrt{E + i\hbar\omega_0 + \Delta E_{iv}} \qquad [8.40]$$

where Z_{iv} is the number of possible final valleys for intervalley processes and 1 for an intravalley process. When a zero-order process is forbidden by selection rules

[STR 70, LAX 72], it may be permitted at first order in the phonon wave vector [FER 76, DOL 97], which yields the scattering rates:

$$s_1(\mathbf{k},\mathbf{k}') = \frac{V}{8\pi^3}\frac{\pi}{\rho V \omega_1} D_1^2 q^2 \left[N_{\mathbf{q}}(\omega_1) + \frac{1}{2} - \frac{i}{2} \right] \delta(E' - i\hbar\omega_1 - E) \quad [8.41]$$

and

$$\begin{aligned} \Gamma_1(E) = Z_{iv} \frac{\sqrt{2}}{\pi} \frac{m^{5/2}}{\hbar^5 \rho \omega_1} D_1^2 \left[N_{\mathbf{q}}(\omega_1) + \frac{1}{2} - \frac{i}{2} \right] \\ \times \sqrt{E + i\hbar\omega_0 + \Delta E_{iv}} \, (2E + i\hbar\omega_0 + \Delta E_{iv}) \end{aligned} \quad [8.42]$$

where the deformation potential D_1 has the unit of energy.

8.3.1.1.3. Electrostatic interaction – Intravalley acoustic piezoelectric phonons

An acoustic wave propagating in a piezoelectric material generates an electric field likely to act on the charged carriers. Considering this interaction as a perturbation leads to the scattering rate from state **k** to state $\mathbf{k}'$ in the form [RID 82]

$$\begin{aligned} s_{pz}(\mathbf{k},\mathbf{k}') = \frac{V}{8\pi^3} \frac{4\pi^2 e^2 p^2}{\rho V q u \varepsilon^2} \left(\frac{q^2}{q^2 + \beta^2} \right)^2 \\ \times \left[N_{\mathbf{q}} + \frac{1}{2} - \frac{i}{2} \right] \mathcal{G}(\mathbf{k},\mathbf{k}') \, \delta(E' - i\hbar q u - E) \end{aligned} \quad [8.43]$$

where p is the appropriate component of the piezoelectric tensor, ε is the dielectric constant, u is the average sound velocity and β is the inverse screening length. For a non-degenerate electron gas of density n, β is given at equilibrium by the Debye theory as $\beta = \left(e^2 n / \varepsilon \varepsilon_0 k_B T \right)^{1/2}$. In the elastic approximation, emission and absorption can be merged in a single process, the rate of which for a parabolic band simplifies in:

$$s_{pz}(\mathbf{k},\mathbf{k}') = \frac{e^2 p^2 k_B T}{8\pi^2 \hbar \rho u^2 \varepsilon^2 \varepsilon_0^2} \frac{1}{q^2} \left(\frac{q^2}{q^2 + \beta^2} \right)^2 \delta(E' - E) \quad [8.44]$$

The scattering probability per unit of time is finally:

$$\Gamma_{pz}(E)=\frac{e^2 K_{pz}^2\, k_B T\sqrt{m}}{2\pi\sqrt{2}\,\hbar^2\ \varepsilon\varepsilon_0}\sqrt{E}\left[\log\left(1+\frac{8mE}{\hbar^2\beta}\right)-\frac{1}{1+\left(\hbar^2\beta/8mE\right)}\right] \qquad [8.45]$$

where the dimensionless piezoelectric coupling constant K_{pz} is given by:

$$K_{pz}^2=\frac{p^2}{\varepsilon\varepsilon_0\,\rho u^2} \qquad [8.46]$$

This scattering mechanism is strongly anisotropic [HES 84] though an isotropic approximation is sometimes used when it is not the predominant source of scattering.

8.3.1.1.4. Electrostatic interaction – Polar optical phonons

In a polar material, when the two atoms of the unit cell vibrate in a longitudinal optical mode, they generate an oscillating dipole. The resulting electric field can scatter the electrons and holes, according to the theory of Froelich [FRO 47]. The corresponding Hamiltonian interaction has been derived by Ziman [ZIM 60] and in the form of [8.34], the scattering rate writes:

$$\begin{aligned} s_{po}(\mathbf{k},\mathbf{k}')=&\frac{2\pi}{\hbar}\frac{V}{8\pi^3}F(\mathbf{q})\left[N_{\mathbf{q}}\left(\omega_{op}\right)+\frac{1}{2}-\frac{i}{2}\right]\\ &\times\mathcal{G}(\mathbf{k},\mathbf{k}')\,\delta\left(E'-i\hbar\omega_{op}-E\right)\end{aligned} \qquad [8.47]$$

where ω_{op} is the angular frequency of the longitudinal optical phonon. The function $F(\mathbf{q})$ is given by [MOG 93]

$$F(\mathbf{q})=\frac{e^2}{2\varepsilon_0 V}\hbar\omega_{op}\frac{1}{q^2}\left(\frac{1}{\varepsilon_\infty}-\frac{1}{\varepsilon}\right) \qquad [8.48]$$

where ε is the dielectric constant, ε_∞ is the high-frequency dielectric constant and ε_0 is the permittivity of vacuum. Considering the final energy $E'=E+i\hbar\omega_{op}$ (with $i=1$ for an absorption and $i=-1$ for an emission), for a parabolic and isotropic band of mass *m*, the probability of scattering per unit of time is given by:

$$\Gamma_{po}(E) = \frac{1}{\sqrt{2}\,4\pi}\frac{m^{1/2}e^2}{\hbar\varepsilon_0}\omega_{op}\left(\frac{1}{\varepsilon_\infty}-\frac{1}{\varepsilon}\right) \times\left[N_{\mathbf{q}}\left(\omega_{op}\right)+\frac{1}{2}-\frac{i}{2}\right]\frac{1}{\sqrt{E}}\log\left|\frac{\sqrt{E}+\sqrt{E'}}{\sqrt{E}-\sqrt{E'}}\right| \qquad [8.49]$$

However, this intravalley scattering mechanism is important at low field in III–V semiconductors where the main conduction band (in Γ valley) is usually non-parabolic and the hole bands (heavy and light holes) are strongly anisotropic. In this case, appropriate equations of the overlap factor $\mathcal{G}(\mathbf{k},\mathbf{k}')$ must be taken into account and the energy dependence of the scattering probability is more complicated [MOG 93].

While previous scattering mechanisms may be considered as isotropic (at least, an isotropic approximation is possible), the interaction with polar optical phonons is strongly anisotropic. Due to the $1/q^2$ term in [8.48], the mechanism is strongest for small q, that is for $\mathbf{k}=\mathbf{k}'$, which means that it weakly deviates the trajectory of particles. For a spherical parabolic band the angular dependence of this scattering is given by:

$$P(\theta)d\theta = \frac{E\,E'}{E+E'-2\sqrt{E\,E'}\cos\theta}\sin\theta\,d\theta \qquad [8.50]$$

Regarding all electron–phonon scattering mechanisms, the appropriate phonon energies, coupling constants and other material parameters may be found in many articles and textbooks, such as [BLA 82] for GaAs, [JAC 83] for Si, [JAC 81] for Ge. See also discussions in [MOG 93]. It should be noted that for many years, deformation potentials have been considered as empirical parameters adjusted to fit with experimental velocity-field characteristics or other experimental transport data. However, the choice of this set of parameters is not unique and significant discrepancies occur between values proposed by different authors. As an example, the deformation potential associated with the intervalley phonon g3 of energy 63.2 meV in silicon varies from 1.75×10^8 eV/cm [FIS 88] to 11.0×10^8 eV/cm [JAC 83]. The modern approach now consists of deducing these parameters from first principles simulation [WAN 11]. See also Chapter 2 of this book for details.

8.3.1.2. *Impurity scattering*

The Coulomb interaction due to an ionized impurity is often described in the Brooks-Herring approach where the screening effect of other carriers is simply introduced through the screening length β^{-1} [RID 82]. In the simplest approximation at thermal equilibrium, the inverse Debye screening length $\beta = \left(e^2 n / \varepsilon \varepsilon_0 k_B T\right)^{1/2}$ may be considered for an electron gas of density n. The scattering potential is then:

$$V(r) = \frac{Z e^2}{4\pi \varepsilon \varepsilon_0} \exp(-\beta r) \qquad [8.51]$$

Where Z is the number of charge units of the impurity and ε the dielectric constant of the material. Inserting the Fourier transform of the screened potential in the Fermi golden rule, we obtain the scattering rate due to a single impurity:

$$s_{imp}(\mathbf{k}, \mathbf{k}') = \frac{1}{4\pi^3 V} \frac{Z^2 e^4}{\hbar \varepsilon \varepsilon_0} \frac{1}{\left(q^2 + \beta^2\right)^2} \mathcal{G}(\mathbf{k}, \mathbf{k}') \, \delta(E' - E) \qquad [8.52]$$

To improve the computing efficiency, this scattering is commonly approximated by an elastic–isotropic process that has the same relaxation time. It is derived by inserting the factor $1 - \cos\theta$ in the integration over all final states, where θ is the scattering angle. For an ionized impurity density n_I and a parabolic band with an overlap factor $\mathcal{G} = 1$ the electron–impurity relaxation time is finally in the Brooks-Herring theory [RID 82]:

$$\frac{1}{\tau_{imp}(E)} = \frac{Z^2}{16\sqrt{2}\,\pi} \frac{n_I e^4}{\sqrt{m}\,(\varepsilon \varepsilon_0)^2} \frac{1}{E^{3/2}} \left[\log\left(1 + \frac{8 m E}{\hbar^2 \beta^2}\right) - \frac{1}{1 + \hbar^2 \beta^2 / 8 m E} \right] \qquad [8.53]$$

In this simple model, it is assumed that the interacting electron does not perturb the Coulomb field of the impurity by its presence. To improve the description of screening effects, more sophisticated models have been developed, for example [TAK 59, HAL 62, CHA 81, RID 82].

8.3.1.3. *Alloy scattering*

In an alloyed compound such as $Si_{1-x}Ge_x$ or $Ga_{1-x}Al_xAs$, with a fraction $1 - x$ of atoms of type A and x of atoms of type B, the random distribution of A and B atoms gives rise to an additional scattering process that is usually considered in the formulation developed by Harrison and Hauser [HAR 76]. It is based on the simple assumption that when an electron moves from a site occupied by an atom of type A to a site occupied by an atom of type B, its potential energy changes by a quantity E_{all} defined by:

$$E_{all} = \begin{cases} U_{all} & \text{for } r \leq r_0 \\ 0 & \text{for } r > r_0 \end{cases} \qquad [8.54]$$

where r_0 is the space extension of the alloy potential U_{all}. It is arbitrarily defined as the distance between nearest-neighbor atoms, that is $r_0 = \sqrt{3}/4\, a_0$, where a_0 is the lattice parameter. Considering the number N_{site} of possible sites for atoms of type B, the alloy scattering rate is given by:

$$\begin{aligned} s_{all}(\mathbf{k},\mathbf{k}') = &\frac{2\pi}{\hbar}\frac{V}{8\pi^3 V}\frac{16\pi^2}{q^6} x(1-x)\left(-qr_0 \cos qr_0 + \sin qr_0\right)^2 \\ &\times U_{all}^2\, \mathcal{G}(\mathbf{k},\mathbf{k}')\, \delta(E'-E) \end{aligned} \qquad [8.55]$$

In MC simulation, this scattering mechanism is usually treated through its isotropic relaxation rate. In the case of a binary alloy such as SiGe, the number of sites in the crystal of volume V is $N_{site} = 8V/a_0^3$. For a spherical parabolic band the relaxation rate is:

$$\frac{1}{\tau_{all}(\varepsilon)} = \frac{3\pi\sqrt{2}}{16}\frac{m^{3/2}\, a_0^3}{\hbar^4} x(1-x)\, U_{all}^2 \sqrt{\varepsilon} \qquad [8.56]$$

In the case of SiGe alloys, the alloy potentials $U_{all} = 0.8$ eV (for electrons) and $U_{all} = 0.6$ eV (for holes) are commonly used, as determined in [VEN 93] and [LI 93].

8.3.1.4. *Surface roughness scattering*

In devices such as MOSFETs where the carrier transport takes place near an oxide/semiconductor interface, the roughness of the interface is an important additional source of scattering that degrades the effective mobility, especially in a strong inversion regime. To describe this effect within MC simulation, it is possible to derive scattering rates on the basis of fluctuations of potential generated by the fluctuations of the interface position characterized by a root-mean-square Δ and a correlation length Λ that can be determined experimentally [FIS 93, ESS 11]. Standard values $\Delta = 0.5$ nm and $\Lambda = 1.5$ nm of these technology-dependent parameters are commonly used for the Si/SiO$_2$ interface. However, in this approach the scattering rate is dependent on the parallel component of the particle energy that is not easy to decouple from the total kinetic energy in the general case and to include in the MC algorithm of scattering processes for a 3D electron gas. It is very well suited to the case of a 2D gas as in multi-subband MC simulation that will be described later. For a 3D gas, it is usually preferred to describe this elastic process using an empirical combination of specular and diffusive scattering for carriers that hit the interface [SAN 92]. Usually, the fraction of diffusive scattering N_{diff} is considered as a constant value that may vary significantly according to the authors: for example, N_{diff} can be equal to 6% [SAN 92], 8.5% [FIE 92], 15% [BUF 03] or 50% [FIS 91]. Alternatively, N_{diff} can be chosen to vary with the effective vertical field E_{eff} and described as a unique function of E_{eff} to accurately reproduce the universal mobility laws for electrons in unstrained Si [AUB 05] or strained Si [AUB 06].

8.3.2. *Phonon and impurity scattering in the Wigner formalism*

Can the classical Boltzmann scattering and the quantum Wigner potential operators be considered within the same equation? The answer is not trivial: derivations from first principles and analysis of the assumptions and approximations have been provided only recently for interactions with ionized impurities [QUE 09] and with phonons [NED 05], by two different approaches. The full derivations and discussions of the assumptions leading to the classical limit of scattering can be found in [NED 11].

Consider for instance the short-range Coulomb potential created by an ionized impurity $e^2 \exp\left(-\beta\left|\mathbf{r}-\mathbf{r}_i\right|\right)/4\pi\varepsilon\left|\mathbf{r}-\mathbf{r}_i\right|$, where ε is the semiconductor permittivity and β is the screening factor in the static screening approximation. The demonstration starts with the derivation of the Wigner potential associated with this Coulomb potential, from which a quantum evolution term is derived. After some tedious but straightforward calculations, considering a large number of dopants and within the fast collision approximation, the electron–impurity collision term finally takes exactly the same form as commonly derived for the Boltzmann collision operator with continuous doping density [NED 11].

The semiclassical phonon collision rate is derived from the equation for the generalized Wigner function [ROS 94, BOR 99]. In addition to the electron coordinates, the function depends on the occupation number of the phonon states in the system. Of interest is the electron, or reduced, Wigner function obtained from the generalized Wigner function by a trace over the phonon coordinates. A closed equation for the reduced Wigner function can be derived after a hierarchy of approximations, which includes the weak scattering limit and assumes that the phonon system is in equilibrium. These approximations concern only the interaction with the phonons, while the potential operator [8.17] remains exact. In the resulting equation, the phonon interaction, being nonlocal in both space and time, is still quantum. The Wigner–Boltzmann equation is obtained after a classical limit in the phonon term, leading to the instantaneous and local Boltzmann collision operator that neglects some quantum features of scattering processes as CB and CR or ICFE, which we discuss now.

Effects of time-dependent CB and CR of phonon replicas have been investigated theoretically and experimentally in homogeneous semiconductors [SCH 94, FUE 97, BOR 96, KUH 92, ROS 02]. These effects are related to the lack of energy conservation and the memory character of the electron–phonon dynamics, and are a consequence of the finite duration of the interaction process. The effect of the electric field during the process of collision (ICFE) has attracted scientific attention for quite some time [THO 78, BAR 80, FIS 84]. Numerical studies have investigated the CB, CR and ICFE effects in the case of ultrafast and/or high-field transport in semiconductors and insulators [BOR 99, JAC 01, BOR 03, LIP 91, GUR 02, NED 07] and in the case of photo-excited semiconductors [KUH 92, ROS 02]. It has been shown that the classical energy-conserving delta-function establishes at long times. Semiclassically forbidden states are occupied only at early evolution times [ROS 02, NED 05]. The first experimental evidence of memory effects and energy nonconserving transitions in the relaxation of hot carrier distributions was reported in 1997 [FUE 97]. At longer times, typically above a few hundred femtoseconds for GaAs, the semiclassical (Boltzmann) limit dominates in the carrier evolution. Finally, it appears that the ICFE is not important in stationary high-field transport in semiconductors [LIP 91]. This effect must be sought in the early-time evolution, which precedes the formation of the classical energy conserving δ-function [GUR 02, HEB 03]. We note that the above considerations hold in the weak collision limit, where the next interaction begins well after the completion of the current interaction.

8.3.3. *Wigner–Boltzmann equation*

From the previous sections, we can conclude that, under some conditions, the collision term widely used in the BTE can also be used in the WTE. It is a strong result and one of the main advantages of the Wigner function with regard to device modeling. All the knowledge acquired in the past in the treatment of scattering in semiclassical transport can still be reused for quantum transport in the Wigner formalism. It makes it possible to study new problems such as scattering-induced decoherence and the transition from quantum to semiclassical transport regimes. The WTE including the Boltzmann collision term is usually called the Wigner-Boltzmann transport equation (WBTE) [NED 04]. It gives, finally:

$$\frac{\partial}{\partial t} f_w(\mathbf{r},\mathbf{k},t) + \frac{\hbar}{m}\mathbf{k}\cdot\nabla_{\mathbf{r}} f_w(\mathbf{r},\mathbf{k},t) - \frac{1}{\hbar}\nabla V_{slow}\cdot\nabla_{\mathbf{k}} f_w = Q_{V_{rapid}} f_w(\mathbf{r},\mathbf{k},t) + C f_w(\mathbf{r},\mathbf{k},t) \quad [8.57]$$

If, as in most cases, the potential V is not separated in rapidly and slowly varying parts, [8.57] reduces to:

$$\frac{\partial}{\partial t} f_w(\mathbf{r},\mathbf{k},t) + \frac{\hbar}{m}\mathbf{k}\cdot\nabla_{\mathbf{r}} f_w(\mathbf{r},\mathbf{k},t) = Q f_w(\mathbf{r},\mathbf{k},t) + C f_w(\mathbf{r},\mathbf{k},t) \quad [8.58]$$

where the Boltzmann collision term derives from the transition probabilities per unit of time $s_i(\mathbf{k},\mathbf{k}')$ of each scattering process calculated in the first-order perturbation theory of the Fermi golden rule, that is:

$$C f_w(\mathbf{r},\mathbf{k},t) = \sum_i \int d\mathbf{k}'\left[s_i(\mathbf{k},\mathbf{k}') f_w(\mathbf{r},\mathbf{k}',t) - s_i(\mathbf{k}',\mathbf{k}) f_w(\mathbf{r},\mathbf{k},t)\right] \quad [8.59]$$

The quantum evolution term $Q f_w(\mathbf{r},\mathbf{k},t)$ due to the potential V is given by [8.17], including the Wigner potential $V_w(\mathbf{r},\mathbf{k})$ defined by [8.18].

8.4. The MC method for solving the BTE

8.4.1. *Main features of the MC algorithm*

To develop a statistical particle (MC) approach to solving the BTE, the Boltzmann function is described as a sum of Dirac excitations, that is:

$$f_b(\mathbf{r},\mathbf{k},t) = \sum_j f_j(\mathbf{r},\mathbf{k},t) = \sum_j \delta(\mathbf{r}-\mathbf{r}_j(t))\,\delta(\mathbf{k}-\mathbf{k}_j(t)) \quad [8.60]$$

These excitations $f_j(\mathbf{r},\mathbf{k},t)$ are thus localized in the phase-space at positions $\mathbf{r}_j$ and $\mathbf{k}_j$. They are equivalent to semiclassical particles. Substituting the excitation $f_j(\mathbf{r},\mathbf{k},t)$ into the BTE [8.24] leads to the semiclassical motion equations for the *j*th particle of effective mass *m* in a potential $V=-eU$

$$\begin{cases} \dfrac{d}{dt}\mathbf{r}_j = \dfrac{1}{\hbar}\nabla_{\mathbf{k}}E(\mathbf{k}) = \dfrac{\hbar\mathbf{k}_j}{m} \\ \dfrac{d}{dt}\mathbf{k}_j = \dfrac{e}{\hbar}\nabla_{\mathbf{r}}U(\mathbf{r}) = -\dfrac{1}{\hbar}\nabla_{\mathbf{r}}V(\mathbf{r}) \end{cases} \qquad [8.61]$$

The MC technique consists of statistically describing the motion of each particle as a succession of free flights interrupted by instantaneous scattering events on the basis of appropriate probability laws. The probability $\Gamma_i(\mathbf{k})$ per unit of time for a particle to scatter from initial state $\mathbf{k}$ to any state $\mathbf{k}'$ by the *i*-th scattering process is given by:

$$\Gamma_i(\mathbf{k}) = \int s_i(\mathbf{k},\mathbf{k}')d\mathbf{k}' \qquad [8.62]$$

The total scattering rate is thus:

$$\Gamma_{tot}(\mathbf{k}) = \sum_i \Gamma_i(\mathbf{k}) \qquad [8.63]$$

The duration of free flights is directly related to the total scattering rate $\Gamma_{tot}(\mathbf{k})$. The type of scattering experienced by the particle after a free flight is related to the weights $\Gamma_i(\mathbf{k})$ of the scattering mechanisms allowed at state **k** and the effect of a scattering mechanism *i* is linked to the angular scattering probability contained in $\Gamma_i(\mathbf{k})$. The MC method is based on the random selection of these three steps of each elemental fraction of motion, from the scattering rates calculated prior to the simulation and usually stored in a look-up table. The principles of selection processes are briefly described below. More details may be found in textbooks fully devoted to the MC method [JAC 89, MOG 93, JUN 03].

8.4.1.1. *Selection of free-flight duration*

First, assume the total scattering rate to be independent of the energy (or the wave vector) of the particle. Hence,

$$\Gamma_{tot}(E) = \sum_i \Gamma_i(E) = \text{Const} = \Gamma_0 \qquad [8.64]$$

Given $n_{ball}(t)$ the number of electrons at time t have experienced a ballistic (collision-free) flight from time $t = 0$. Since the total scattering rate is constant and the same for all particles, the time evolution of $n_{ball}(t)$ is simply given by:

$$\frac{d\,n_{ball}}{dt} = -\Gamma_0\, n_{ball} \quad [8.65]$$

which trivially leads to an exponential function of time. In other words, the probability of a particle having a ballistic free flight of duration t is

$$\frac{n_{ball}(t)}{n_{ball}(0)} = \exp(-\Gamma_0 t) \quad [8.66]$$

Additionally, the probability of a particle experiencing a scattering event during the small time step dt is just $\Gamma_0\, dt$. The probability of a particle having a free flight between 0 and t, and experiencing a scattering event between t and $t+dt$ is thus the product of the two previous probabilities, assumed to be independent, that is:

$$P(t)dt = \exp(-\Gamma_0 t) \times \Gamma_0\, dt \quad [8.67]$$

$P(t)$ is thus the probability per unit of time of having a free flight between t and $t+dt$. The probability of having a free flight of duration smaller than t_f is thus

$$R_f = R(t_f) = \int_0^{t_f} P(t)dt \quad [8.68]$$

It is easy to check that this probability is normalized to unity, that is $R(\infty) = 1$. We can associate to the probability on time $P(t)$ a probability $P_R(R)$ on real numbers R lying between 0 and 1, such as:

$$P_R(R)dR = P(t)dt\text{, with } \int_0^1 P_R(R)dR = 1 \quad [8.69]$$

Of course, practically, a flat probability $P_R(R)=1$ is chosen to make the generation of random numbers easier. Finally, the random number R_f is associated with the time of free flight t_f according to $\int_0^{R_f} dR = \int_0^{t_f} P(t)\,dt$, that is:

$$t_f = -\frac{\ln(R_f)}{\Gamma_0} \quad \text{with} \quad R_f \in [0,1] \qquad [8.70]$$

This simple procedure is based on the assumption that the total scattering rate is independent of the state of the particle and thus on time. It is of course far from the truth. The rigorous determination of the free-flight duration is actually more numerically complicated and time consuming [MOG 93]. To circumvent this difficulty, a simple way consists of introducing an additional fictitious scattering mechanism, called "self-scattering". The scattering rate $\Gamma_{sc}(E)$ of this mechanism is adjusted in such a way that the new total scattering rate is actually constant and time independent, that is:

$$\Gamma_{sc}(E)+\Gamma_{tot}(E)=\Gamma_{sc}(E)+\sum_i \Gamma_i(E)=\text{Const}=\Gamma_0 \qquad [8.71]$$

Then the equation [8.70] of the free-flight duration is correct. The self-scattering must of course be included in the MC procedure as with any other scattering mechanism. If a self-scattering is selected after a free flight, it has no effect on the particle state, which does not modify the statistics of real events experienced by the electron gas. This fictitious event just slows down the simulation since it generates additional numerical operations.

8.4.1.2. *Selection of scattering events*

The choice of the type of scattering experienced after a free flight may be done by using a flat random number R_s and the cumulative scattering probability $P_i(E)$ defined as:

$$P_i(E)=\sum_{j=1}^{i}\frac{\Gamma_j(E)}{\Gamma_0} \qquad [8.72]$$

The scattering process i is selected if the random number R_s satisfies:

$$P_{i-1}(E)<R_s\le P_i(E) \qquad [8.73]$$

8.4.1.3. *Selection of scattering effects*

The scattering-induced change in energy of the particle is usually fully determined by the type of scattering selected. In the case of scattering with a phonon of energy, $\hbar\omega_q$, the electron energy after scattering is $E' = E \pm \hbar\omega_q$ (with + for an absorption, – for an emission). In most other cases the interaction is elastic, that is $E' = E$. To calculate the scattering angles that determine the initial direction of the next free flight, it is usually preferred to substitute in [8.62] the integral over the energy *E*, the polar angle θ and the azimuth φ for the integral over the wave vector, that is:

$$\begin{aligned} \Gamma_i(E) &= \iiint s_i(E,E',\theta,\varphi) J(E,\theta,\varphi) dE' \sin\theta \; d\theta \; d\varphi \\ &= A(E) \int_0^{\pi} d\theta \sin\theta \int_0^{2\pi} d\varphi \, B_E(\theta,\varphi) \end{aligned} \quad [8.74]$$

where $J(E,\theta,\varphi)$ is the Jacobian of the change in variables, $A(E)$ is the angle-dependent part of the scattering rate and $B_E(\theta,\varphi)$ is the angle-dependent part of the integrand of the first integral in [8.74]. For each type of scattering, we can define the angular function:

$$I(\theta_s,\varphi_s) = \int_0^{\theta_s} d\theta \sin\theta \int_0^{\varphi_s} d\varphi \, B_E(\theta,\varphi) \quad [8.75]$$

From flat random numbers R_θ and R_φ the angles θ_s and φ_s are selected as the angles that satisfy:

$$\frac{I(\theta_s,2\pi)}{I(\pi,2\pi)} = R_\theta \quad \text{and} \quad \frac{I(\theta_s,\varphi_s)}{I(\theta_s,2\pi)} = R_\varphi \quad [8.76]$$

In the case of isotropic effective mass (spherical energy–wave vector relationship, as in the Γ valley of GaAs) and isotropic scattering process, we have $B_E(\theta,\varphi) = 1$ and the integrals over θ_s and φ_s are independent. The equations [8.76] simplify to:

$$\cos\theta_s = 1 - 2R_\theta \quad \text{and} \quad \varphi_s = 2\pi R_\varphi \quad [8.77]$$

In the case of the conduction band of silicon with ellipsoidal bands of longitudinal and transverse effective masses m_l and m_t, it is convenient to make the Herring–Vogt transform of the **k**-space that reduces the ellipsoidal equi-energy surfaces to spheres and is defined by $k_i^* = T_{ij}\, k_j$, where T is a diagonal matrix of elements $T_{ii} = \left(m_{l,t}\right)^{-1/2}$. In the starred space, the relationships [8.77] apply.

8.4.1.4. *Device simulation: Ensemble MC method*

To simulate a device connected to external contacts, it is necessary to couple the MC algorithm of particle motion to the Poisson's equation. To describe the time behavior of a device, the ensemble MC (EMC) technique must be implemented. This consists of simultaneously simulating a large number of particles over very small time steps (typically 0.1–10 fs depending on the timescales of relevant phenomena, such as the dielectric relaxation time and plasma frequency). The Poisson equation is solved at the end of each time step with updated carrier density distribution and possibly new electrostatic potential at contacts. To extract stationary values of physical quantities, the corresponding instantaneous quantities are time-averaged once the transient regime has elapsed.

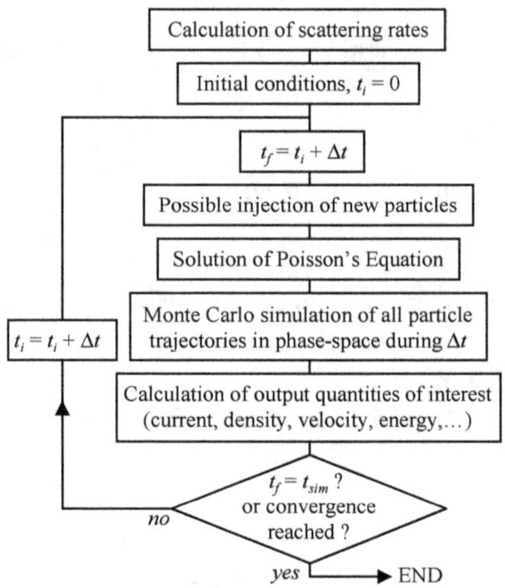

Figure 8.1. *Flow chart of the ensemble MC simulation of bulk devices; the iterative loop ends once the total simulation time has reached t_{sim}*

The EMC method is the most widely used for device simulation. The flow chart of the EMC method for bulk device simulation is presented in Figure 8.1. The problem of particle injection at contacts will be considered in detail in section 8.5.2.2, which is devoted to the MC solution of the WBTE. However, in the semiclassical case of the BTE, for conventional devices where particles enter and leave the device through Ohmic contact, the only condition of carrier injection is the electrical neutrality of cells adjacent to the contact assumed to always be under thermal equilibrium. At each timestep, the appropriate number of particles is injected in these cells to fill this condition of neutrality.

8.4.2. *Interaction with single impurities*

Statistical fluctuations of the channel dopant number are known to be a serious physical limitation of MOSFET downscaling. Entering into the nanometer regime indeed results in a decreasing number of channel impurities whose random distribution leads to significant fluctuations of threshold voltage and off-state leakage current. To include these effects in device simulation, the doping atoms can no longer be considered as a continuum of charges, but must be treated as discrete charges.

When treating simultaneously ionized impurities and carriers as point-charge particles, the electron–impurity interaction is to be considered in the carrier motion calculation directly. Simulation methods have initially been developed by Hockney and Eastwood [HOC 81]. The most important are the particle–particle (PP) and the particle–mesh (PM) models. The former consists of calculating the Coulomb force $\mathbf{F}_i$ applied to particle i of charge q_i by summing the contributions of all other particles of charge q_j at position $\mathbf{r}_j$ as $\mathbf{F}_i = q_i \sum_{j \neq i} q_j \left(\mathbf{r}_i - \mathbf{r}_j\right) / \left|\mathbf{r}_i - \mathbf{r}_j\right|^3$. This approach is accurate but computationally demanding and not suitable for device simulation. The PM method is less time consuming but gives a Coulomb less accurate force estimation [HOC 81, GRO 99, WOR 00]. The local electric field acting on a particle is considered to be the gradient of the potential distribution V as it results from the solution of Poisson's equation in a predefined meshing. The force $\mathbf{F}_i$ is then given by $\mathbf{F}_i = -q_i \nabla V(\mathbf{r})$. In principle, this mesh force should fully include the electron–impurity interaction through a long-range approach. However, once the grid has been defined, the particle charges have to be assigned to the grid nodes before solving Poisson's equation and we can no longer consider particles as point particles [BAR 02]. At very short distances from the impurity, the mesh force experienced by an electron approaches zero due to the mesh-induced charge smoothing, while the real Coulomb force tends to infinity. The validity of the PM approach is thus limited by inadequacies between the concepts of point charge and spatial discretization. A correction to the mesh force used in the PM method is thus necessary in the vicinity of impurities, that is in the cells where the mesh force calculation is dubious.

A possible method consists of defining a short-range zone (SRZ) around each impurity [BAR 02]. Inside an SRZ, the field applied to electrons is deduced from the potential at the zone boundaries resulting from Poisson's equation solution. The mesh force thus vanishes and the electron–impurity interaction is described by a short-range scattering mechanism included in the MC scattering procedure, similar to the standard method used for continuous doping. Outside the SRZ, the long-range mesh force is still applied without impurity scattering, as in the pure PM method. Provided that the space extension of SRZs is correctly defined within an appropriate meshing to take care of screening effects [BAR 02], this method gives good results, for example in terms of mobility of both minority and majority carriers, and has been successfully used to investigate the effect of random dopant distribution in nanoscale MOSFETs [DOL 04].

8.4.3. *Quantization effects: the multi-subband approach*

In many current nanodevices (such as nano-MOSFETs, HEMTs or structures based on quantum point contact), carriers are strongly confined in one direction. In such devices, two different kinds of quantum effects are observed: (i) a quantization of energy levels in the confinement (gate-to-substrate) direction that makes the electron gas two-dimensional and (ii) possible coherent transport effects in the source-to-drain direction. It appears that these effects may often be decoupled. In many cases the simulation of these structures can be considered as a quasi-2D problem where the confinement direction and the transport direction are treated separately. This approach is called "*mode–space*" approximation. Within this approximation the transport may be described either semiclassically or in a quantum way. Here, we consider the case where the Boltzmann formalism can still be used in the 2D plane to describe transport for the general description of the principle and technique of the mode–space approach, which will be straightforwardly extended to the Wigner formalism in section 8.5.

To illustrate the *mode–space* approximation, we consider a typical nanostructure with a confined electron gas: the double gate MOSFET. This type of transistor is schematized in Figure 8.2, where x is the source-to-drain axis and z is the gate-to-gate confinement axis. The width along y axis is assumed to be infinite. The mass tensor is assumed to be diagonal and characterized by the effective masses m_x, m_y and m_z. In the mode–space approximation, the 2D electron wave function is written in the form:

$$\Psi(x,z)=\sum_{n}\varphi_n(x)\xi_n(x,z) \qquad [8.78]$$

where $\xi_n(x,z)$ is the wave function associated with the nth eigen-state, or mode, of energy $E_n(z)$ in a slice of position x. These quantities are thus a solution of the Schrödinger equation:

$$-\frac{\hbar^2}{2m_z}\frac{\partial^2}{\partial z^2}\xi_n(x,z) - eV(x,z)\xi_n(x,z) = E_n(x)\xi_n(x,z) \quad [8.79]$$

The wave functions $\xi_n(x,z)$ form an orthogonal basis. In this formalism, the quantity $|\varphi_n(x)|^2$ represents the probability of an electron being on the n-th mode in the slice of position x. Thanks to this choice of wave functions, under some conditions it is possible to decouple the 2D stationary Schrödinger equation:

$$-\frac{\hbar^2}{2m_x}\frac{\partial^2\Psi(x,z)}{\partial x^2} - \frac{\hbar^2}{2m_z}\frac{\partial^2\Psi(x,z)}{\partial z^2} - eV(x,z)\Psi(x,z) = E\,\Psi(x,z) \quad [8.80]$$

into two 1D equations along z axis (equation [8.79]) and along x axis, respectively. In the latter case, the 1D equation to be solved is:

$$-\frac{\hbar^2}{2m_x}\frac{d^2}{dx^2}\varphi_n(x) - E_n(x)\varphi_n(x) = E\,\varphi_n(x) \quad [8.81]$$

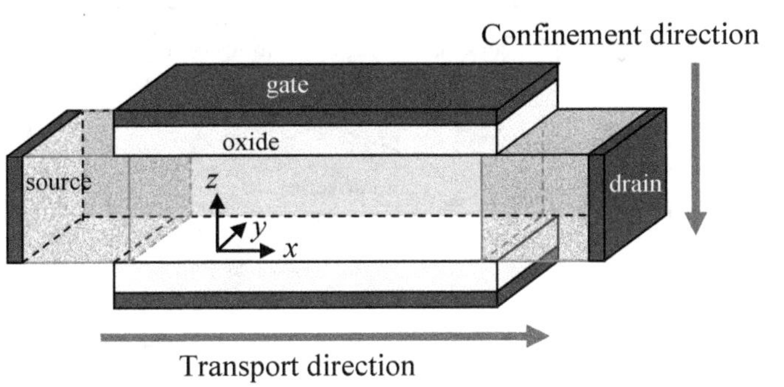

Figure 8.2. *Typical planar DG MOSFET. For a color version of this figure, see www.iste.co.uk/dollfus/nanodevices.zip*

The transport equation [8.81] along the x-axis and the confinement equation for the calculation of modes along the z-axis are fully decoupled. The discussion on the validity of this approximation is beyond the scope of this chapter. The reader may find detailed information on this point in more specialized textbooks, such as [QUE 10a]. In brief, this approximation is justified if the wave functions associated with modes do not evolve too rapidly along the device, that is $\partial\xi_m(x_0,z)/\partial x << \xi_n(x_0,z)/L_x$ and $\partial^2\xi_m(x_0,z)/\partial x^2 << \xi_n(x_0,z)/L_x^2$ (L_x being a relevant length in the source-to-drain direction). The mode–space approximation has been validated for ultrathin double-gate (DG) structures ($T_{Si} < 2$ nm) in the ballistic limit by comparing full 2D and mode–space quantum NEGF calculations [VEN 02]. It has also been validated for a DG–MOSFET of body thickness $T_{Si} < 2$ nm and of gate length $L_G = 10$ nm from MC simulation [QUE 10a].

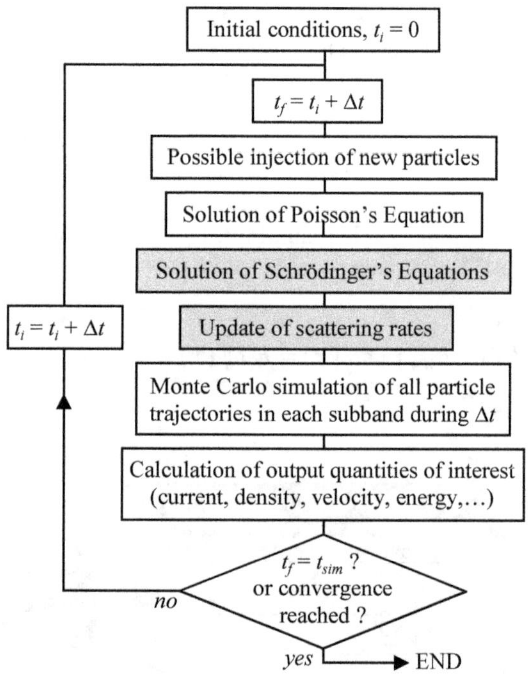

Figure 8.3. *Flow chart of the multi-subband MC device simulation; the shaded boxes are the main changes with respect to the flowchart of bulk simulation (Figure 8.1). For a color version of this figure, see www.iste.co.uk/dollfus/nanodevices.zip*

The flowchart of multi-subband simulation is significantly modified when compared to that of bulk simulation. The main change is the introduction in the time-loop of the solution of Schrödinger's equation in each vertical slice normal to the transport direction, as shown in Figure 8.3. Another important change that further weighs down the computation is a consequence of the dependence of scattering rates on eigen energies $E_n(x)$ and wave functions $\xi_n(x_0,z)$. They are thus time and space dependent and must be updated at each time step. This update is now in the loop too. Finally, scattering rates adapted to the mode–space approach need to be used. More detailed information on these points may be found in [SAI 06, LUC 07, QUE 10a, ESS 11].

8.4.4. *Quantization effects: the effective potential approach*

To reduce the computational cost of multi-subband simulation, it is possible to capture a part of quantization effects using “bulk” simulation in an unconfined 3D **k**-space. Indeed, in a MOSFET, quantum corrections may be applied to mimic the shift of the peak density from the oxide interface, typically of 1 nm. This shift increases the effective oxide thickness and reduces the transconductance, which contributes to enhance the threshold voltage [DOR 91].

The basic idea of quantum corrections consists of correcting the description and the dynamics of the electron wave packet by keeping the exact formulation [8.24] of its energy [FER 00]. If the wave packet is assumed to be Gaussian, a *Gaussian effective potential* (GEP) can be used as potential energy instead of the electrostatic potential. This effective potential writes:

$$V_{eff}^{GEP}(x) = \frac{1}{\sqrt{\pi}\,\sigma}\int dx'V(x')\exp\left(-\frac{(x-x')^2}{\sigma^2}\right) \qquad [8.82]$$

However, in addition to the important question on the natural size σ of a wave packet [FER 02, QUE 10a], the validity of Gaussian effective models is limited. In particular, it has been shown that the GEP cannot correctly reproduce both the total inversion charge and the carrier density profile in the MOSFET channel because of overestimation of the repulsive effect at oxide interface [JAU 06]. It has also been shown that the GEP is not able to satisfactorily describe the tunneling effect [FER 02]. Actually, the limitation of this approach is related to two important approximations: the extension σ is considered as a universal parameter for all electrons, and all wave packets are assumed to be Gaussian. Some improvements to this approach make it possible to obtain better results. To describe the effect of interfaces on the wave packet properly, a nonsymmetric Pearson IV distribution function can be used in place of the Gaussian function [JAU 07, JAU 08]. The resulting Pearson effective potential (PEP) correction provides a density profile in MOS structures in much better agreement with Schrödinger–Poisson simulation than the conventional GEP correction. Other correction techniques consist of considering either a quantum force formulation based on the Wigner formalism [TSU 01], or a direct solution of Schrödinger's equation [WIN 03].

8.4.5. *Full-band MC simulation*

All models and algorithms presented above are based on the effective mass approximation of the band structure and can be extended straightforwardly to non-parabolic bands with analytical dispersion relation. This approximation works very well in many cases of electron transport at reasonably high energy and in many materials such as Si, SiGe, GaAs and most usual III-V semiconductors. However, this approximation fails to properly describe the transport of holes due to the strong non-parabolicity and anisotropy of hole bands. It fails also to describe the hot electron transport effects since at high energy the conduction bands are far from being parabolic. To overcome this limitation, it is necessary to consider a full-band description of energy bands. In this approach, the energy bands are calculated numerically using appropriate methods based for instance on the $\mathbf{k} \cdot \mathbf{p}$ theory [CAR 66, RID 06] or the empirical pseudo-potentials [CHE 76, RIE 93]. Instead of using analytical relations of energy dispersion, the information on the band structure is discretized in the **k**-space and stored in a look-up table, which significantly increases the computational requirements. The discretization of the reciprocal space also requires appropriate algorithms of interpolation to obtain a smooth description of particle trajectories, which contributes to the additional computational cost.

The description of this approach is beyond the scope of this book. Advanced readers may refer to some pioneering works and reviews, for instance [TAN 83, FIS 88, HES 91]. Numerical details on the implementation may be found in [JUN 03].

8.5. Extension of the MC method for solving the WBTE

Since it can take negative values, it may look nonsensical to represent the Wigner function with particles that cannot have a "negative presence". Classically, electrons either do or do not exist. To solve this apparent inconsistency with a view to developing a statistical particle approach to the solution of the WBTE, it is necessary to give simulated particles the strange property of carrying negative contributions. With this in mind, it has been suggested that the Wigner function be described as a sum of Dirac excitations of the form:

$$f_j(\mathbf{r},\mathbf{k},t) = \delta(\mathbf{r} - \mathbf{r}_j(t))\,\delta(\mathbf{k} - \mathbf{k}_j(t))\,A_i(t) \qquad [8.83]$$

which are still localized in the phase-space but weighted by an amplitude A_i, called affinity in [SHI 02]. It was not necessary to consider this amplitude in the solution of the BTE since the Boltzmann function is always positive, but it is required to reconstruct the Wigner function that can take negative values. Accordingly, this amplitude may also be negative. The particle amplitudes will contain all information on the quantum state of the electron system. Depending on the technique used, the affinity can take integer or real values, as explained below.

Consistent with the Heisenberg inequalities, such excitations, that we will call pseudo-particles, cannot represent physical particles. They should just be considered as mathematical tools for the solution of the WBTE. The advantage of this formulation is that the evolution of these excitations follows simple equations. Three techniques of particle simulation of the WBTE have been developed. We now briefly review them.

8.5.1. *Overview of the different methods*

The first technique, developed by Jacoboni's group at University of Modena, makes it possible to go beyond the Wigner–Boltzmann equation since it can solve the Wigner equation exactly without approximation on the electron/phonon coupling [PAS 98, BER 99a, BER 99b, BER 03, BOR 99, JAC 01, JAC 04]. It is based on the integral form of the Wigner equation and on a Neumann expansion of the Wigner function. Then, a subtle evolution of particle affinities and wave vectors is performed by following paths induced by the Wigner potential and phonon scattering. It is a rich tool for fundamental physics investigation but difficult to implement for device simulation. An important result obtained using the Wigner paths technique is the fact that quantum collision effects play a reduced role inside devices, which tends to reinforce the validity of the Wigner–Boltzmann approach [BER 99b]. Indeed, the multiple collision processes tend to average the advanced effect of collisions, which may explain the successful use of the Fermi golden rule beyond its expected theoretical limit.

The second method is the "most MC" approach to solving the WBTE. It has been developed by Nedjalkov and collaborators of the Technical University of Vienna. Indeed, in this method, even the effect of the potential on the Wigner function is treated stochastically, in the same way as a scattering process. Indeed, the quantum evolution term of the WBTE may be rewritten in the form [NED 04]:

$$Qf_w(\mathbf{r},\mathbf{k}) = \int d\mathbf{k}'\left[\frac{1}{3}V_w^+(\mathbf{r},\mathbf{k}-\mathbf{k}') - \frac{1}{3}V_w^-(\mathbf{r},\mathbf{k}-\mathbf{k}') + \frac{1}{3}\gamma(\mathbf{r})\delta(\mathbf{k}-\mathbf{k}')\right] 3f_w(\mathbf{r},\mathbf{k}') - f_w(\mathbf{r},\mathbf{k})\gamma(\mathbf{r}) \qquad [8.84]$$

where the Wigner potential has been separated into positive and negative parts V_w^+ and V_w^-, respectively, that is $V_w(\mathbf{r},\mathbf{k}) = V_w^+(\mathbf{r},\mathbf{k}) - V_w^-(\mathbf{r},\mathbf{k})$, and an out-scattering term $\gamma(\mathbf{r},\mathbf{k})$ has been defined as $\gamma(\mathbf{r},\mathbf{k}) = \gamma(\mathbf{r}) = \int d\mathbf{k}' V_w^+(\mathbf{r},\mathbf{k}-\mathbf{k}')$. The term [8.84] can be interpreted as a collision term, similar to the actual collision term [8.59], with collisions that may occur with a probability per unit of time $\gamma(\mathbf{r})$. There are then three possible events of equal probability (1/3), that is

– the final wave vector is determined by the transition rate $V_w^+(\mathbf{r},\mathbf{k}-\mathbf{k}')$. The carrier affinity is multiplied by 3, due to the fact that the function $3f_w(\mathbf{r},\mathbf{k}')$ appears in the in-scattering term instead of only $f_w(\mathbf{r},\mathbf{k}')$;

– the final wave vector is determined by the transition rate $V_w^-(\mathbf{r},\mathbf{k}-\mathbf{k}')$ and the carrier affinity is multiplied by -3;

– the wave vector is unchanged. The carrier affinity is multiplied by 3.

Hence, we have here a particle MC interpretation of the Wigner equation very similar to that of the Boltzmann equation. The real-space coordinates of pseudo-particles evolve like those of classical particles, that is $d\mathbf{r}_j/dt = \hbar\mathbf{k}_j/m$. During a free flight the **k**-space coordinates are constant unless the potential has been separated into a slowly varying part and a rapidly varying part, the former being treated classically (see [8.29] and [8.30]). The effect of the collision term Cf_w [8.59] on the excitations is still interpreted as a scattering event, as in the BTE. Each scattering event instantaneously changes the wave vector of pseudo-particles without changing their affinity. The quantum evolution term Qf_w [8.84] also induces collisions as described above, which change both the wave vector and the affinity of pseudo-particles.

In this approach, called "signed method", the affinity of pseudo-particles can only take integer values, which makes the technique naturally instable since it yields exponential growth of the affinities [NED 04]. However, an algorithm has been proposed to reach convergence, including generation and recombination of particles [SVE 06]. While the Wigner MC technique has always been implemented for 1D transport problems, the possibility of extending the technique to 2D transport problems using supercomputers has been suggested [NED 08]. Recently, it has been extended also to treat many-body problems [SEL 14].

The third method, called "affinity method", is less demanding in terms of computational resources and probably easier to quickly make the convergence. In this approach, the interpretation of the quantum evolution term is different from that of the previous "signed" technique. Here, this term does not generate events determined by an MC technique, but it is considered to act continuously on the particle affinity at each time step. Only the collision term is assumed to generate random events, as in the Boltzmann simulation. This method, inspired by the works of Shifren, Ringhofer and Ferry [SHI 04], has been developed by the authors and successfully used to study several transport problems and devices [QUE 06a, QUE 06b, QUE 07, QUE 08, QUE 09, QUE 10a, QUE 11, QUE 13]. It will be described in more details in what follows.

8.5.2. *The "continuous affinity" method*

8.5.2.1. *Principles*

The method is thus very similar to the Boltzmann MC technique. The main change consists of adding the update of the Wigner function and of the particle affinities at each time step. In a mesh of the phase-space $M(x,k)$ the quantum evolution term $Q f_w$ induces the change in the affinity of particles in the mesh according to:

$$\sum_{i \in M(x,k)} \frac{d A_i}{dt} = Q f_w(x,k) \qquad [8.85]$$

which means that at each time step the affinity of all pseudo-particles in a mesh of the phase-space is updated as a function of $Q f_w$. The simple idea on which the quantum EMC simulation is based now appears clearly. During a time step, the real and reciprocal space coordinates of pseudo-particles evolve as in conventional EMC simulation [8.61]. In other words if the potential has been separated in slowly and rapidly varying parts [8.29] the coordinates of the *j*-th obey, in the effective mass approximation,

$$\begin{cases} \dfrac{d}{dt}\mathbf{r}_j = \dfrac{\hbar \mathbf{k}_j}{m} \\ \dfrac{d}{dt}\mathbf{k}_j = -\dfrac{1}{\hbar}\nabla_{\mathbf{r}} V_{slow}(\mathbf{r}) \end{cases} \qquad [8.86]$$

If the whole potential is included in the quantum evolution term, that is $V_{slow} = 0$, the wave vector of each particle is constant during a free flight and can take a new value only after scattering. At the end of a time step the Poisson equation is solved, the Wigner potential and the quantum evolution term $Q f_w$ are updated and the particle affinities are recalculated according to [8.85]. The main steps of the algorithm are summarized in the flowchart of Figure 8.4, to be compared to that of Figure 8.1 for bulk semiclassical simulation. The particle injection process and the boundary conditions will be discussed in the next section. Similar changes may be applied to the multi-subband algorithm of Figure 8.3 for the self-consistent simulation of quantum transport in 1D or 2D electron gas.

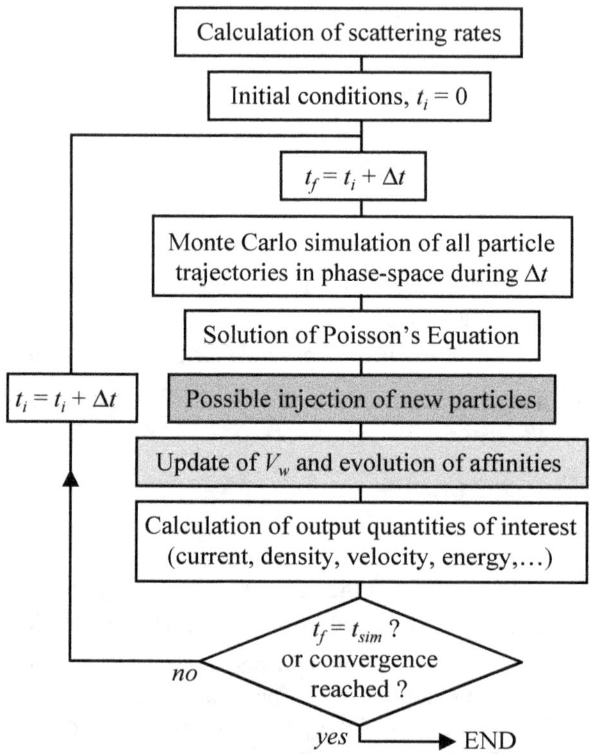

Figure 8.4. *Flow chart of the Wigner–Boltzmann MC device simulation; with respect to the flowchart of bulk simulation (Figure 8.1), shaded boxes are either new (Wigner potential of affinities) or modified (injection of particles, see next section). For a color version of this figure, see www.iste.co.uk/dollfus/nanodevices.zip*

We should keep in mind that by using this technique, the potential may have two effects. If one part of the potential is treated as a slowly varying potential V_{slow}, it induces an evolution of the pseudo-particle wave vectors as in semi-classical simulation. The rest of the potential V_{rapid} induces the evolution of affinities at each time step.

In the semi-classical limit, that is if the full potential is treated as a slowly varying quantity $V = V_{slow}$, the quantum evolution term $Q f_w$ is zero and the particle affinity is constant. The method turns out to be equivalent to the semi-classical EMC. It should be noted that the strong similarity and even compatibility of this technique with the conventional MC solution of the BTE is one of its greatest advantages.

8.5.2.2. *Pseudo-particle injection in device simulation*

In semi-classical (Boltzmann) device simulation with Ohmic contacts, the only condition of particle injection is the neutrality of real-space cells adjacent to the Ohmic contacts. After each time step, if particles are missing in some "Ohmic" cells with respect to the charge neutrality, the appropriate number of carriers (of affinity equal to 1) is injected in these cells to recover the neutrality. Assuming these "Ohmic" regions to be in thermal equilibrium, an equilibrium distribution is used to randomly select their wave vector components. In this way, consistency between the distribution of potential and the average number and distribution of particles in the device is reached. Obviously, this condition of particle injection should still be used in quantum simulation of Ohmic contacts if the transport in the contact region is assumed to be essentially semi-classical. Very good results are obtained by implementing this semi-classical condition in Wigner–Boltzmann device simulation. However, it is not enough to ensure the conservation of total affinity and charge within an algorithm in which the particle affinity evolves continuously.

Actually, it has been shown that it is not enough to inject particles at the contacts [QUE 06a]. Indeed, one of the most important difficulties in this MC method lies in the fact that even a particle with zero affinity may gain finite affinity through the quantum evolution term $Q f_w$ according to [8.86]. This means that, if there is no particle in a particular region of the phase-space where the affinity should evolve, a significant error may occur with possible non-conservation of charge, since the contribution of each particle to the total charge in the device is weighted by its affinity. This problem is very important for device simulation and should be fixed by implementing an appropriate algorithm to inject particles of convenient affinity.

This problem can be solved as follows. After each time step, the quantum evolution term $Q f_w(x,k)$ is calculated in the full phase-space. If in a mesh $M(x,k)$ of the phase-space, even inside the device, the quantity $\left|Q_{V_{rapid}} f_w(x,k)\right|$ is finite or greater than a threshold value, a pseudo-particle of zero affinity is injected in the mesh. This technique is proved to be very effective.

In summary, in the case of Wigner simulation it is necessary to combine the "semi-classical injection" of particles of affinity equal to 1 at Ohmic contacts to guarantee the electrical neutrality near the contacts and the "quantum injection" of pseudo-particles of 0 affinity in all regions of the phase-space where particles are missing and $Q\,f_w$ takes significant values [QUE 06a, 10a].

8.5.2.3. *Calculation of the Wigner potential*

Another fundamental problem lies in the choice of the limits of integration for the calculation of the Wigner potential [8.18]. There are two approaches, depending on whether the contacts are assumed to be coherent or non-coherent. In the former case the integration is cut at a maximum size from the contact corresponding to the "coherence length" beyond which no quantum effect may occur [NED 04], which raises the question of the relevant choice of the coherence length in the contact. In the latter case, the integration should be limited to positions x' such that both $x+x'/2$ and $x-x'/2$ belong to the device [BER 03]. This approach is used in the model the authors have developed. However, we have checked that all studied devices such as MOSFETs and RTDs, all limits of integration larger than that corresponding to the hypothesis of decoherent contacts yield the same results. This insensitivity is certainly due to the fact that in these cases contact regions, or access regions, have a semiclassical behavior dominated by scattering [QUE 10a]. Interesting recent discussions on this important question may be found for instance in [ROS 13, JIA 14].

8.5.2.4. *Affinity evolution*

To describe the time evolution of pseudo-particle affinities, a very stable discretization scheme is required. Indeed, we observed that due to the noise inherent to the technique, the MC simulation acts as a stiff problem that tends to make the solution of [8.85] unstable. In our model, an implicit backward Euler scheme was finally implemented, that is:

$$A_i(t+dt)-A_i(t)=\frac{1}{N}dt\times Q f_w(x,k,t+dt) \qquad [8.87]$$

where N is here the number of pseudo-particles in the mesh $M(x,k)$ of the phase-space. This backward Euler scheme is implicit. It may be implemented by matrix inversion of the quantum evolution operator Q, or by using a predictor/corrector technique of high order, at least fourth order. The two techniques give the same results but the predictor/corrector one is faster.

All higher precision schemes were proved to be detrimental to the simulation stability and required longer simulation time to obtain good average quantities. In particular, Cayley's scheme, known to be the best technique for the evaluation of the time derivative in the deterministic solution of the WBTE, leads to instable results in MC simulation.

Further details on the numerical implementation of the "affinity" MC technique may be found in [QUE 10a].

8.5.3. *About the validity of the method*

To validate the WBMC method described above, we just take two simple but typical examples. The goal is to show briefly that it can describe very well the two extreme cases, that is (1) quantum effects in a purely coherent configuration and (2) classical transport in a structure where quantum and coherent effects are not expected to occur.

8.5.3.1. *Quantum mechanical case: Interaction of a wave packet with a tunneling barrier*

Here, we consider the typical situation previously suggested in [SHI 01]. It consists of simulating the interaction of a ballistic Gaussian wave packet with a tunnel barrier. The Gaussian wave packet is thus in the form:

$$\psi(x) = N\exp\left[-\frac{(x-x_0)^2}{2\sigma^2}\right]\exp\left[ik_0x\right] \qquad [8.88]$$

The Wigner function of which is:

$$f_w(x,k) = N'\exp\left[-\frac{(x-x_0)^2}{\sigma^2}\right]\exp\left[-(k-k_0)^2\sigma^2\right] \qquad [8.89]$$

Such a simple situation without any collision term can be solved easily from the Schrödinger equation. Though there is no space to show the result here, an impressive agreement was obtained between the particle approach of the WTE and the solution of the Schrödinger equation [QUE 10a, QUE 10b]. In particular, the oscillations of the wave-function are very well reproduced. This shows that the method can accurately model the existence of space correlations and interferences.

8.5.3.2. *Semi-classical case: The $N^+/N/N^+$ diode*

In the MC simulation technique described above, the potential felt by electrons may be treated in two different ways with, *a priori*, two very different effects. It may induce the evolution of the electron wave vector according to [8.86] if it is treated semi-classically, or it may be considered to act only on the electron affinity through [8.85] if it is treated in a quantum way. These two treatments seem mathematically very different. It is interesting and relevant to check that they coincide in a semi-classical situation to prove that the quantum model is able to describe it properly.

A simple GaAs $N^+/N/N^+$ structure was considered, with N^+ regions doped to 10^{18} cm^{-3} and with the 50-nm-long central region doped to 10^{16} cm^{-3}. The conduction band structure was limited to the Γ valley with an effective mass of 0.6 m_0. The scattering mechanisms included in the collision operator are interactions with acoustic phonons (in the elastic approximation), polar optical phonons and ionized impurity at room temperature. The transport simulation was self-consistently coupled with the 1D Poisson equation. It is performed using both the semi-classical EMC algorithm and the quantum EMC model. In the latter case, the Poisson potential is fully included in the quantum evolution term, that is $V = V_{rapid}$, which means that the electron wave vector is constant during free flights and may change only when the electron is scattered by a phonon or an impurity. After each time step the Wigner potential associated with the Poisson potential is recalculated and the affinity of particles are updated according to [2.29].

For this structure with a 50-nm-long central region, no quantum effect is expected to occur. Accordingly, the *I–V* characteristics (not shown here) are identical for semiclassical and quantum treatment of the potential [QUE 10b]. The results are superimposed on three orders of magnitude of current and voltage. The conduction band profiles are also identical. These results clearly show that, in a semiclassical situation, that is for slowly varying potential and in the presence of scattering mechanisms, the semiclassical evolution of wave vectors and the quantum evolution of affinities are fully equivalent. It is an important validation step for the Wigner MC technique.

8.6. Bibliography

[AUB 06] AUBRY-FORTUNA V., BOURNEL A., DOLLFUS P. *et al.*, "Ultra-short n MOSFETs with strained Si: Device performance and effect of ballistic transport using Monte Carlo simulation", *Semiconductor Science and Technology*, vol. 21, pp. 422–428, 2006.

[AUB 10] AUBRY-FORTUNA V., DOLLFUS P., "Electron transport properties in high-purity Ge down to cryogenic temperatures", *Journal of Applied Physics*, vol. 108, p. 123706, 2010.

[BAR 80] BARKER J., FERRY D., "On the physics and modeling of small semiconductor devices – I", *Solid-State Electronics*, vol. 23, pp. 519–530, 1980.

[BAR 02] BARRAUD S., DOLLFUS P., GALDIN S. *et al.*, "Short-range and long-range Coulomb interactions for 3D Monte Carlo device simulation with discrete impurity distribution", *Solid-State Electronics*, vol. 46, pp. 1061–1067, 2002.

[BER 99a] BERTONI A., BORDONE P., BRUNETTI R. *et al.*, "The Wigner function for electron transport in mesoscopic systems", *Journal of Physics: Condensed Matter*, vol. 11, pp. 5999–6012, 1999.

[BER 99b] BERTONI A., BORDONE P., BRUNETTI R. *et al.*, "Quantum versus classical scattering in semiconductor charge transport: A quantitative comparison", *Physica B: Condensed Matter*, vol. 272, pp. 299–301, 1999.

[BER 02] BERTET P., AUFFEVES A., MAIOLI P. *et al.*, "Direct measurement of the Wigner function of a one-photon fock state in a cavity", *Physical Review Letters*, vol. 89, no. 20, p. 200402, 2002.

[BER 03] BERTONI A., BORDONE P., FERRARI G. *et al.*, "Proximity effect of the contacts on electron transport in mesoscopic devices", *Journal of Computational Electronics*, vol. 2, pp. 137–140, 2003.

[BLA 82] BLAKEMORE J.S., "Semiconducting and other major properties of gallium arsenide", *Journal of Applied Physics*, vol. 53, pp. R123–R181, 1982.

[BOR 96] BORDONE P., VASILESKA D., FERRY D., "Collision-duration time for optical-phonon emission in semiconductors", *Physical Review B*, vol. 53, no. 7, pp. 3846–3855, 1996.

[BOR 99] BORDONE P., PASCOLI M., BRUNETTI R. *et al.*, "Quantum transport of electrons in open nanostructures with the Wigner-function formalism", *Physical Review B*, vol. 59, no. 4, pp. 3060-3069, 1999.

[BOR 03] BORDONE P., BERTONI A., BRUNETTI R. *et al.*, "Monte Carlo simulation of quantum electron transport based on Wigner paths", *Mathematics and Computers in Simulation*, vol. 62, nos. 3–6, pp. 307–314, 2003.

[BRU 89] BRUNETTI R., JACOBONI C., ROSSI F., "Quantum theory of transient transport in semiconductors: A Monte Carlo approach", *Physical Review B*, vol. 39, no. 15, pp. 10781–10790, 1989.

[BUF 03] BUFLER F.M., FICHTNER W., "Scaling of strained-Si n-MOSFETs into the ballistic regime and associated anisotropic effects", *IEEE Transactions on Electron Devices*, vol. 50, pp. 278–284, 2003.

[CAL 88] CALZETTA E., HU B.L., "Nonequilibrium quantum fields: Closed-time-path effective action, Wigner function, and Boltzmann equation", *Physical Review D*, vol. 37, no. 10, pp. 2878–2900, 1988.

[CAR 66] CARDONA M., POLLACK F.H., "Energy-band structure of germanium and silicon: The k·p method", *Physical Review*, vol. 142, pp. 530–543, 1966.

[CHA 81] CHATTOPADHYAY D., QUEISSER H.J., "Electron scattering by ionized impurities in semiconductors", *Reviews of Modern Physics*, vol. 53, pp. 745–768, 1981.

[CHE 84] CHELIKOWSKY J.R., COHEN M.L., "Nonlocal pseudopotential calculations for the electronic structure of eleven diamond and zinc-blende semiconductors", *Physical Review B*, vol. 14, pp. 556–582, 1976; *Physical Review B*, vol. 30, p. 4828, 1984.

[COH 06] COHEN-TANNOUDJI C., DIU B., LALOË F., *Quantum Mechanics*, Wiley Interscience, New York, 2006.

[DEL 08] DELEGLISE S., DOTSENKO I., SAYRIN C. *et al.*, "Reconstruction of non-classical cavity field states with snapshots of their decoherence", *Nature*, vol. 455, no. 7212, pp. 510–514, 2008.

[DIN 12] DINH T.V., VALAVANIS A., LEVER L.J.M. *et al.*, "Extended density-matrix model applied to silicon-based terahertz quantum cascade lasers", *Physical Review B*, vol. 85, no. 23, p. 235427, 2012.

[DOL 97] DOLLFUS P., "Si/$Si_{1-x}Ge_x$ heterostructures: Electron transport and field-effect transistor operation using Monte Carlo simulation", *Journal of Applied Physics*, vol. 82, pp. 3911–3916, 1997.

[DOL 04] DOLLFUS P., BOURNEL A., GALDIN-RETAILLEAU S. *et al.*, "Effect of discrete impurities on electron transport in ultra-short MOSFET using 3D Monte Carlo simulation", *IEEE Transactions on Electron Devices*, vol. 51, pp. 749–756, 2004.

[DOR 91] VAN DORT M.J., WOERLEE P.H., WALKER A.J. *et al*, "Quantum-mechanical threshold voltage shifts of MOSFETs caused by high levels of channel doping", in *IEEE International Electron Devices Meeting (IEDM) Technical Digest*, Washington, pp. 495–498, 1991.

[DUP 10] DUPONT E., FATHOLOLOUMI S., LIU H.C., "Simplified density-matrix model applied to three-well terahertz quantum cascade lasers", *Physical Review B*, vol. 81, no. 20, p. 205311, 2010.

[ESS 11] ESSENI D., PALESTRI P., SELMI L., *Nanoscale MOS Transistors*, Cambridge University Press, Cambridge, 2011.

[FAW 70] FAWCET W., BOARDMAN A.D., SWAIN S., "Monte Carlo determination of electron transport properties in gallium arsenide", *Journal of Physics and Chemistry of Solids*, vol. 31, pp. 1963–1990, 1970.

[FER 76] FERRY D.K., "First order optical and intervalley scattering in semiconductors", *Physical Review B*, vol. 14, pp. 1605–1609, 1976.

[FER 00a] FERRY D.K., "Effective potentials and the onset of quantization in ultrasmall MOSFETs", *Superlattices and Microstructures*, vol. 28, pp. 419–423, 2000.

[FER 00b] FERRY D.K., AKIS R., VASILESKA D., "Quantum effects in MOSFETs: Use of an effective potential in 3D Monte Carlo simulation of ultra-short channel devices", in *IEEE International Electron Devices Meeting (IEDM) Technical Digest*, San Francisco, pp. 871–874, 2000.

[FER 02] FERRY D.K., RAMEY S., SHIFREN L. *et al.*, "The effective potential in device modeling: The good, the bad and the ugly", *Journal of Computational Electronics*, vol. 1, pp. 59–65, 2002.

[FIE 93] FIEGNA C., SANGIORGI E., "Modeling of high-energy electrons in MOS devices at the microscopic level", *IEEE Transactions on Electron Devices*, vol. 40, pp. 619–627, 1993.

[FIS 84] FISCHETTI M.V., "Monte Carlo solution to the problem of high-field electron heating in SiO_2", *Physical Review Letters*, vol. 53, pp. 1755–1758, 1984.

[FIS 88] FISCHETTI M.V., LAUX S.E., "Monte Carlo analysis of electron transport in small semiconductor devices including band-structure and space-charge effects", *Physical Review B*, vol. 38, pp. 9721–9745, 1988.

[FIS 91] FISCHETTI M.V., LAUX S.E., "Monte Carlo simulation of transport in technologically significant semiconductors of the diamond and zinc-blende structures – Part II: Submicrometer MOSFET's", *IEEE Transactions on Electron Devices*, vol. 38, pp. 650–660, 1991.

[FIS 93] FISCHETTI M.V., LAUX S.E. "Monte Carlo study of electron transport in silicon inversion layers", *Physical Review B*, vol. 48, pp. 2244–2274, 1993.

[FIS 98] FISCHETTI M.V., "Theory of electron transport in small semiconductor devices using the Pauli master equation", *Journal of Applied Physics*, vol. 83, no. 1, pp. 270–291, 1998.

[FIS 99] FISCHETTI M.V., "Master-equation approach to the study of electronic transport in small semiconductor devices", *Physical Review B*, vol. 59, no. 7, pp. 4901–4917, 1999.

[FRO 47] FROEHLICH H., "On the theory of dielectric breakdown in solids", *Proceedings of the Royal Society of London* A, vol. 188, pp. 521–532, 1947.

[FUE 97] FUERST C., LEITENSTORFER A., LAUBEREAU A. *et al.*, "Quantum kinetic electron-phonon interaction in GaAs: Energy nonconserving scattering events and memory effects", *Physical Review Letters*, vol. 78, pp. 3733–3736, 1997.

[GEB 04a] GEBAUER R., CAR R., "Kinetic theory of quantum transport at the nanoscale", *Physical Review B,* vol. 70, pp. 12, p. 125324, 2004.

[GEB 04b] GEBAUER R., CAR R., "Current in open quantum systems", *Physical Review Letters*, vol. 93, no. 16, p. 160404, 2004.

[GRO 99] GROSS W.J., VASILESKA D., FERRY D. K., "A novel approach for introducing the electron–electron and electron–impurity interactions in particle-based simulations", *IEEE Electron Device Letters*, vol. 20, pp. 463–465, 1999.

[GUR 02] GUROV T., NEDJALKOV M., WHITLOCK P. *et al.*, "Femtosecond relaxation of hot electrons by phonon emission in presence of electric field", *Physica B*, vol. 314, pp. 301–304, 2002.

[HAL 62] HALL G.L., "Ionized impurity scattering in semiconductors", *Journal of Physics and Chemistry of Solids*, vol. 23, pp. 1147–1151, 1962.

[HAR 76] HARRISON J.W., HAUSER J.R., "Alloy scattering in ternary III-V compounds", *Physical Review B*, vol. 13, pp. 5347–5350, 1976.

[HAU 76] HAUSER J.R., LITTLEJOHN M.A., GLISSON T.H., "Velocity-field relationship of InAs-InP alloys including the effects of alloy scattering", *Applied Physics Letters*, vol. 28, pp. 458–461, 1976.

[HER 03] HERBST M., GLANEMANN M., AXT V. *et al.*, "Electron-phonon quantum kinetics for spatially inhomogeneous excitations", *Physical Review B*, vol. 67, p. 195305, 2003.

[HES 84] HESTO P., PELOUARD J.L., CASTAGNÉ R. *et al.*, "Nonelastic acoustic-phonon-electron interactions in Monte Carlo simulations at low fields", *Applied Physics Letters*, vol. 45, pp. 641–643, 1984.

[HES 91] HESS K., *Monte Carlo Device Simulation: Full Band and Beyond*, Kluwer Academic Publishers, Boston, 1991.

[HOC 81] HOCKNEY R.W., EASTWOOD J.W., *Computer Simulation Using Particles*, McGraw-Hill, New York, 1981.

[JAC 81] JACOBONI C., NAVA F., CANALI C. *et al.*, "Electron drift velocity and diffusivity in germanium", *Physical Review B*, vol. 24, pp. 1014–1026, 1981.

[JAC 83] JACOBONI C., REGGIANI L., "The Monte Carlo method for the solution of charge transport in semiconductors with applications to covalent materials", *Reviews of Modern Physics*, vol. 55, pp. 645–705, 1983.

[JAC 89] JACOBONI C., LUGLI P., *The Monte Carlo Method for Semiconductor Device Simulation*, Springer-Verlag, 1989.

[JAC 92] JACOBONI C., "Comparison between quantum and classical results in hot-electron transport", *Semiconductor Science and Technology*, vol. 7, no. 3B, pp. B6–B11, 1992.

[JAC 01a] JACOBONI C., BERTONI A., BORDONE P. *et al.*, "Wigner-function formulation for quantum transport in semiconductors: Theory and Monte Carlo approach", *Mathematics and Computers in Simulations*, vol. 55, nos. 1–3, pp. 67–78, 2001.

[JAC 01b] JACOBONI C., BRUNETTI R., BORDONE P. *et al.*, "Quantum transport and its simulation with the Wigner-function approach," *International Journal of High Speed Electronics and Systems*, vol. 11, pp. 387–423, 2001.

[JAC 04] JACOBONI C., BORDONE P., "The Wigner-function approach to non-equilibrium electron transport", *Reports on Progress in Physics*, vol. 67, pp. 1033–1071, 2004.

[JAU 06] JAUD M.A., BARRAUD S., DOLLFUS P. *et al.*, "Validity of the effective potential approach for the simulation of quantum confinement effects: A Monte-Carlo study", *Journal of Computational Electronics*, vol. 5, pp. 171–175, 2006.

[JAU 07] JAUD M.A., BARRAUD S., DOLLFUS P. *et al.*, "Pearson versus Gaussian effective potentials for quantum-corrected Monte-Carlo simulation", *Journal of Computational Electronics*, vol. 6, pp. 19–22, 2007.

[JAU 08] JAUD M.A., BARRAUD S., SAINT-MARTIN J. *et al.*, "A Pearson effective potential for Monte-Carlo simulation of quantum confinement effects in nMOSFETs", *IEEE Transactions on Electron Devices*, vol. 55, pp. 3450–3458, 2008.

[JIA 14] JIANG H., LU T., CAI W., "A device adaptive inflow boundary condition for Wigner equations of quantum transport", *Journal of Computational Physics*, vol. 258, pp. 773–786, 2014.

[JUN 03] JUNGEMANN C., MEINERZHAGEN B., *Hierarchical Device Simulation, the Monte Carlo Perspective*, Springer, New York, 2003.

[KIT 96] KITTEL C., *Introduction to Solid State Physics*, Wiley, New York, 1996.

[KUH 92] KUHN T., ROSSI F., "Monte Carlo simulation of ultrafast processes in photoexcited semiconductors: Coherent and incoherent dynamics", *Physical Review B*, vol. 46, pp. 7496–7514, 1992.

[LAX 72] LAX M., BIRMAN J.L., "Intervalley scattering selection rules for Si and Ge", *Physica Status Solidi*, vol. 49, pp. K153–K154, 1972.

[LI 93] LI S.H., HINCKLEY J.M., SINGH J. *et al.*, "Carrier velocity-field characteristics and alloy scattering potential in Si_{1-x}Gex/Si", *Applied Physics Letters*, vol. 63, pp. 1393–1395, 1993.

[LIP 91] LIPAVSKI P., KHAN F., ABDOLSALAMI F. *et al.*, "High-field transport in semiconductors. I. Absence of the intra-collisional field effect", *Physical Review B*, vol. 43, no. 6, pp. 4885–4896, 1991.

[LIU 05] LIU X.Y., JOVANOVIC D., STUMPF R., "First-principles study of Si–SiO_2 interface and the impact on mobility", *Applied Physics Letters*, vol. 86, no. 8, p. 082104, 2005.

[LUC 07] LUCCI L., PALESTRI P., ESSENI D. *et al.*, "Multisubband Monte Carlo study of transport, quantization, and electron-gas degeneration in ultrathin SOI n-MOSFETs", *IEEE Transactions on Electron Devices*, vol. 54, pp. 1156–1164, 2007.

[LUN 00] LUNDSTROM M., *Fundamentals of Carrier Transport*, Cambridge University Press, Cambridge, 2000.

[LUS 05] LUSAKOWSKI J., KNAP W., DYAKONOVA N. *et al.*, "Voltage tunable terahertz emission from a ballistic nanometer InGaAs/InAlAs transistor", *Journal of Applied Physics*, vol. 97, no. 6, p. 064307, 2005.

[LUT 97] LUTTERBACH L.G., DAVIDOVICH L., "Method for Direct Measurement of the Wigner Function in Cavity QED and Ion Traps", *Physical Review Letters*, vol. 78, no. 13, pp. 2547–2550, 1997.

[MIC 11] De MICHIELIS M., CONZATTI F., ESSENI D. *et al.*, "On the surface-roughness scattering in biaxially strained n- and p-MOS transistors", *IEEE Transactions on Electron Devices*, vol. 58, no. 9, pp. 3219–3223, 2011.

[MIL 08] MILLITHALLER J.F., REGGIANI L., POUSSET J. *et al.*, "Monte Carlo investigation of terahertz plasma oscillations in ultrathin layers of n-type $In_{0.53}Ga_{0.47}As$", *Applied Physics Letters*, vol. 92, no. 4, p. 042113, 2008.

[MOG 93] MOGLESTUE C., *Monte Carlo Simulation of Semiconductor Devices*, Chapman & Hall, London, 1993.

[MUR 11] MURPHY-ARMANDO F., FAHY S., "First principles calculation of electron-phonon and alloy scattering in strained SiGe", *Journal of Applied Physics*, vol. 110, p. 123706, 2011.

[NED 04] NEDJALKOV M., KOSINA H., SELBERHERR S. *et al.*, "Unified particle approach to Wigner-Boltzmann transport in small semiconductor devices", *Physical Review B*, vol. 70, no. 11, p. 115319, 2004.

[NED 05] NEDJALKOV M., "Wigner transport in presence of phonons: Particle models of the electron kinetics", *From Nanostructures to Nanosensing Applications,* Societa Italiana Di Fisica, IOS Press, Amsterdam, vol. 160, pp. 55–103, 2005.

[NED 07] NEDJALKOV M., VASILESKA D., ATANASSOV E. *et al.*, "Ultrafast Wigner transport in quantum wires", *Journal of Computational Electronics*, vol. 6, pp. 235–238, 2007.

[NED 08] NEDJALKOV M., VASILESKA D., "Semi-discrete 2D Wigner-particle approach", *Journal of Computational Electronics*, vol. 7, pp. 222–225, 2008.

[NED 11] NEDJALKOV M., QUERLIOZ D., DOLLFUS P. *et al.*, "Wigner function approach", in VASILESKA D., GOODNICK S. (eds), *Nano-Electronic Devices: Semiclassical and Quantum Transport Modeling*, Springer, pp. 289–358, 2011.

[NGU 11] NGUYEN H.N., QUERLIOZ D., GALDIN-RETAILLEAU S. *et al.*, "Semi-classical and quantum transport in CNTFETs using Monte Carlo simulation", *IEEE Transactions on Electron Devices*, vol. 58, pp. 798–804, 2011.

[PAL 05] PALESTRI P., ESSENI D., EMINENTE S. *et al.*, "Understanding quasi-ballistic transport in nano-MOSFETs: Part I – Scattering in the channel, and in the drain", *IEEE Transactions on Electron Devices*, vol. 52, no. 12, pp. 2727–2735, 2005.

[PAS 98] PASCOLI M., BORDONE P., BRUNETTI R. *et al.*, "Wigner paths for electrons interacting with phonons", *Physical Review B*, vol. 58, pp. 3503–3506, 1998.

[QUE 06a] QUERLIOZ D., DOLLFUS P., DO V.N. *et al.*, "An improved Wigner Monte Carlo technique for the self-consistent simulation of RTDs", *Journal of Computational Electronics*, vol. 5, pp. 443–446, 2006.

[QUE 06b] QUERLIOZ D., SAINT-MARTIN J., DO V.N. *et al.*, "A study of quantum transport in end-of-roadmap DG-MOSFETs using a fully self-consistent Wigner Monte Carlo approach", *IEEE Transactions on Nanotechnology*, vol. 5, pp. 737–744, 2006.

[QUE 07] QUERLIOZ D., SAINT-MARTIN J., HUET K. *et al.*, "On the ability of the particle Monte Carlo technique to include quantum effects in nano-MOSFET simulation", *IEEE Transactions on Electron Devices*, vol. 54, pp. 2232–2242, 2007.

[QUE 08] QUERLIOZ D., SAINT-MARTIN J., BOURNEL A. *et al.*, "Wigner Monte Carlo simulation of phonon induced electron decoherence in semiconductor nanodevices", *Physical Review B*, vol. 78, no. 16, p. 165306, 2008.

[QUE 09] QUERLIOZ D., NGUYEN H.N., SAINT-MARTIN J. *et al.*, "Wigner-Boltzmann Monte Carlo approach to nanodevice simulation: From quantum to semiclassical transport", *Journal of Computational Electronics*, vol. 8, nos. 3–4, pp. 324–335, 2009.

[QUE 10a] QUERLIOZ D., DOLLFUS P, *The Wigner Monte Carlo Method for Nanoelectronic Devices*, ISTE Ltd., London, and John Wiley & Sons, New York, 2010.

[QUE 10b] QUERLIOZ D., SAINT-MARTIN J., DOLLFUS P., "Implementation of the Wigner-Boltzmann transport equation within particle Monte Carlo simulation", *Journal of Computational Electronics*, vol. 9, pp. 224–231, 2010.

[RID 82] RIDLEY B.K., *Quantum Processes in Semiconductors*, Clarendon Press, Oxford, 1982.

[RID 06] RIDEAU D., FERAILLE M., CIAMPOLINI L. *et al.*, "Strained Si, Ge, and $Si_{1-x}Ge_x$ alloys modeled with a first-principles-optimized full-zone k · p method", *Physical Review B*, vol. 74, p. 195208, 2006.

[RIE 93] RIEGER M.M., VOGL P., "Electronic-band parameters in strained $Si_{1x}Ge_x$ alloys on $Si_{1-y}Ge_y$ substrates", *Physical Review B*, vol. 48, pp. 14276–14287, 1993.

[ROS 92] ROSSI F., JACOBONI C., "Self-scattering in Monte Carlo simulation of quantum transport", *Europhysics Letters*, vol. 18, no. 2, pp. 169–173, 1992.

[ROS 94] ROSSI F., JACOBONI C., NEDJALKOV M., "A Monte Carlo solution of the Wigner transport equation", *Semiconductor Science and Technology*, vol. 9, pp. 934–936, 1994.

[ROS 02] ROSSI F., KUHN T., "Theory of ultrafast phenomena in photoexcited semiconductors", *Reviews of Modern Physics*, vol. 74, pp. 895–950, 2002.

[ROS 13] ROSATI R., DOLCINI F., IOTTI R.C. *et al.*, "Wigner-function formalism applied to semiconductor quantum devices: Failure of the conventional boundary condition scheme", *Physical Review B*, vol. 88, p. 035401, 2013.

[SAI 04] SAINT-MARTIN J., BOURNEL A., DOLLFUS P., "On the ballistic transport in nanometer-scaled DG MOSFETs", *IEEE Transactions on Electron Devices*, vol. 51, no. 7, pp. 1148–1155, 2004.

[SAI 06] SAINT-MARTIN J., BOURNEL A., MONSEF F. *et al.*, "Multi sub-band Monte Carlo simulation of an ultra-thin double gate MOSFET with 2D electron gas", *Semiconductor Science and Technology*, vol. 21, pp. L29–L31, 2006.

[SAN 92] SANGIORGI E., PINTO M.R., "A semi-empirical model of surface scattering for Monte Carlo simulation of silicon n-MOSFET's", *IEEE Transactions on Electron Devices*, vol. 39, pp. 356–361, 1992.

[SCH 94] SCHILP J., KUHN T., MAHLER G., "Electron-phonon quantum kinetics in pulse-excited semiconductors: Memory and renormalization effects", *Physical Review B*, vol. 50, no. 8, pp. 5435–5447, 1994.

[SCH 13] SCHWAHA P., QUERLIOZ D., DOLLFUS P. *et al.*, "Decoherence effects in the Wigner function formalism", *Journal of Computational Electronics*, vol. 12, pp. 388–396, 2013.

[SEL 14] SELLIER J.M., DIMOV I., "The many-body Wifner Monte Carlo method for time-dependent ab-initio quantum simulations", *Journal of Computational Physics*, vol. 273, pp. 589–597, 2014.

[SHI 01] SHIFREN L., FERRY D.K., "Particle Monte Carlo simulation of Wigner function tunneling", *Physics Letters A*, vol. 285, pp. 217–221, 2001.

[SHI 02] SHIFREN L., FERRY D.K., "A Wigner function based ensemble Monte Carlo approach for accurate incorporation of quantum effects in device simulation", *Journal of Computational Electronics*, vol. 1, pp. 55–58, 2002.

[SHI 03] SHIFREN L., RINGHOFER C., FERRY D.K., "A Wigner function-based quantum ensemble Monte Carlo study of a resonant tunneling diode", *IEEE Transactions on Electron Devices*, vol. 50, pp. 769–773, 2003.

[STR 70] STREITWOLF H.W., "Intervalley scattering selection rules for Si and Ge", *Physica Status Solidi*, vol. 37, pp. K47–K49, 1970.

[SVE 06] SVERDLOV V., GRASSER T., KOSINA H. *et al.*, "Scattering and space-charge effects in Wigner Monte Carlo simulations of single and double barrier devices", *Journal of Computational Electronics*, vol. 5, pp. 447–450, 2006.

[TAK 59] TAKIMOTO N., "On the screening of impurity potential by conduction electrons", *Journal of the Physical Society of Japan*, vol. 14, pp. 1142–1158, 1959.

[TAN 83] TANG J.Y., HESS K., "Impact ionization of electrons in silicon (steady state)", *Journal of Applied Physics*, vol. 54, pp. 5139–5144, 1983.

[THO 78] THORNBER K., "High-field electronic conduction in insulators", *Solid-State Electronics*, vol. 21, pp. 259–266, 1978.

[TSU 01] TSUCHIYA H., RAVAIOLI U., "Particle Monte Carlo simulation of quantum phenomena in semiconductor nanostructures", *Journal of Applied Physics*, vol. 89, pp. 4023–4029, 2001.

[VEN 93] VENKATARAMAN V., LIU C.W., STURM J.C., "Alloy scattering limited transport of two-dimensional carriers in strained $Si_{1-x}Ge_x$ quantum wells", *Applied Physics Letters*, vol. 63, pp. 2795–2797, 1993.

[VEN 02] VENUGOPAL R., REN Z., DATTA S. *et al.*, "Simulating quantum transport in nanoscale transistors: Real versus mode-space approaches", *Journal of Applied Physics*, vol. 92, pp. 3730–3739, 2002.

[WAN 11] WANG Z., WANG S., OBUKHOV S. *et al.*, "Thermoelectric transport properties of silicon: Toward an ab initio approach", *Physical Review B*, vol. 83, p. 205208, 2011.

[WIG 32] WIGNER E., "On the quantum correction for thermodynamic equilibrium", *Physical Review*, vol. 40, no. 5, pp. 749–759, 1932.

[WIL 71] WILEY J.D., "Polar mobility of holes in III-V compounds", *Physical Review B*, vol. 4, pp. 2485–2493, 1971.

[WIN 03] WINSTEAD B., RAVAIOLI U., "A quantum correction based on Schrodinger equation applied to Monte Carlo device simulation", *IEEE Transactions on Electron Devices*, vol. 50, pp. 440–446, 2003.

[WOR 00] WORDELMAN C.J., RAVAIOLI U., "Integration of a particle-particle-particle-mesh algorithm with the ensemble MC method for the simulation of ultrasmall semiconductor devices", *IEEE Transactions on Electron Devices*, vol. 47, pp. 410–416, 2000.

[ZIM 60] ZIMAN J. M., *Electrons and Phonons*, Oxford University Press, Oxford, 1960.

List of Authors

Alfonso ALARCÓN
Universitat Autònoma de Barcelona (UAB)
Barcelona
Spain

Guillermo ALBAREDA
Universitat de Barcelona (UB)
Barcelona
Spain

Abdelilah BENALI
Universitat Autònoma de Barcelona (UAB)
Barcelona
Spain

Marc BESCOND
Institut Matériaux Microélectronique Nanosciences de Provence (IM2NP)
CNRS
Marseille
France

Philippe DOLLFUS
Centre de Nanosciences et Nanotechnologies (C2N)
CNRS
Orsay

Gaston KANÉ
Laboratoire des Solides Irradiés,
École Polytechnique - CEA/DRF
CNRS
University of Paris-Saclay
Palaiseau
France

Michel LANNOO
Institut Matériaux Microélectronique Nanosciences de Provence (IM2NP)
CNRS
Marseille
France

Damiano MARIAN
Universitat Autònoma de Barcelona (UAB)
Barcelona
Spain

Simeon MOISES
Universitat Autònoma de Barcelona (UAB)
Barcelona
Spain

Yann-Michel Niquet
Institut Nanosciences et Cryogénie (INAC)
CEA
Grenoble
France

Xavier Oriols
Universitat Autònoma de Barcelona (UAB)
Barcelona
Spain

Damien Querlioz
Centre de Nanosciences et Nanotechnologies (C2N)
CNRS
Orsay
France

Denis Rideau
STMicroelectronics
Crolles
France

Stephan Roche
Catalan Institute of Nanoscience and Nanotechnology (ICN2)
CSIC and Barcelona Institute of Science and Technology
and
ICREA - Institucio Catalana de Recerca i Estudis Avancats
Barcelona
Spain

Jérôme Saint-Martin
Centre de Nanosciences et Nanotechnologies (C2N)
University of Paris-Sud
Orsay
France

Jelena Sjakste
Laboratoire des Solides Irradiés, École Polytechnique - CEA/DRF
CNRS
University of Paris-Saclay
Palaiseau
France

Virginie Trinité
III-V Lab, TRT THALES
Ecole Polytechnique
Palaiseau
France

François Triozon
Laboratoire d'Electronique et de Technologie de l'Information (LETI)
CEA
Grenoble
France

Nathalie Vast
Laboratoire des Solides Irradiés, École Polytechnique - CEA/DRF
CNRS
University of Paris-Saclay
Palaiseau
France

Sebastian Volz
EM2C
CNRS
CentraleSupélec
Châtenay-Malabry
France

Index

G, H, I

K, L, M

N, P, Q

S, T

W, Z

Other titles from

iSTE

in

Nanoscience and Nanotechnology

2016

BASSET Philippe, BLOKHINA Elena, GALAYKO Dimitri
Electrostatic Kinetic Energy Harvesting
(Nanotechnologies for Energy Recovery Set – Volume 3)

LAFONTAINE Eric, COMET Marc
Nanothermites

2015

LEPRINCE-WANG Yamin
Piezoelectric ZnO Nanostructure for Energy Harvesting
(Nanotechnologies for Energy Recovery Set – Volume 1)

ROSSI Carole
Al-based Energetic Nanomaterials
(Nanotechnologies for Energy Recovery Set – Volume 2)

2014

BALESTRA Francis
Beyond CMOS Nanodevices 1
Beyond CMOS Nanodevices 2

LACAZE Pierre Camille, LACROIX Jean-Christophe
Non-volatile Memories

LATU-ROMAIN Laurence, OLLIVIER Maelig
Silicon Carbide One-dimensional Nanostructures

NICU Liviu, LEÏCHLE Thierry
Micro- and Nanoelectromechanical Biosensors

SENEZ Vincent, THOMY Vincent, DUFOUR Renaud
Nanotechnologies for Synthetic Super Non-wetting Surfaces

TRIPON-CANSELIET Charlotte, CHAZELAS Jean
Nanoscale Microwave Engineering: Optical Control of Nanodevices

2012

BELLEVILLE Marc, CONDEMINE Cyril
Energy Autonomous Micro and Nano Systems

CLAVERIE Alain
Transmission Electron Microscopy in Micro-nanoelectronics

LACAZE PIERRE Camille
Nanotechnologies: Concepts, Processing and Applications

2010

BALESTRA Francis
Nanoscale CMOS / Innovative Materials, Modeling and Characterization

QUERLIOZ Damien, DOLLFUS Philippe
The Wigner Monte Carlo Method for Nanoelectronic Devices: Particle Description of Quantum Transport and Decoherence

TAKADOUM Jamal
Nanomaterials and Surface Engineering

VODA Alina
Micro, Nanosystems and Systems on Chips: Modeling, Control and Estimation

2008

DAVIM J. Paulo, JACKSON J.
Nano and Micromachining

NOUAILLHAT Alain
An Introduction to Nanoscience and Nanotechnology

OUISSE Thierry
Electron Transport in Nanostructures and Mesoscopic Devices

2006

RIGNEAULT Hervé *et al.*
Nanophotonics

www.ingramcontent.com/pod-product-compliance
Lightning Source LLC
LaVergne TN
LVHW011906200926
848484LV00002B/541
9781848215665